Study and Solutions Guide for

COLLEGE ALGEBRA

THIRD EDITION

Larson/Hostetler

Dianna L. Zook

Indiana University
Purdue University at Fort Wayne, Indiana

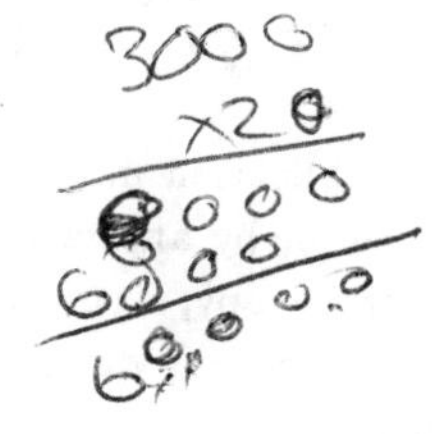

D. C. Heath and Company

Lexington, Massachusetts Toronto

Published simultaneously in Canada.

Printed in the United States of America.

International Standard Book Number: 0–669–28306–1

10 9 8 7 6 5 4 3 2

TO THE STUDENT

The *Study and Solutions Guide for College Algebra* is a supplement to the text by Roland E. Larson and Robert P. Hostetler.

As a mathematics instructor, I often have students come to me with questions about the assigned homework. When I ask to see their work, the reply often is "I didn't know where to start." The purpose of the *Study Guide* is to provide brief summaries of the topics covered in the textbook and enough detailed solutions to problems so that you will be able to work the remaining exercises.

A special thanks to Patricia S. Larson for typing this guide. Also I would like to thank my husband Edward L. Schlindwein for his support during the several months I worked on this project.

If you have any corrections or suggestions for improving this *Study Guide*, I would appreciate hearing from you.

Good luck with your study of algebra.

Dianna L. Zook
Indiana University,
Purdue University at
Fort Wayne, Indiana 46805

CONTENTS

STUDY STRATEGIES

- Attend all classes and come prepared. Have your homework completed. Bring the text, paper, pen or pencil, and a calculator (scientific or graphing) to each class.
- Read the section in the text that is to be covered before class. Make notes about any questions that you have and, if they are not answered during the lecture, ask them at the appropriate time.
- Participate in class. As mentioned above, ask questions. Also, do not be afraid to answer questions.
- Take notes on all definitions, concepts, rules, formulas and examples. After class, read your notes and fill in any gaps, or make notations of any questions that you have.
- DO THE HOMEWORK!!! You learn mathematics by doing it yourself. Allow **at least** two hours outside of each class for homework. Do not fall behind.
- Seek help when needed. Visit your instructor during office hours and come prepared with specific questions; check with your school's tutoring service; find a study partner in class; check additional books in the library for more examples—just do something before the problem becomes insurmountable.
- Do not cram for exams. Each chapter in the text contains a chapter review and this study guide contains a practice test at the end of each chapter. (The answers are at the back of the study guide.) Work these problems a few days before the exam and review any areas of weakness.

CHAPTER 1

Review of Fundamental Concepts of Algebra

SECTION 1.1

The Real Number System

- You should know the following sets.

  (a) The set of real numbers includes the rational numbers and the irrational numbers.

  (b) The set of rational numbers includes all real numbers that can be written as the ratio p/q of two integers, where $q \neq 0$.

  (c) The set of irrational numbers includes all real numbers which are not rational.

  (d) The set of integers: $\{\cdots, -3, -2, -1, 0, 1, 2, 3, \cdots\}$

  (e) The set of whole numbers: $\{0, 1, 2, 3, 4, \cdots\}$

  (f) The set of natural numbers: $\{1, 2, 3, 4, \cdots\}$

- The real number line is used to represent the real numbers.

- Know the inequality symbols.

  (a) $a < b$ means a is less than b.

  (b) $a \leq b$ means a is less than or equal to b.

  (c) $a > b$ means a is greater than b.

  (d) $a \geq b$ means a is greater than or equal to b.

- You should know that

  $$|a| = \begin{cases} a & \text{, if } a \geq 0 \\ -a & \text{, if } a < 0. \end{cases}$$

- Know the properties of absolute value.

  (a) $|a| \geq 0$ (b) $|-a| = |a|$ (c) $|ab| = |a|\,|b|$ (d) $\left|\frac{a}{b}\right| = \frac{|a|}{|b|}$

- The distance between a and b on the real line is $|b - a| = |a - b|$.

Solutions to Selected Exercises

5. Determine which numbers in the set $\left\{-\pi, -\frac{1}{3}, \frac{6}{3}, \frac{1}{2}\sqrt{2}, -7.5\right\}$ are (a) natural numbers, (b) integers, (c) rational numbers, and (d) irrational numbers.

Solution:

(a) Natural numbers: $\{\frac{6}{3}\}$

(b) Integers: $\{\frac{6}{3}\}$

(c) Rational numbers: $\{-\frac{1}{3}, \frac{6}{3}, -7.5\}$

(d) Irrational numbers: $\{-\pi, \frac{1}{2}\sqrt{2}\}$

11. Plot the two real numbers $\frac{5}{6}$ and $\frac{2}{3}$ on the real number line and place the appropriate inequality sign between them.

Solution:

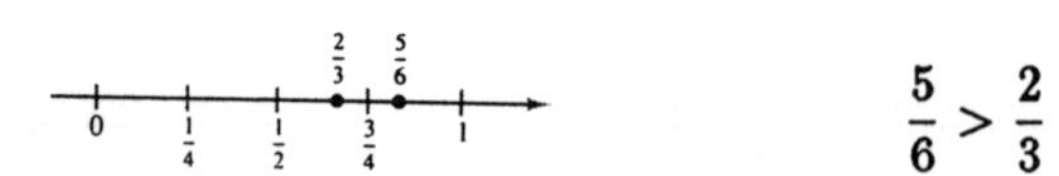

$$\frac{5}{6} > \frac{2}{3}$$

15. Describe the subset of real numbers represented by the inequality $x < 0$. Then sketch the subset on the real number line.

Solution:

The inequality $x < 0$ denotes all real numbers that are less than zero.

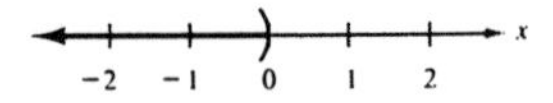

19. Describe the subset of real numbers represented by the inequality $-2 < x < 2$. Then sketch the subset on the real number line.

Solution:

The inequality $-2 < x < 2$ denotes all real numbers between -2 and 2, not including -2 or 2.

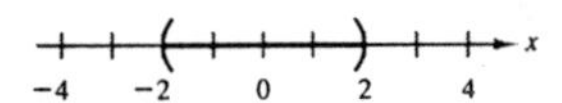

23. Use inequality notation to describe the expression "x is negative."

Solution:

"x is negative" can be written as $x < 0$.

27. Use inequality notation to describe the expression "The person's age, A, is at least 30."

Solution:

The expression "The person's age, A, is at least 30" can be written as $A \geq 30$.

29. Use inequality notation to describe the expression "the annual rate of inflation, r, is expected to be at least 3.5% but no more than 6%".

Solution:

The expression "the annual rate of inflation, r, is expected to be at least 3.5% but no more than 6%" can be written as

$$3.5\% \leq r \leq 6\% \qquad \text{or} \qquad 0.035 \leq r \leq 0.06.$$

31. Evaluate the expression $|-10|$.

Solution:

$$|-10| = -(-10) = 10$$

35. Evaluate the expression $\dfrac{-5}{|-5|}$.

Solution:

$$\frac{-5}{|-5|} = \frac{-5}{-(-5)} = \frac{-5}{5} = -1$$

39. Evaluate the expression $-|16.25| + 20$.

Solution:

$$-|16.25| + 20 = -16.25 + 20 = 3.75$$

43. Place the correct symbol (<, >, or =) between the numbers -5 and $-|5|$.

Solution:

Since $-|5| = -5$, we have $-5 \boxed{=} -|5|$.

49. Find the distance between the points $-\frac{5}{2}$ and 0.

Solution:

$$d\left(-\frac{5}{2},\ 0\right) = \left|0 - \left(-\frac{5}{2}\right)\right| = \left|\frac{5}{2}\right| = \frac{5}{2}$$

53. Find the distance between the points 9.34 and -5.65.

Solution:

$$d(9.34,\ -5.65) = |9.34 - (-5.65)| = |9.34 + 5.65| = |14.99| = 14.99$$

55. Use absolute value notation to describe the expression "The distance between x and 5 is no more than 3."

Solution:

Since $d(x,\ 5) = |x - 5|$ and $d(x,\ 5) \le 3$, we have $|x - 5| \le 3$.

59. Use absolute value notation to describe the expression "y is at least six units from 0."

Solution:

Since $d(y,\ 0) = |y - 0| = |y|$ and $d(y,\ 0) \ge 6$, we have $|y| \ge 6$.

63. The accounting department of a company is checking to see whether the actual expenses of a department differ from the budgeted expenses by more than \$500 or by more than 5%. Complete the table (shown in the textbook) and determine whether the actual expense passes the "budget variance test."

Solution:

	Budgeted expense, b	Actual expense, a	$\lvert a-b\rvert$	$0.05b$
Taxes	\$37,640.00	\$37,335.80	\$304.20	\$1882.00

$$|a-b| = |37{,}335.80 - 37{,}640.00| = |-304.20| = \$304.20$$
$$0.05b = 0.05(37{,}640.00) = \$1882.00$$

Since the difference is less than \$500 and 5% of the budgeted expenses, it passes the "budget variance test."

67. The bar graph shows the receipts of the federal government (in billions of dollars) for selected years from 1960 through 1989. Use the bar graph to find the income y of the Federal Government in 1990. Then, find the absolute value of the surplus or deficit for the same year.

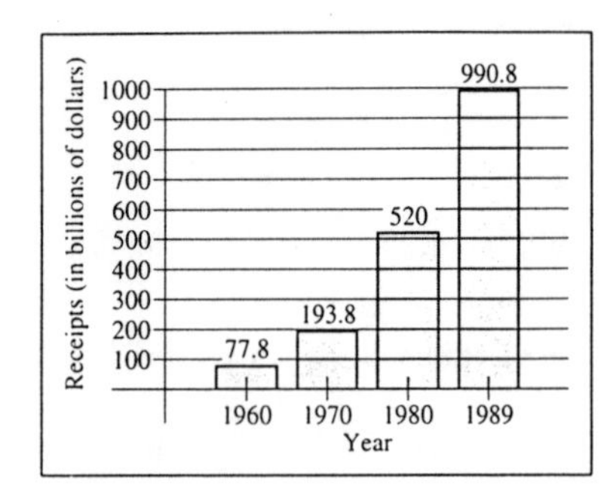

Solution:

	Income, y	Expense, x	$\lvert y-x\rvert$	Surplus or deficit
1980	\$520.0 billion	\$590.2 billion	70.2	Deficit of \$70.2 billion

The income for 1980 was \$520.0 billion.

$$|y-x| = |520.0 - 590.2| = |-70.2| = 70.2$$

There was a deficit of \$70.2 billion.

69. Use a calculator to order the following real numbers, from smallest to largest.

$\frac{7071}{5000}, \frac{584}{413}, \sqrt{2}, \frac{47}{33}, \frac{127}{90}$

Solution:

$$\frac{7071}{5000} = 1.4142$$

$$\frac{584}{413} \approx 1.414043584$$

$$\sqrt{2} \approx 1.414213562$$

$$\frac{47}{33} = 1.42\overline{42}$$

$$\frac{127}{90} = 1.41\overline{1}$$

$$\frac{127}{90} < \frac{584}{413} < \frac{7071}{5000} < \sqrt{2} < \frac{47}{33}$$

73. Use a calculator to find the decimal form of $\frac{41}{333}$. If it is a nonterminating decimal, write the repeating pattern.

Solution:

$$\frac{41}{333} = 0.123123\cdots$$

77. Determine whether the statement "The reciprocal of a nonzero rational number is rational" is true or false.

Solution:

True. Since the number is a nonzero rational number, it can be written as p/q where $p \neq 0$, $q \neq 0$. Its reciprocal is q/p, which is also a nonzero rational number.

SECTION 1.2

The Basic Rules of Algebra

- You should be able to identify the terms in an algebraic expression.
- You should know and be able to use the basic rules of algebra.
- Commutative Property
  (a) Addition: $a + b = b + a$
  (b) Multiplication: $a \cdot b = b \cdot a$
- Associative Property
  (a) Addition: $(a + b) + c = a + (b + c)$
  (b) Multiplication: $(ab)c = a(bc)$
- Identity Property
  (a) Addition: 0 is the identity; $a + 0 = 0 + a = a$.
  (b) Multiplication: 1 is the identity; $a \cdot 1 = 1 \cdot a = a$.
- Inverse Property
  (a) Addition: $-a$ is the inverse of a; $a + (-a) = -a + a = 0$.
  (b) Multiplication: $1/a$ is the inverse of a, $a \neq 0$; $a(1/a) = (1/a)a = 1$.
- Distributive Property
  (a) Left: $a(b + c) = ab + ac$
  (b) Right: $(a + b)c = ac + bc$
- Properties of Negatives
  (a) $(-1)a = -a$
  (b) $-(-a) = a$
  (c) $(-a)b = a(-b) = -ab$
  (d) $(-a)(-b) = ab$
  (e) $-(a + b) = (-a) + (-b) = -a - b$
- Properties of Zero
  (a) $a \pm 0 = a$
  (b) $a \cdot 0 = 0$
  (c) $0 \div a = 0/a = 0, \; a \neq 0$
  (d) If $ab = 0$, then $a = 0$ or $b = 0$.
  (e) $a/0$ is undefined.

- Properties of Fractions ($b \neq 0$, $d \neq 0$)
  (a) Equivalent Fractions: $a/b = c/d$ if and only if $ad = bc$.
  (b) Rule of Signs: $-a/b = a/-b = -(a/b)$ and $-a/-b = a/b$
  (c) Equivalent Fractions: $a/b = ac/bc$, $c \neq 0$
  (d) Addition and Subtraction
    1. Like Denominators: $(a/b) \pm (c/b) = (a \pm c)/b$
    2. Unlike Denominators: $(a/b) \pm (c/d) = (ad \pm bc)/bd$
  (e) Multiplication: $(a/b) \cdot (c/d) = ac/bd$
  (f) Division: $(a/b) \div (c/d) = (a/b) \cdot (d/c) = ad/bc$ if $c \neq 0$.

- Properties of Equality
  (a) Reflexive: $a = a$
  (b) Symmetric: If $a = b$, then $b = a$.
  (c) Transitive: If $a = b$ and $b = c$, then $a = c$.
  (d) Substitution: If $a = b$, then a can be replaced by b in any statement involving a or b.
    1. If $a = b$, then $a + c = b + c$.
    2. If $a = b$, then $ac = bc$.

- Cancellation Laws
  (a) If $a + c = b + c$, then $a = b$.
  (b) If $ac = bc$, then $a = b$, $c \neq 0$.

- You should know the properties of exponents.
  (a) $a^1 = a$
  (b) $a^0 = 1$, $a \neq 0$
  (c) $a^m a^n = a^{m+n}$
  (d) $a^m/a^n = a^{m-n}$, $a \neq 0$
  (e) $a^{-n} = 1/a^n$, $a \neq 0$
  (f) $(a^m)^n = a^{mn}$
  (g) $(ab)^n = a^n b^n$
  (h) $(a/b)^n = a^n/b^n$, $b \neq 0$
  (i) $(a/b)^{-n} = (b/a)^n$, $a \neq 0$, $b \neq 0$
  (j) $|a^2| = |a|^2 = a^2$

- You should be able to write numbers in scientific notation, $c \times 10^n$, where $1 \leq c < 10$ and n is an integer.

- You should be able to use your calculator to evaluate expressions involving exponents.

Solutions to Selected Exercises

3. Identify the terms of the expression $x^2 - 4x + 8$.

Solution:

$$x^2 - 4x + 8 = x^2 + (-4x) + 8$$

Terms: x^2, $-4x$, 8

7. Evaluate the expression $4x - 6$ for (a) $x = -1$, and (b) $x = 0$.

Solution:

(a) $4(-1) - 6 = -4 - 6 = -10$

(b) $4(0) - 6 = 0 - 6 = -6$

11. Evaluate the expression $\dfrac{x+1}{x-1}$ for (a) $x = 1$, and (b) $x = -1$.

Solution:

(a) $\dfrac{1+1}{1-1} = \dfrac{2}{0}$ which is undefined. You cannot divide by zero.

(b) $\dfrac{-1+1}{-1-1} = \dfrac{0}{-2} = 0$

17. Identify the rule illustrated in the equation $2(x+3) = 2x + 6$.

Solution:

By the Distributive Property, we have

$$2(x+3) = 2 \cdot x + 2 \cdot 3 = 2x + 6.$$

21. Identify the rules illustrated in the equation $x(3y) = (x \cdot 3)y = (3x)y$.

Solution:

$$\begin{aligned} x(3y) &= (x \cdot 3)y && \text{by the Associative Property of Multiplication} \\ &= (3x)y && \text{by the Commutative Property of Multiplication} \end{aligned}$$

25. Evaluate the following expression. If this is not possible, state the reason.

$$\frac{8-8}{-9+(6+3)}$$

Solution:

$$\frac{8-8}{-9+(6+3)} = \frac{8-8}{-9+9} = \frac{0}{0}$$

which is undefined since the denominator is zero.

27. Perform the indicated operations:

$(4-7)(-2)$.

Solution:

$$(4-7)(-2) = (-3)(-2) = 6$$

33. Perform the indicated operations:

$\frac{4}{5} \cdot \frac{1}{2} \cdot \frac{3}{4}$.

Solution:

$$\frac{4}{5} \cdot \frac{1}{2} \cdot \frac{3}{4} = \frac{1}{5} \cdot \frac{1}{2} \cdot \frac{3}{1} = \frac{3}{10}$$

35. Perform the indicated operation:

$12 \div \frac{1}{4}$.

Solution:

$$12 \div \frac{1}{4} = 12 \times \frac{4}{1} = 12 \times 4 = 48$$

39. Use a calculator to evaluate

$$\frac{11.46-5.37}{3.91}.$$

Round your answer to two decimal places.

Solution:

$$\frac{11.46-5.37}{3.91} = \frac{6.09}{3.91} \approx 1.56$$

41. Evaluate the expression

$$\frac{5^5}{5^2}.$$

Solution:

$$\frac{5^5}{5^2} = 5^{5-2} = 5^3 = 125$$

45. Evaluate the expression $(2^3 \cdot 3^2)^2$.

Solution:

$$(2^3 \cdot 3^2)^2 = (8 \cdot 9)^2 = 72^2 = 5184$$

47. Evaluate the expression

$$\frac{4 \cdot 3^{-2}}{2^{-2} \cdot 3^{-1}}.$$

Solution:

$$\frac{4 \cdot 3^{-2}}{2^{-2} \cdot 3^{-1}} = \frac{4 \cdot 2^2 \cdot 3^1}{3^2} = \frac{4 \cdot 4}{3} = \frac{16}{3}$$

51. Evaluate $6x^0 - (6x)^0$ when $x = 10$.

Solution:

$$6(10)^0 - (6 \cdot 10)^0 = 6 \cdot 1 - (60)^0 = 6 - 1 = 5$$

55. Simplify $5x^4(x^2)$.

Solution:

$$5x^4(x^2) = 5x^{4+2} = 5x^6$$

57. Simplify $6y^2(2y^4)^2$.

Solution:

$$\begin{aligned} 6y^2(2y^4)^2 &= 6y^2(2)^2(y^4)^2 \\ &= 6y^2(4)(y^8) = 24y^{10} \end{aligned}$$

63. Simplify

$$\frac{12(x+y)^3}{9(x+y)}.$$

Solution:

$$\begin{aligned} \frac{12(x+y)^3}{9(x+y)} &= \frac{3 \cdot 4(x+y)^{3-1}}{3 \cdot 3} \\ &= \frac{4(x+y)^2}{3} \end{aligned}$$

67. Simplify

$$(2x^2)^{-2}.$$

Solution:

$$(2x^2)^{-2} = \frac{1}{(2x^2)^2} = \frac{1}{4x^4}$$

69. Simplify $(-2x^2)^3(4x^3)^{-1}$.

Solution:

$$\begin{aligned} (-2x^2)^3(4x^3)^{-1} &= \frac{(-2x^2)^3}{4x^3} \\ &= \frac{-8x^6}{4x^3} = -2x^3 \end{aligned}$$

73. Simplify $(4a^{-2}b^3)^{-3}$.

Solution:

$$\begin{aligned} (4a^{-2}b^3)^{-3} &= (4)^{-3}(a^{-2})^{-3}(b^3)^{-3} \\ &= 4^{-3}a^6b^{-9} = \frac{a^6}{4^3b^9} = \frac{a^6}{64b^9} \end{aligned}$$

77. Simplify $3^n \cdot 3^{2n}$.

Solution:

$$3^n \cdot 3^{2n} = 3^{n+2n} = 3^{3n}$$

79. Simplify

$$\left(\frac{a^{-2}}{b^{-2}}\right)\left(\frac{b}{a}\right)^3.$$

Solution:

$$\left(\frac{a^{-2}}{b^{-2}}\right)\left(\frac{b}{a}\right)^3 = \left(\frac{b^2}{a^2}\right)\left(\frac{b^3}{a^3}\right) = \frac{b^5}{a^5}$$

83. Write the number in scientific notation.

Relative density of hydrogen: 0.0000899.

Solution:

$$0.0000899 = 8.99 \times 10^{-5}$$

87. Write the number in decimal form.

Charge of electron: 4.8×10^{-10} electrostatic units.

Solution:

$$4.8 \times 10^{-10} = 0.00000000048$$

89. Use a calculator to evaluate the following. Round your answers to three decimal places.

(a) $750\left(1+\dfrac{0.11}{365}\right)^{800}$

(b) $\dfrac{67{,}000{,}000+93{,}000{,}000}{0.0052}$

Solution:

(a) $750\left(1+\dfrac{0.11}{365}\right)^{800} \approx 954.448$

750 $\boxed{\times}$ $\boxed{(}$ 1 $\boxed{+}$ .11 $\boxed{\div}$ 365 $\boxed{)}$ $\boxed{y^x}$ 800 $\boxed{=}$

(b) $\dfrac{67{,}000{,}000+93{,}000{,}000}{0.0052} \approx 3.077 \times 10^{10}$

$\boxed{(}$ 67000000 $\boxed{+}$ 93000000 $\boxed{)}$ $\boxed{\div}$ 0.0052 $\boxed{=}$

91. The speed of light is 11,160,000 miles per minute. The distance from the sun to the earth is 93,000,000 miles. Find the time it takes for light to travel from the sun to the earth.

Solution:

$$\frac{93{,}000{,}000 \text{ miles}}{11{,}160{,}000 \text{ miles/minute}} = 8\tfrac{1}{3} \text{ minutes}$$

95. Write the expression that corresponds to the following keystrokes.

5 $\boxed{\times}$ $\boxed{(}$ 2.7 $\boxed{-}$ 9.4 $\boxed{)}$ $\boxed{=}$

Solution:

5 $\boxed{\times}$ $\boxed{(}$ 2.7 $\boxed{-}$ 9.4 $\boxed{)}$ $\boxed{=}$ corresponds to $5(2.7-9.4)$.

SECTION 1.3

Radicals and Rational Exponents

- You should know the properties of radicals.
  - (a) $\sqrt[n]{a^m} = (\sqrt[n]{a})^m$
  - (b) $\sqrt[n]{a} \cdot \sqrt[n]{b} = \sqrt[n]{ab}$
  - (c) $\dfrac{\sqrt[n]{a}}{\sqrt[n]{b}} = \sqrt[n]{\dfrac{a}{b}}$
  - (d) $\sqrt[m]{\sqrt[n]{a}} = \sqrt[mn]{a}$
  - (e) $(\sqrt[n]{a})^n = a$
  - (f) For n even, $\sqrt[n]{a^n} = |a|$
For n odd, $\sqrt[n]{a^n} = a$
  - (g) $a^{1/n} = \sqrt[n]{a}$
  - (h) $a^{m/n} = (\sqrt[n]{a})^m = \sqrt[n]{a^m}$
- You should be able to simplify radicals.
  - (a) All possible factors have been removed from the radical sign.
  - (b) All fractions have radical-free denominators.
  - (c) The index for the radical has been reduced as far as possible.
- You should be able to use your calculator to evaluate radicals.

Solutions to Selected Exercises

3. Find the radical form for $32^{1/5} = 2$.

Solution:

Radical form: $\sqrt[5]{32} = 2$

7. Find the rational exponent form for

$$\sqrt[3]{-216} = -6.$$

Solution:

Rational exponent form: $(-216)^{1/3} = -6$

11. Find the rational exponent form for

$$\sqrt[4]{81^3} = 27.$$

Solution:

Rational exponent form: $(81)^{3/4} = 27$

15. Evaluate $\sqrt[3]{8}$ without using a calculator.

Solution:

$$\sqrt[3]{8} = \sqrt[3]{2^3} = 2$$

17. Evaluate $-\sqrt[3]{-27}$ without using a calculator.

Solution:

$$-\sqrt[3]{-27} = -\sqrt[3]{(-3)^3} = -(-3) = 3$$

19. Evaluate $4/\sqrt{64}$ without using a calculator.

Solution:

$$\frac{4}{\sqrt{64}} = \frac{4}{8} = \frac{1}{2}$$

23. Evaluate $36^{3/2}$ without using a calculator.

Solution:

$$36^{3/2} = (\sqrt{36})^3 = (6)^3 = 216$$

27. Evaluate $\left(\frac{16}{81}\right)^{-3/4}$ without using a calculator.

Solution:

$$\left(\frac{16}{81}\right)^{-3/4} = \left(\frac{81}{16}\right)^{3/4} = \left(\sqrt[4]{\frac{81}{16}}\right)^3 = \left(\frac{3}{2}\right)^3 = \frac{27}{8}$$

31. Simplify $\sqrt{8}$ by removing all possible factors from the radical.

Solution:

$$\sqrt{8} = \sqrt{4 \cdot 2} = \sqrt{4}\sqrt{2} = 2\sqrt{2}$$

35. Simplify $\sqrt{72x^3}$ by removing all possible factors from the radical.

Solution:

$$\sqrt{72x^3} = \sqrt{36 \cdot 2 \cdot x^2 \cdot x} = 6x\sqrt{2x}$$

41. Simplify $\sqrt{75x^2y^{-4}}$ by removing all possible factors from the radical.

Solution:

$$\sqrt{75x^2y^{-4}} = \sqrt{\frac{75x^2}{y^4}} = \sqrt{\frac{25 \cdot 3x^2}{y^4}} = \frac{5|x|\sqrt{3}}{y^2}$$

45. Rewrite $8/\sqrt[3]{2}$ by rationalizing the denominator. Simplify your answer.

Solution:

$$\frac{8}{\sqrt[3]{2}} = \frac{8}{\sqrt[3]{2}} \cdot \frac{\sqrt[3]{2^2}}{\sqrt[3]{2^2}} = \frac{8\sqrt[3]{2^2}}{\sqrt[3]{2^3}} = \frac{8\sqrt[3]{4}}{2} = 4\sqrt[3]{4}$$

49. Rewrite the following by rationalizing the denominator. Simplify your answer.

$$\frac{3}{\sqrt{5}+\sqrt{6}}$$

Solution:

$$\frac{3}{\sqrt{5}+\sqrt{6}} = \frac{3}{\sqrt{5}+\sqrt{6}} \cdot \frac{\sqrt{5}-\sqrt{6}}{\sqrt{5}-\sqrt{6}} = \frac{3(\sqrt{5}-\sqrt{6})}{5-6} = -3(\sqrt{5}-\sqrt{6}) = 3(\sqrt{6}-\sqrt{5})$$

53. Rewrite the following by rationalizing the numerator. Simplify your answer.

$$\frac{\sqrt{5}+\sqrt{3}}{3}$$

Solution:

$$\frac{\sqrt{5}+\sqrt{3}}{3} = \frac{\sqrt{5}+\sqrt{3}}{3} \cdot \frac{\sqrt{5}-\sqrt{3}}{\sqrt{5}-\sqrt{3}} = \frac{5-3}{3(\sqrt{5}-\sqrt{3})} = \frac{2}{3(\sqrt{5}-\sqrt{3})}$$

59. Reduce the index of $\sqrt[6]{(x+1)^4}$.

Solution:

$$\sqrt[6]{(x+1)^4} = (x+1)^{4/6}$$
$$= (x+1)^{2/3} = \sqrt[3]{(x+1)^2}$$

63. Write $\sqrt{\sqrt[4]{2x}}$ as a single radical.

Solution:

$$\sqrt{\sqrt[4]{2x}} = \left((2x)^{1/4}\right)^{1/2}$$
$$= (2x)^{1/8} = \sqrt[8]{2x}$$

67. Combine and simplify $2\sqrt{50}+12\sqrt{8}$.

Solution:

$$2\sqrt{50}+12\sqrt{8} = 2\sqrt{25 \cdot 2}+12\sqrt{4 \cdot 2} = 2(5)\sqrt{2}+12(2)\sqrt{2} = 10\sqrt{2}+24\sqrt{2} = 34\sqrt{2}$$

71. Simplify $5^{4/3} \cdot 5^{8/3}$.

Solution:

$$5^{4/3} \cdot 5^{8/3} = 5^{4/3\,+\,8/3} = 5^{12/3} = 5^4 = 625$$

75. Simplify $\dfrac{x^{-3}x^{1/2}}{x^{3/2}x^{-1}}$, $x > 0$.

Solution:

$$\frac{x^{-3}x^{1/2}}{x^{3/2}x^{-1}} = \frac{x^{1/2}x^{1}}{x^{3}x^{3/2}} = \frac{x^{3/2}}{x^{9/2}} = \frac{1}{x^{9/2\,-\,3/2}} = \frac{1}{x^{6/2}} = \frac{1}{x^3}$$

79. Use a calculator to approximate $\sqrt[6]{125}$. Round your answer to three decimal places.

Solution:

$$\sqrt[6]{125} \approx 2.236$$

125 [y^x] [(] 1 [÷] 6 [)] [=]

Note: You may get an error if you raise a negative number to a power on your calculator. To use the [y^x] key when your base is negative, omit the sign and then give the result the appropriate sign.

85. Fill in the blank with <, >, or = by finding the decimal approximation.

$5 \ \_\_\_\_ \ \sqrt{3^2 + 2^2}$

Solution:

$$\sqrt{3^2 + 2^2} = \sqrt{9 + 4} = \sqrt{13} \approx 3.6056$$

Therefore, $5 > \sqrt{3^2 + 2^2}$.

89. Find the annual depreciation rate r by the <u>declining balances method</u> using the formula

$$r = 1 - \left(\frac{S}{C}\right)^{1/n}$$

where $n = 8$ years, $S = \$1500$ and $C = \$10{,}400$.

Solution:

$$r = 1 - \left(\frac{1500}{10{,}400}\right)^{1/8} \approx 0.215 = 21.5\%$$

1 [−] [(] 1500 [÷] 10,400 [)] [y^x] [(] 1 [÷] 8 [)] [=]

91. Find the dimensions of a cube that has a volume of 13,824 cubic inches.

Solution:

$$\text{Volume} = x^3$$
$$13{,}824 = x^3$$
$$\sqrt[3]{13{,}824} = x$$
$$x = 24$$

The dimensions are 24 in. × 24 in. × 24 in.

95. Enter any positive real number in your calculator and repeatedly take the square root. What real number does the display appear to be approaching?

Solution:

$$x^{1/2 \cdot 1/2 \cdot 1/2 \cdot 1/2 \cdot 1/2 \cdots} \approx 1$$

SECTION 1.4

Polynomials and Special Products

- Given a polynomial in x, $a_nx^n + a_{n-1}x^{n-1} + \ldots + a_1x + a_0$, where $a_n \neq 0$, you should be able to identify the following:

  (a) Degree: n
  (b) Terms: a_nx^n, $a_{n-1}x^{n-1}$, $\ldots$, a_1x, a_0
  (c) Coefficients: a_n, a_{n-1}, $\ldots$, a_1, a_0
  (d) Leading coefficient: a_n
  (e) Constant term: $a.$

- You should be able to add and subtract polynomials.

- You should be able to multiply polynomials by either

  (a) The Distributive Law or
  (b) The Vertical Method.

- You should know the special binomial products.

  (a) $(ax + b)(cx + d) = acx^2 + adx + bcx + bd$ FOIL
  $\qquad = acx^2 + (ad + bc)x + bd$
  (b) $(u \pm v)^2 = u^2 \pm 2uv + v^2$
  (c) $(u + v)(u - v) = u^2 - v^2$
  (d) $(u \pm v)^3 = u^3 \pm 3u^2v + 3uv^2 \pm v^3$

Solutions to Selected Exercises

5. Find the degree and the leading coefficient of $4x^5 + 6x^4 - x - 1$.

Solution:

For $4x^5 + 6x^4 - x - 1$, we have the following.

Degree: 5 Leading coefficient: 4

9. Determine whether $\dfrac{3x+4}{x}$ is a polynomial. If it is, write the polynomial in standard form.

Solution:

$$\frac{3x+4}{x} = \frac{3x}{x} + \frac{4}{x} = 3 + 4x^{-1}$$

This is not a polynomial because of the negative exponent.

13. Simplify $(6x+5)-(8x+15)$.

Solution:

$$(6x+5)-(8x+15) = 6x+5-8x-15 = 6x-8x+5-15 = -2x-10$$

17. Simplify $(15x^2-6)-(-8x^3-14x^2-17)$.

Solution:

$$(15x^2-6)-(-8x^3-14x^2-17) = 15x^2-6+8x^3+14x^2+17 = 8x^3+29x^2+11$$

19. Simplify $5z-[3z-(10z+8)]$.

Solution:

$$\begin{aligned} 5z-[3z-(10z+8)] &= 5z-[3z-10z-8] \\ &= 5z-[-7z-8] \\ &= 5z+7z+8 = 12z+8 \end{aligned}$$

25. Simplify $(-2x)(-3x)(5x+2)$.

Solution:

$$(-2x)(-3x)(5x+2) = 6x^2(5x+2) = 6x^2(5x)+6x^2(2) = 30x^3+12x^2$$

29. Find the product.

$$(3x-5)(2x+1)$$

Solution:

$$(3x-5)(2x+1) = 6x^2+3x-10x-5 = 6x^2-7x-5$$

33. Find the product.

$$(2x-5y)^2$$

Solution:

$$(2x-5y)^2 = (2x)^2-2(2x)(5y)+(5y)^2 = 4x^2-20xy+25y^2$$

35. Find the product.

$$[(x-3)+y]^2$$

Solution:

$$\begin{aligned} [(x-3)+y]^2 &= (x-3)^2+2(x-3)y+y^2 \\ &= x^2-6x+9+2xy-6y+y^2 \\ &= x^2+2xy+y^2-6x-6y+9 \end{aligned}$$

41. Find the product.

$$(m-3+n)(m-3-n)$$

Solution:

$$\begin{aligned}(m-3+n)(m-3-n) &= [(m-3)+n][(m-3)-n] \\ &= (m-3)^2 - n^2 \\ &= m^2 - 6m + 9 - n^2 = m^2 - n^2 - 6m + 9\end{aligned}$$

47. Find the product.

$$(2x-y)^3$$

Solution:

$$(2x-y)^3 = (2x)^3 - 3(2x)^2y + 3(2x)y^2 - y^3 = 8x^3 - 12x^2y + 6xy^2 - y^3$$

51. Find the product.

$$(4x^3-3)^2$$

Solution:

$$(4x^3-3)^2 = (4x^3)^2 - 2(4x^3)(3) + (3)^2 = 16x^6 - 24x^3 + 9$$

55. Find the product.

$$(x^2-x+1)(x^2+x+1)$$

Solution:

By the Vertical Method we have

$$\begin{array}{ccccc} x^2 & -\,x & +\,1 & & \\ x^2 & +\,x & +\,1 & & \\ \hline x^4 & -\,x^3 & +\,x^2 & & \\ & x^3 & -\,x^2 & +\,x & \\ & & x^2 & -\,x & +\,1 \\ \hline x^4 & +\,0x^3 & +\,x^2 & +\,0x & +\,1 \end{array}$$

which equals $x^4 + x^2 + 1$

59. Find the product.

$$(x+\sqrt{5})(x-\sqrt{5})(x+4)$$

Solution:

$$\begin{aligned}(x+\sqrt{5})(x-\sqrt{5})(x+4) &= (x^2-5)(x+4) && \text{Special Product} \\ &= x^3 + 4x^2 - 5x - 20 && \text{FOIL Method}\end{aligned}$$

63. An open box is made by cutting squares out of the corners of a piece of metal that is 18 inches by 26 inches (see figure). If the edge of each cut-out square is x inches, what is the volume of the box? Find the volume when $x = 1$, $x = 2$, and $x = 3$.

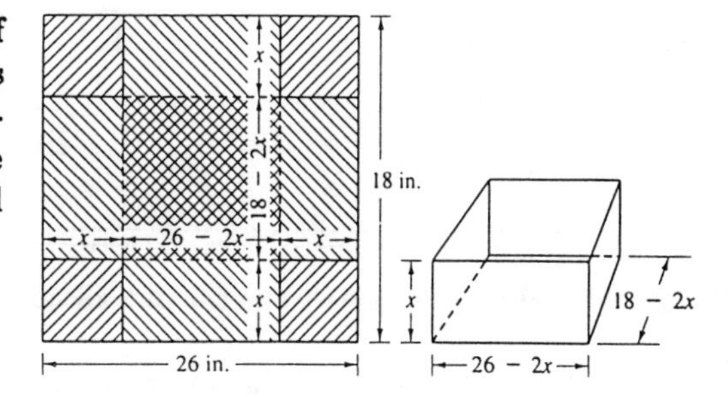

Solution:

Volume = (length)(width)(height)

$$V = (26 - 2x)(18 - 2x)x = (468 - 88x + 4x^2)x = 4x^3 - 88x^2 + 468x$$

When $x = 1 : V = 4(1)^3 - 88(1)^2 + 468(1) = 384$ cubic inches

When $x = 2 : V = 4(2)^3 - 88(2)^2 + 468(2) = 616$ cubic inches

When $x = 3 : V = 4(3)^3 - 88(3)^2 + 468(3) = 720$ cubic inches

67. The stopping distance of an automobile is the distance traveled during the driver's reaction time plus the distance traveled after the brakes are applied. In an experiment these distances were measured (in feet) when the automobile was traveling at a speed of x miles per hour (see figure). The distance traveled during the reaction time is $R = 1.1x$ and the braking distance is $B = 0.14x^2 - 4.43x + 58.40$. Determine the polynomial that represents the total stopping distance. Use this polynomial to estimate the total stopping distance when $x = 30$, $x = 40$ and $x = 55$.

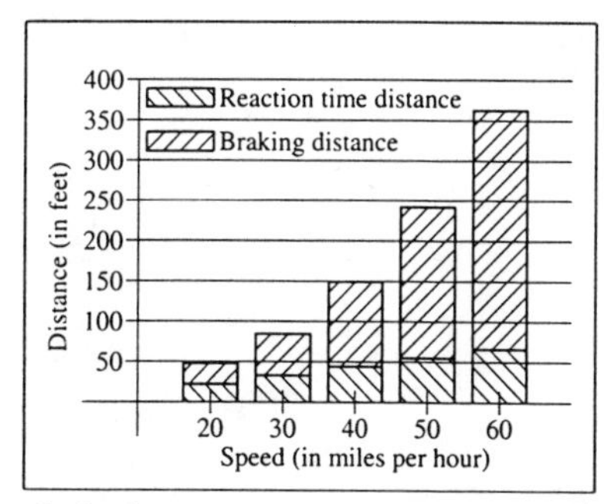

Solution:

Stopping distance = Reaction time distance + Braking distance

$$S = R + B = 1.1x + (0.14x^2 - 4.43x + 58.40) = 0.14x^2 - 3.33x + 58.40$$

When $x = 30 :$ $S = 84.5$ feet

When $x = 40 :$ $S = 149.2$ feet

When $x = 55 :$ $S = 298.75$ feet

SECTION 1.5

Factoring

- You should be able to factor out all common factors, the first step in factoring.
- You should be able to factor the following special polynomial forms.

  (a) $u^2 - v^2 = (u + v)(u - v)$
  (b) $u^2 \pm 2uv + v^2 = (u \pm v)^2$
  (c) $mx^2 + nx + r = (ax + b)(cx + d)$, where $m = ac,\ r = bd,\ n = ad + bc$
  Note: Not all trinomials can be factored (using real coefficients).
  (d) $u^3 \pm v^3 = (u \pm v)(u^2 \mp uv + v^2)$
- You should be able to factor by grouping.

Solutions to Selected Exercises

5. Remove the common factor: $(x - 1)^2 + 6(x - 1)$.

Solution:

$$(x - 1)^2 + 6(x - 1) = (x - 1)[(x - 1) + 6] = (x - 1)(x + 5)$$

9. Factor $16y^2 - 9$.

Solution:

$$16y^2 - 9 = (4y)^2 - (3)^2 = (4y + 3)(4y - 3)$$

11. Factor $(x - 1)^2 - 4$.

Solution:

$$(x - 1)^2 - 4 = [(x - 1) + 2][(x - 1) - 2] = (x + 1)(x - 3)$$

15. Factor $4t^2 + 4t + 1$.

Solution:

$$4t^2 + 4t + 1 = (2t)^2 + 2(2t)(1) + (1)^2 = (2t + 1)^2$$

21. Factor $s^2 - 5s + 6$.

Solution:

$$s^2 - 5s + 6 = (s-2)(s-3) \text{ since } (-2)(-3) = 6 \text{ and } (-2) + (-3) = -5$$

25. Factor $x^2 - 30x + 200$.

Solution:

$$x^2 - 30x + 200 = (x-10)(x-20) \text{ since } (-10)(-20) = 200 \text{ and } (-10) + (-20) = -30$$

29. Factor $9z^2 - 3z - 2$.

Solution:

$$9z^2 - 3z - 2 = (3z+1)(3z-2)$$

35. Factor $y^3 + 64$.

Solution:

$$\begin{aligned} y^3 + 64 &= y^3 + 4^3 \\ &= (y+4)(y^2 - 4y + 16) \end{aligned}$$

37. Factor $8t^3 - 1$.

Solution:

$$\begin{aligned} 8t^3 - 1 &= (2t)^3 - (1)^3 \\ &= (2t-1)(4t^2 + 2t + 1) \end{aligned}$$

39. Factor $x^3 - x^2 + 2x - 2$ by grouping.

Solution:

$$\begin{aligned} x^3 - x^2 + 2x - 2 &= x^2(x-1) + 2(x-1) \\ &= (x-1)(x^2+2) \end{aligned}$$

43. Factor $6 + 2x - 3x^3 - x^4$ by grouping.

Solution:

$$6 + 2x - 3x^3 - x^4 = 2(3+x) - x^3(3+x) = (3+x)(2-x^3)$$

47. Factor $6x^2 + x - 2$ by grouping.

Solution:

We have $a = 6$ and $c = -2$ which implies that the product $ac = -12$. Now, since $-12 = 4(-3)$ and $4 - 3 = 1 = b$, we rewrite the middle term as $x = 4x - 3x$. Thus, we have

$$\begin{aligned} 6x^2 + x - 2 &= 6x^2 + 4x - 3x - 2 \\ &= (6x^2 - 3x) + (4x - 2) \\ &= 3x(2x-1) + 2(2x-1) = (2x-1)(3x+2) \end{aligned}$$

49. Factor $15x^2 - 11x + 2$ by grouping.

Solution:

We have $a = 15$ and $c = 2$ which implies that the product $ac = 30$. Now, since $30 = (-6)(-5)$ and $-6 - 5 = -11 = b$, we rewrite the middle term as $-11x = -6x - 5x$. Thus, we have

$$\begin{aligned} 15x^2 - 11x + 2 &= 15x^2 - 6x - 5x + 2 \\ &= (15x^2 - 6x) - (5x - 2) \\ &= 3x(5x - 2) - (5x - 2) = (5x - 2)(3x - 1) \end{aligned}$$

51. Completely factor $x^3 - 4x^2$.

Solution:

$$x^3 - 4x^2 = x^2(x - 4)$$

53. Completely factor $1 - 4x + 4x^2$.

Solution:

$$1 - 4x + 4x^2 = (1 - 2x)^2$$

57. Completely factor $9x^2 + 10x + 1$.

Solution:

$$9x^2 + 10x + 1 = (9x + 1)(x + 1)$$

61. Completely factor $x^4 - 4x^3 + x^2 - 4x$.

Solution:

$$\begin{aligned} x^4 - 4x^3 + x^2 - 4x &= x^3(x - 4) + x(x - 4) \\ &= (x - 4)(x^3 + x) \\ &= (x - 4)x(x^2 + 1) \\ &= x(x - 4)(x^2 + 1) \end{aligned}$$

65. Completely factor $(x^2 + 1)^2 - 4x^2$.

Solution:

$$\begin{aligned} (x^2 + 1)^2 - 4x^2 &= [(x^2 + 1) + 2x][(x^2 + 1) - 2x] \\ &= (x^2 + 2x + 1)(x^2 - 2x + 1) = (x + 1)^2(x - 1)^2 \end{aligned}$$

69. Completely factor $4x(2x - 1) + (2x - 1)^2$.

Solution:

$$4x(2x - 1) + (2x - 1)^2 = (2x - 1)[4x + (2x - 1)] = (2x - 1)(6x - 1)$$

71. Completely factor $2(x + 1)(x - 3)^2 - 3(x + 1)^2(x - 3)$.

Solution:

$$\begin{aligned} 2(x + 1)(x - 3)^2 - 3(x + 1)^2(x - 3) &= (x + 1)(x - 3)[2(x - 3) - 3(x + 1)] \\ &= (x + 1)(x - 3)[2x - 6 - 3x - 3] \\ &= (x + 1)(x - 3)(-x - 9) \\ &= -(x + 1)(x - 3)(x + 9) \end{aligned}$$

75. Completely factor $2x(x-5)^4 - x^2(4)(x-5)^3$.

Solution:

$$\begin{aligned}2x(x-5)^4 - x^2(4)(x-5)^3 &= 2x(x-5)^3[(x-5) - x(2)]\\ &= 2x(x-5)^3[x-5-2x]\\ &= 2x(x-5)^3(-x-5) = -2x(x+5)(x-5)^3\end{aligned}$$

79. Completely factor $x^2(x^2+1)^{-5} - (x^2+1)^{-4}$ and express the answer with no negative exponents.

Solution:

$$\begin{aligned}x^2(x^2+1)^{-5} - (x^2+1)^{-4} &= (x^2+1)^{-5}[x^2 - (x^2+1)]\\ &= (x^2+1)^{-5}[x^2 - x^2 - 1]\\ &= -(x^2+1)^{-5} = -\frac{1}{(x^2+1)^5}\end{aligned}$$

85. Match $a^2 + 2a + 1$ with the correct factoring formula.

Solution:

$a^2 + 2a + 1 = (a+1)^2$ matches model (a)

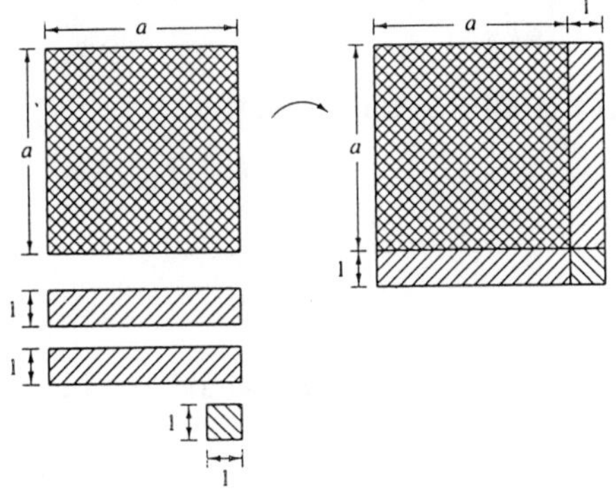

89. Make a "geometric factoring model" to represent $2x^2 + 7x + 3 = (2x+1)(x+3)$.

Solution:

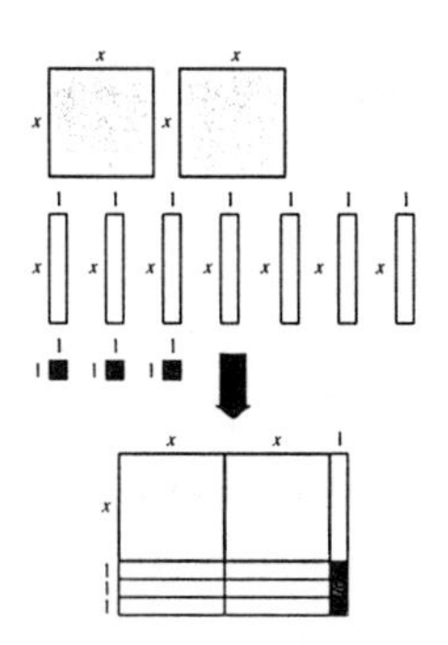

SECTION 1.6

Fractional Expressions

- You should know that a rational expression is the quotient of two polynomials.
- You should be able to simplify rational expressions by reducing them to lowest terms. This may involve factoring both the numerator and the denominator.
- You should be able to add, subtract, multiply, and divide rational expressions.
- You should be able to simplify compound fractions.

Solutions to Selected Exercises

1. Find the domain of $3x^2 - 4x + 7$.

Solution:

The domain of the **polynomial** $3x^2 - 4x + 7$ is the set of all real numbers.

5. Find the domain of $\dfrac{1}{x-2}$.

Solution:

The domain of $\dfrac{1}{x-2}$ is the set of all real numbers except $x = 2$, which would produce an undefined division by zero.

9. Find the domain of $\sqrt{x+1}$.

Solution:

The domain of $\sqrt{x+1}$ is the set of real numbers greater than or equal to -1 since $x + 1 \geq 0$ when $x \geq -1$.

13. Find the factor that makes the two fractions equivalent.

$$\frac{x+1}{x} = \frac{(x+1)(\qquad)}{x(x-2)}$$

Solution:

$$\frac{x+1}{x} = \frac{x+1}{x} \cdot \frac{x-2}{x-2} = \frac{(x+1)(x-2)}{x(x-2)}, \quad x \neq 2$$

19. Reduce to lowest terms.

$$\frac{3xy}{xy+x}$$

Solution:

$$\frac{3xy}{xy+x} = \frac{3xy}{x(y+1)}$$
$$= \frac{3y}{y+1}, \quad x \neq 0$$

23. Reduce to lowest terms.

$$\frac{x^3+5x^2+6x}{x^2-4}$$

Solution:

$$\frac{x^3+5x^2+6x}{x^2-4} = \frac{x(x+2)(x+3)}{(x+2)(x-2)}$$
$$= \frac{x(x+3)}{x-2}, \quad x \neq -2$$

27. Reduce to lowest terms.

$$\frac{2-x+2x^2-x^3}{x-2}$$

Solution:

$$\frac{2-x+2x^2-x^3}{x-2} = \frac{(2-x)+x^2(2-x)}{x-2}$$
$$= \frac{(2-x)(1+x^2)}{x-2}$$
$$= \frac{-(x-2)(x^2+1)}{x-2} = -(x^2+1), \quad x \neq 2$$

29. Reduce to lowest terms.

$$\frac{z^3-8}{z^2+2z+4}$$

Solution:

$$\frac{z^3-8}{z^2+2z+4} = \frac{(z-2)(z^2+2z+4)}{z^2+2z+4} = z-2$$

33. Simplify

$$\frac{(x-9)(x+7)}{x+1} \cdot \frac{x}{9-x}.$$

Solution:

$$\frac{(x-9)(x+7)}{x+1} \cdot \frac{x}{9-x} = -\frac{(9-x)(x+7)x}{(x+1)(9-x)} = -\frac{x(x+7)}{x+1}, \quad x \neq 9$$

37. Simplify

$$\frac{t^2-t-6}{t^2+6t+9}\cdot\frac{t+3}{t^2-4}.$$

Solution:

$$\frac{t^2-t-6}{t^2+6t+9}\cdot\frac{t+3}{t^2-4}=\frac{(t+2)(t-3)}{(t+3)(t+3)}\cdot\frac{t+3}{(t+2)(t-2)}=\frac{t-3}{(t+3)(t-2)},\quad t\neq-2$$

39. Simplify

$$\frac{x^2+xy-2y^2}{x^3+x^2y}\cdot\frac{x}{x^2+3xy+2y^2}.$$

Solution:

$$\begin{aligned}\frac{x^2+xy-2y^2}{x^3+x^2y}\cdot\frac{x}{x^2+3xy+2y^2}&=\frac{(x+2y)(x-y)}{x^2(x+y)}\cdot\frac{x}{(x+2y)(x+y)}\\&=\frac{x-y}{x(x+y)^2},\quad x\neq-2y\end{aligned}$$

43. Simplify

$$\frac{\left(\dfrac{x^2}{(x+1)^2}\right)}{\left(\dfrac{x}{(x+1)^3}\right)}.$$

Solution:

$$\frac{\left(\dfrac{x^2}{(x+1)^2}\right)}{\left(\dfrac{x}{(x+1)^3}\right)}=\frac{x^2}{(x+1)^2}\div\frac{x}{(x+1)^3}=\frac{x^2}{(x+1)^2}\cdot\frac{(x+1)^3}{x}=x(x+1),\quad x\neq0,\ -1$$

47. Simplify

$$6-\frac{5}{x+3}.$$

Solution:

$$6-\frac{5}{x+3}=\frac{6(x+3)}{x+3}-\frac{5}{x+3}=\frac{6(x+3)-5}{x+3}=\frac{6x+18-5}{x+3}=\frac{6x+13}{x+3}$$

53. Simplify

$$\frac{1}{x^2 - x - 2} - \frac{x}{x^2 - 5x + 6}.$$

Solution:

$$\begin{aligned}\frac{1}{x^2 - x - 2} - \frac{x}{x^2 - 5x + 6} &= \frac{1}{(x-2)(x+1)} - \frac{x}{(x-2)(x-3)}\\ &= \frac{(x-3) - x(x+1)}{(x+1)(x-2)(x-3)}\\ &= \frac{-x^2 - 3}{(x+1)(x-2)(x-3)} = -\frac{x^2+3}{(x+1)(x-2)(x-3)}\end{aligned}$$

57. Simplify

$$\frac{\left(\frac{x}{2} - 1\right)}{(x-2)}.$$

Solution:

$$\frac{\left(\frac{x}{2} - 1\right)}{(x-2)} = \frac{\left(\frac{x}{2} - 1\right)}{(x-2)} \cdot \frac{2}{2} = \frac{(x-2)}{2(x-2)} = \frac{1}{2}, \quad x \neq 2$$

61. Simplify

$$\frac{\left(\frac{x+3}{x-3}\right)^2}{\frac{1}{x+3} + \frac{1}{x-3}}.$$

Solution:

(a) Combining method:

$$\begin{aligned}\frac{\left(\frac{x+3}{x-3}\right)^2}{\frac{1}{x+3} + \frac{1}{x-3}} &= \frac{\frac{(x+3)^2}{(x-3)^2}}{\frac{(x-3)+(x+3)}{(x+3)(x-3)}}\\ &= \frac{(x+3)^2}{(x-3)^2} \cdot \frac{(x+3)(x-3)}{2x} = \frac{(x+3)^3}{2x(x-3)}\end{aligned}$$

–CONTINUED ON NEXT PAGE–

61. --CONTINUED--

(b) LCD method:

$$\frac{\dfrac{(x+3)^2}{(x-3)^2}}{\dfrac{1}{x+3}+\dfrac{1}{x-3}} \cdot \frac{(x+3)(x-3)^2}{(x+3)(x-3)^2} = \frac{(x+3)^3}{(x-3)^2+(x+3)(x-3)}$$

$$= \frac{(x+3)^3}{(x^2-6x+9)+(x^2-9)}$$

$$= \frac{(x+3)^3}{2x^2-6x}$$

$$= \frac{(x+3)^3}{2x(x-3)}$$

65. Simplify

$$\frac{\left(\sqrt{x}-\dfrac{1}{2\sqrt{x}}\right)}{\sqrt{x}}.$$

Solution:

$$\frac{\left(\sqrt{x}-\dfrac{1}{2\sqrt{x}}\right)}{\sqrt{x}} = \frac{\left(\sqrt{x}-\dfrac{1}{2\sqrt{x}}\right)}{\sqrt{x}} \cdot \frac{2\sqrt{x}}{2\sqrt{x}} = \frac{2x-1}{2x}, \quad x>0$$

67. Simplify

$$\frac{\dfrac{t^2}{\sqrt{t^2+1}}-\sqrt{t^2+1}}{t^2}.$$

Solution:

$$\frac{\dfrac{t^2}{\sqrt{t^2+1}}-\sqrt{t^2+1}}{t^2} = \frac{\dfrac{t^2}{\sqrt{t^2+1}}-\sqrt{t^2+1}}{t^2} \cdot \frac{\sqrt{t^2+1}}{\sqrt{t^2+1}}$$

$$= \frac{t^2-(t^2+1)}{t^2\sqrt{t^2+1}} = -\frac{1}{t^2\sqrt{t^2+1}}$$

71. Rationalize the numerator of

$$\frac{\sqrt{x+2}-\sqrt{x}}{2}.$$

Solution:

$$\frac{\sqrt{x+2}-\sqrt{x}}{2} = \frac{\sqrt{x+2}-\sqrt{x}}{2} \cdot \frac{\sqrt{x+2}+\sqrt{x}}{\sqrt{x+2}+\sqrt{x}}$$

$$= \frac{(x+2)-x}{2(\sqrt{x+2}+\sqrt{x})} = \frac{2}{2(\sqrt{x+2}+\sqrt{x})}$$

$$= \frac{1}{\sqrt{x+2}+\sqrt{x}}$$

75. Find the average of $\frac{x}{3}$ and $\frac{2x}{5}$.

Solution:

$$\text{Average} = \frac{\left(\frac{x}{3}+\frac{2x}{5}\right)}{2} = \frac{\left(\frac{x}{3}+\frac{2x}{5}\right)}{2} \cdot \frac{15}{15}$$

$$= \frac{5x+6x}{30} = \frac{11x}{30}$$

79. When two resistors are connected in parallel, the total resistance is given by

$$\frac{1}{\frac{1}{R_1}+\frac{1}{R_2}}.$$

Simplify this compound fraction.

Solution:

$$\frac{1}{\frac{1}{R_1}+\frac{1}{R_2}} = \frac{1}{\frac{1}{R_1}+\frac{1}{R_2}} \cdot \frac{R_1R_2}{R_1R_2} = \frac{R_1R_2}{R_2+R_1}$$

SECTION 1.7

Algebraic Errors and Some Algebra of Calculus

- You should be able to recognize and avoid the common algebraic errors listed in this section.
- You should be able to "unsimplify" algebraic expressions by the following methods:

  (a) Unusual Factoring
  (b) Inserting Factors or Terms
  (c) Rewriting with Negative Exponents
  (d) Writing a Fraction as a Sum of Terms.

Solutions to Selected Exercises

3. Find and correct any errors in $5z + 3(x - 2) = 5z + 3x - 2$.

Solution:

$$\begin{aligned} 5z + 3(x-2) &= 5z + 3x - 6 \qquad \text{By the Distributive Law} \\ &\neq 5z + 3x - 2 \end{aligned}$$

5. Find and correct any errors in $-\dfrac{x-3}{x-1} = \dfrac{3-x}{1-x}$.

Solution:

$$\begin{aligned} -\frac{x-3}{x-1} &= \frac{-(x-3)}{x-1} \\ &= \frac{3-x}{x-1} \qquad \text{Only the numerator is multiplied by } (-1). \\ &\neq \frac{3-x}{1-x} \end{aligned}$$

11. Find and correct any errors in $\sqrt{x+9} = \sqrt{x} + 3$.

Solution:

$$\sqrt{x+9} \neq \sqrt{x} + 3$$

The root of a sum does not equal the sum of the roots.

$$\sqrt{x} + 3 = \sqrt{(\sqrt{x}+3)^2} = \sqrt{x + 6\sqrt{x} + 9} \neq \sqrt{x+9}$$

$\sqrt{x+9}$ cannot be simplified.

15. Find and correct any errors in

$$\frac{1}{x+y^{-1}} = \frac{y}{x+1}.$$

Solution:

$$\frac{1}{x+y^{-1}} = \frac{1}{x+(1/y)} \bullet \frac{y}{y} = \frac{y}{xy+1} \neq \frac{y}{x+1}$$

19. Find and correct any errors in $\sqrt[3]{x^3+7x^2} = x^2\sqrt[3]{x+7}$.

Solution:

$$\sqrt[3]{x^3+7x^2} = \sqrt[3]{x^2(x+7)} = \sqrt[3]{x^2}\sqrt[3]{x+7} \neq x^2\sqrt[3]{x+7}$$

Radicals apply to every factor of the radicand.

23. Find and correct any errors in

$$\frac{1}{2y} = \left(\frac{1}{2}\right)y.$$

Solution:

$$\frac{1}{2y} = \frac{1}{2} \bullet \frac{1}{y} = \left(\frac{1}{2}\right)\frac{1}{y} \neq \left(\frac{1}{2}\right)y$$

Use the definition for multiplying fractions.

27. Insert the required factor in the parentheses.

$$\frac{1}{3}x^3 + 5 = (\qquad)(x^3+15)$$

Solution:

$$\frac{1}{3}x^3 + 5 = \frac{1}{3}x^3 + \frac{1}{3}(15) = \left(\frac{1}{3}\right)(x^3+15)$$

31. Insert the required factor in the parentheses.

$$\frac{1}{\sqrt{x}(1+\sqrt{x})^2} = (\quad)\frac{1}{(1+\sqrt{x})^2}\left(\frac{1}{2\sqrt{x}}\right)$$

Solution:

$$\frac{1}{\sqrt{x}(1+\sqrt{x})^2} = \frac{1}{\sqrt{x}} \bullet \frac{1}{(1+\sqrt{x})^2} = (2)\left(\frac{1}{2\sqrt{x}}\right)\frac{1}{(1+\sqrt{x})^2}$$

$$= (2)\frac{1}{(1+\sqrt{x})^2}\left(\frac{1}{2\sqrt{x}}\right)$$

35. Insert the required factor in the parentheses.

$$\frac{3}{x}+\frac{5}{2x^2}-\frac{3}{2}x=(\quad)(6x+5-3x^3)$$

Solution:

$$\frac{3}{x}+\frac{5}{2x^2}-\frac{3}{2}x=\frac{6x}{2x^2}+\frac{5}{2x^2}-\frac{3x^3}{2x^2}=\left(\frac{1}{2x^2}\right)(6x+5-3x^3)$$

37. Insert the required factor in the parentheses.

$$\frac{x^2}{1/12}-\frac{y^2}{2/3}=\frac{12x^2}{(\ )}-\frac{3y^2}{(\ )}$$

Solution:

$$\frac{x^2}{1/12}-\frac{y^2}{2/3}=x^2\left(\frac{12}{1}\right)-y^2\left(\frac{3}{2}\right)=\frac{12x^2}{1}-\frac{3y^2}{2}$$

41. Insert the required factor in the parentheses.

$$\frac{x^2}{\sqrt{x^2+1}}-\sqrt{x^2+1}=\frac{1}{\sqrt{x^2+1}}(\quad)$$

Solution:

$$\begin{aligned}\frac{x^2}{\sqrt{x^2+1}}-\sqrt{x^2+1}&=\frac{x^2}{\sqrt{x^2+1}}-\frac{\sqrt{x^2+1}}{1}\cdot\frac{\sqrt{x^2+1}}{\sqrt{x^2+1}}\\&=\frac{x^2-(x^2+1)}{\sqrt{x^2+1}}=\frac{-1}{\sqrt{x^2+1}}\\&=\frac{1}{\sqrt{x^2+1}}(-1)\end{aligned}$$

43. Insert the required factor in the parentheses.

$$\frac{1}{10}(2x+1)^{5/2}-\frac{1}{6}(2x+1)^{3/2}=\frac{(2x+1)^{3/2}}{15}(\quad)$$

Solution:

$$\begin{aligned}\frac{1}{10}(2x+1)^{5/2}-\frac{1}{6}(2x+1)^{3/2}&=\frac{3}{30}(2x+1)^{3/2}(2x+1)^1-\frac{5}{30}(2x+1)^{3/2}\\&=\frac{1}{30}(2x+1)^{3/2}[3(2x+1)-5]\\&=\frac{1}{30}(2x+1)^{3/2}(6x-2)\\&=\frac{1}{30}(2x+1)^{3/2}2(3x-1)\\&=\frac{1}{15}(2x+1)^{3/2}(3x-1)\end{aligned}$$

47. Write the following as a sum of two or more terms.

$$\frac{4x^3 - 7x^2 + 1}{x^{1/3}}$$

Solution:

$$\begin{aligned}\frac{4x^3 - 7x^2 + 1}{x^{1/3}} &= \frac{4x^3}{x^{1/3}} - \frac{7x^2}{x^{1/3}} + \frac{1}{x^{1/3}} \\ &= 4x^{3-1/3} - 7x^{2-1/3} + \frac{1}{x^{1/3}} \\ &= 4x^{8/3} - 7x^{5/3} + \frac{1}{x^{1/3}}\end{aligned}$$

51. Simplify the expression.

$$\frac{-2(x^2-3)^{-3}(2x)(6x+1)^3 - 3(6x+1)^2(6)(x^2-3)^{-2}}{[(6x+1)^3]^2}$$

Solution:

$$\begin{aligned}&\frac{-2(x^2-3)^{-3}(2x)(6x+1)^3 - 3(6x+1)^2(6)(x^2-3)^{-2}}{[(6x+1)^3]^2} \\ &\quad = \frac{2(x^2-3)^{-3}(6x+1)^2[-2x(6x+1) - 9(x^2-3)]}{(6x+1)^6} \\ &\quad = \frac{2[-12x^2 - 2x - 9x^2 + 27]}{(x^2-3)^3(6x+1)^4} \\ &\quad = \frac{2(-21x^2 - 2x + 27)}{(x^2-3)^3(6x+1)^4} \\ &\quad = \frac{-2(21x^2 + 2x - 27)}{(x^2-3)^3(6x+1)^4}\end{aligned}$$

55. Simplify the expression.

$$\frac{(3x+2)^{3/4}(2x+3)^{-2/3}(2) - (2x+3)^{1/3}(3x+2)^{-1/4}(3)}{[(3x+2)^{3/4}]^2}$$

Solution:

$$\begin{aligned}&\frac{(3x+2)^{3/4}(2x+3)^{-2/3}(2) - (2x+3)^{1/3}(3x+2)^{-1/4}(3)}{[(3x+2)^{3/4}]^2} \\ &\quad = \frac{(3x+2)^{-1/4}(2x+3)^{-2/3}[2(3x+2) - 3(2x+3)]}{(3x+2)^{6/4}} \\ &\quad = \frac{6x + 4 - 6x - 9}{(3x+2)^{1/4}(2x+3)^{2/3}(3x+2)^{6/4}} \\ &\quad = \frac{-5}{(2x+3)^{2/3}(3x+2)^{7/4}}\end{aligned}$$

REVIEW EXERCISES FOR CHAPTER 1

Solutions to Selected Exercises

1. Determine which numbers in the set $\left\{11,\ -14,\ -\frac{8}{9},\ \frac{5}{2},\ \sqrt{6},\ 0.4\right\}$ are (a) natural numbers, (b) integers, (c) rational numbers, and (d) irrational numbers.

Solution:

(a) Natural numbers: $\{11\}$

(b) Integers: $\{11,\ -14\}$

(c) Rational numbers: $\{11,\ -14,\ -\frac{8}{9},\ \frac{5}{2},\ 0.4\}$

(d) Irrational numbers: $\{\sqrt{6}\}$

5. Give a verbal description of the subset of real numbers that is represented by the inequality $x \le 7$, and sketch the subset on the real number line.

Solution:

The inequality $x \le 7$ denotes all real numbers that are less than or equal to 7.

9. Use absolute value notation to describe the statement "the distance between x and 7 is at least 4".

Solution:

$$d(x,\ 7) = |x - 7|$$

Since $d(x,\ 7)$ is at least 4, we have $|x - 7| \ge 4$.

11. Use absolute value notation to describe the statement "the distance between y and -30 is less than 5".

Solution:

$$d(y,\ -30) = |y - (-30)| = |y + 30|$$

Since $d(y,\ -30)$ is less than 5, we have $|y + 30| < 5$.

13. Perform the indicated operations for $|-3| + 4(-2) - 6$.

Solution:

$$|-3| + 4(-2) - 6 = 3 - 8 - 6 = -11$$

17. Perform the indicated operations for $6[4-2(6+8)]$.

Solution:

$$6[4-2(6+8)] = 6[4-2(14)] = 6[4-28] = 6[-24] = -144$$

19. Perform the indicated operations for $\left(\frac{3^2}{5^2}\right)^{-3}$.

Solution:

$$\left(\frac{3^2}{5^2}\right)^{-3} = \left(\frac{9}{25}\right)^{-3} = \left(\frac{25}{9}\right)^{3} = \frac{15{,}625}{729}$$

23. Perform the indicated operations for $(3 \times 10^4)^2$.

Solution:

$$(3 \times 10^4)^2 = 3^2 \times (10^4)^2 = 9 \times 10^8$$

27. Write the number in decimal form.

Distance between Sun and Jupiter: 4.883×10^8 miles.

Solution:

$4.833 \times 10^8 = 483{,}300{,}000$ miles

29. Use a calculator to evaluate the given expression. Round your answer to three decimal places.

(a) $1800(1+0.08)^{24}$ (b) $0.0024(7{,}658{,}400)$

Solution:

(a) $1800(1+0.08)^{24} \approx 11{,}414.125$

1800 $\boxed{\times}$ 1.08 $\boxed{y^x}$ 24 $\boxed{=}$

(b) $0.0024\ (7{,}658{,}400) = 18{,}380.160$

0.0024 $\boxed{\times}$ 7,658,400 $\boxed{=}$

33. Describe the *error* and make the necessary correction.

$4(\frac{3}{7}) = \frac{12}{28}$

Solution:

Use the definition for multiplying fractions.

$$4\left(\frac{3}{7}\right) = \frac{4}{1} \cdot \frac{3}{7} = \frac{12}{7} \neq \frac{12}{28}$$

35. Describe the *error* and make the necessary correction.

$$\frac{x-1}{1-x} = 1$$

Solution:

Cancel only common factors.

$$\frac{x-1}{1-x} = \frac{-(1-x)}{1-x} = -1 \neq 1$$

37. Describe the *error* and make the necessary correction for $(2x)^4 = 2x^4$.

Solution:

Apply the exponent to both factors in the parentheses.

$$(2x)^4 = 2^4x^4 = 16x^4 \neq 2x^4$$

39. Describe the *error* and make the necessary correction for $(3^4)^4 = 3^8$.

Solution:

Multiply the exponents.

$$(3^4)^4 = 3^{4\times 4} = 3^{16} \neq 3^8$$

45. Write the rational exponent form of $\sqrt{16} = 4$.

Solution:

$\sqrt{16} = 4$ Radical form

$16^{1/2} = 4$ Rational exponent form

49. Simplify $\sqrt{50} - \sqrt{18}$.

Solution:

$$\sqrt{50} - \sqrt{18} = \sqrt{25 \cdot 2} - \sqrt{9 \cdot 2} = 5\sqrt{2} - 3\sqrt{2} = 2\sqrt{2}$$

55. Perform the indicated operations and/or simplify $(x^2 - 2x + 1)(x^3 - 1)$.

Solution:

$$\begin{aligned}(x^2 - 2x + 1)(x^3 - 1) &= (x^2 - 2x + 1)x^3 - (x^2 - 2x + 1)(1) \\ &= x^5 - 2x^4 + x^3 - x^2 + 2x - 1\end{aligned}$$

57. Perform the indicated operations and/or simplify $(y^2 - y)(y^2 + 1)(y^2 + y + 1)$.

Solution:

$$(y^2 - y)(y^2 + 1)(y^2 + y + 1) = (y^4 - y^3 + y^2 - y)(y^2 + y + 1)$$

By the Vertical Method we have the following.

$$\begin{array}{rrrrrrr}
y^4 & - y^3 & + y^2 & - y & & & \\
 & y^2 & + y & + 1 & & & \\
\hline
y^6 & - y^5 & + y^4 & - y^3 & & & \\
 & y^5 & - y^4 & + y^3 & - y^2 & & \\
 & & y^4 & - y^3 & + y^2 & - y & \\
\hline
y^6 & & + y^4 & - y^3 & & - y &
\end{array}$$

59. Factor completely: $x^3 - x$.

Solution:

$$x^3 - x = x(x^2 - 1) = x(x + 1)(x - 1)$$

63. Factor completely: $x^3 - x^2 + 2x - 2$.

Solution:

$$x^3 - x^2 + 2x - 2 = x^2(x - 1) + 2(x - 1) = (x - 1)(x^2 + 2)$$

67. Insert the missing factor: $\dfrac{t}{\sqrt{t+1}} - \sqrt{t+1} = \dfrac{1}{\sqrt{t+1}}(\quad)$.

Solution:

$$\begin{aligned}
\frac{t}{\sqrt{t+1}} - \sqrt{t+1} &= \frac{t}{\sqrt{t+1}} - \frac{\sqrt{t+1}}{1} \cdot \frac{\sqrt{t+1}}{\sqrt{t+1}} \\
&= \frac{t - (t+1)}{\sqrt{t+1}} \\
&= \frac{-1}{\sqrt{t+1}} \\
&= \frac{1}{\sqrt{t+1}}(-1)
\end{aligned}$$

69. Perform the indicated operations and/or simplify.

$$\frac{x^2-4}{x^4-2x^2-8} \bullet \frac{x^2+2}{x^2}$$

Solution:

$$\frac{x^2-4}{x^4-2x^2-8} \bullet \frac{x^2+2}{x^2} = \frac{x^2-4}{(x^2-4)(x^2+2)} \bullet \frac{x^2+2}{x^2} = \frac{1}{x^2}$$

73. Perform the indicated operations and/or simplify.

$$x-1+\frac{1}{x+2}+\frac{1}{x-1}$$

Solution:

$$\begin{aligned} x-1+\frac{1}{x+2}+\frac{1}{x-1} &= \frac{x(x+2)(x-1)-(x+2)(x-1)+(x-1)+(x+2)}{(x+2)(x-1)} \\ &= \frac{(x^3+x^2-2x)-(x^2+x-2)+(2x+1)}{(x+2)(x-1)} \\ &= \frac{x^3-x+3}{(x+2)(x-1)} \end{aligned}$$

77. Perform the indicated operations and/or simplify.

$$\frac{1}{x-2}+\frac{1}{(x-2)^2}+\frac{1}{x+2}$$

Solution:

$$\begin{aligned} \frac{1}{x-2}+\frac{1}{(x-2)^2}+\frac{1}{x+2} &= \frac{(x-2)(x+2)+(x+2)+(x-2)^2}{(x+2)(x-2)^2} \\ &= \frac{(x^2-4)+(x+2)+(x^2-4x+4)}{(x+2)(x-2)^2} \\ &= \frac{2x^2-3x+2}{(x+2)(x-2)^2} \end{aligned}$$

81. Simplify the compound fraction.

$$\frac{\left(\dfrac{3a}{(a^2/x)-1}\right)}{\left(\dfrac{a}{x}-1\right)}$$

Solution:

$$\frac{\left(\dfrac{3a}{(a^2/x)-1}\right)}{\left(\dfrac{a}{x}-1\right)} = \frac{\left[\dfrac{3a}{(a^2/x)-1}\right]\dfrac{x}{x}}{\dfrac{a-x}{x}} = \frac{3ax}{a^2-x} \bullet \frac{x}{a-x} = \frac{3ax^2}{(a^2-x)(a-x)}, \quad x \neq 0$$

83. Use a calculator to complete the table shown in the textbook.

What number is $5/\sqrt{n}$ approaching as n increases without bound?

Solution:

n	1	10	10^2	10^4	10^6	10^{10}
$\dfrac{5}{\sqrt{n}}$	5	1.5811	0.5	0.05	0.005	0.00005

$5/\sqrt{n}$ approaches 0 as n increases without bound.

85. Let m and n be any two integers. Then $2m$ and $2n$ are even integers and $(2m+1)$ and $(2n+1)$ are odd integers.

(a) Prove that the sum of two even integers is even.

(b) Prove that the sum of two odd integers is even.

(c) Prove that the product of an even integer with *any* integer is even.

Solution:

(a) The sum of two even integers can be written as

$$2m + 2n = 2(m + n)$$

which is even.

(b) The sum of two odd integers can be written as

$$(2m + 1) + (2n + 1) = 2m + 2n + 2 = 2(m + n + 1)$$

which is even.

(c) Let n be any integer. The product of an even integer with n can be written as

$$(2m)n = 2(mn)$$

which is even.

Practice Test for Chapter 1

1. Evaluate $\dfrac{|-42|-20}{15-|-4|}$.

2. Simplify $\dfrac{x}{z}-\dfrac{z}{y}$.

3. The distance between x and 7 is no more than 4. Use absolute value notation to describe this expression.

4. Evaluate $10(-x)^3$ for $x=5$.

5. Simplify $(-4x^3)(-2x^{-5})\left(\dfrac{1}{16}x\right)$.

6. Change 0.0000412 to scientific notation.

7. Evaluate $125^{2/3}$.

8. Simplify $\sqrt[4]{64x^7y^9}$.

9. Rationalize the denominator and simplify $\dfrac{6}{\sqrt{12}}$.

10. Simplify $3\sqrt{80}-7\sqrt{500}$.

11. Simplify $(8x^4-9x^2+2x-1)-(3x^3+5x+4)$.

12. Multiply $(x-3)(x^2+x-7)$.

13. Multiply $[(x-2)-y]^2$.

14. Factor $16x^4-1$.

15. Factor $6x^2+5x-4$.

16. Factor x^3-64.

17. Combine and simplify $-\dfrac{3}{x}+\dfrac{x}{x^2+2}$.

18. Combine and simplify $\dfrac{x-3}{4x}\div\dfrac{x^2-9}{x^2}$.

19. Simplify $\dfrac{1-(1/x)}{1-\dfrac{1}{1-(1/x)}}$.

20. Factor the expression $\frac{1}{3}(x-1)^{5/2}-\frac{1}{6}(x-1)^{1/2}$ so that at least one factor is a polynomial with integer coefficients.

CHAPTER 2

Algebraic Equations and Inequalities

SECTION 2.1

Linear Equations

- You should know how to solve linear equations. $ax + b = 0$
- An identity is an equation whose solution consists of every real number in its domain.
- To solve an equation you can:
  (a) Add or subtract the same quantity from both sides.
  (b) Multiply or divide both sides by the same nonzero quantity.
- To solve an equation that can be simplified to a linear equation:
  (a) Remove all symbols of grouping and all fractions.
  (b) Combine like terms.
  (c) Solve by algebra.
  (d) Check the answer.
- A "solution" that does not satisfy the original equation is called an extraneous solution.

Solutions to Selected Exercises

3. Determine whether the equation

$$-2(x-3)+5=-2x+10$$

is conditional or an identity.

Solution:

$$\begin{aligned} -2(x-3)+5 &= -2x+10 \\ -2x+6+5 &= -2x+10 \\ -2x+11 &= -2x+10 \\ 11 &\neq 10 \end{aligned}$$

No solution; conditional

7. Determine whether the equation

$$x^2-8x+5=(x-4)^2-11$$

is conditional or an identity.

Solution:

$$\begin{aligned} x^2-8x+5 &= (x-4)^2-11 \\ &= x^2-8x+16-11 \\ &= x^2-8x+5 \end{aligned}$$

Identity

9. Determine whether the following equation is conditional or an identity.

$$3 + \frac{1}{x+1} = \frac{4x}{x+1}$$

Solution:

$$3 + \frac{1}{x+1} = \frac{4x}{x+1}$$

$$\frac{3x+4}{x+1} = \frac{4x}{x+1}$$

$$3x + 4 = 4x$$

$$x = 4$$

Conditional

13. Determine whether the given value of x is a solution of the equation $3x^2 + 2x - 5 = 2x^2 - 2$.

(a) $x = -3$

(b) $x = 1$

(c) $x = 4$

(d) $x = -5$

Solution:

(a) $3(-3)^2 + 2(-3) - 5 \stackrel{?}{=} 2(-3)^2 - 2$

$16 = 16$

$x = -3$ is a solution.

(b) $3(1)^2 + 2(1) - 5 \stackrel{?}{=} 2(1)^2 - 2$

$0 = 0$

$x = 1$ is a solution.

(c) $3(4)^2 + 2(4) - 5 \stackrel{?}{=} 2(4)^2 - 2$

$51 \neq 30$

$x = 4$ is not a solution.

(d) $3(-5)^2 + 2(-5) - 5 \stackrel{?}{=} 2(-5)^2 - 2$

$60 \neq 48$

$x = -5$ is not a solution.

17. Determine whether the given value of x is a solution of $(x+5)(x-3) = 20$.

(a) $x = 3$

(b) $x = -5$

(c) $x = 5$

(d) $x = -7$

Solution:

(a) $(3+5)(3-3) \stackrel{?}{=} 20$

$0 \neq 20$

$x = 3$ is not a solution.

(b) $(-5+5)(-5-3) \stackrel{?}{=} 20$

$0 \neq 20$

$x = -5$ is not a solution.

(c) $(5+5)(5-3) \stackrel{?}{=} 20$

$20 = 20$

$x = 5$ is a solution.

(d) $(-7+5)(-7-3) \stackrel{?}{=} 20$

$20 = 20$

$x = -7$ is a solution.

21. Solve the equation $7 - 2x = 15$.

Solution:

$$\begin{aligned} 7 - 2x &= 15 \\ -2x &= 8 && \text{Subtract 7 from both sides.} \\ x &= -4 && \text{Divide both sides by } -2. \end{aligned}$$

25. Solve the equation

$$2(x+5) - 7 = 3(x-2).$$

Solution:

$$\begin{aligned} 2(x+5) - 7 &= 3(x-2) \\ 2x + 10 - 7 &= 3x - 6 \\ 2x + 3 &= 3x - 6 \\ -x + 3 &= -6 \\ -x &= -9 \\ x &= 9 \end{aligned}$$

29. Solve the equation

$$\frac{5x}{4} + \frac{1}{2} = x - \frac{1}{2}.$$

Solution:

$$\begin{aligned} \frac{5x}{4} + \frac{1}{2} &= x - \frac{1}{2} \\ 4\left(\frac{5x}{4}\right) + 4\left(\frac{1}{2}\right) &= 4(x) + 4\left(-\frac{1}{2}\right) \\ 5x + 2 &= 4x - 2 \\ x + 2 &= -2 \\ x &= -4 \end{aligned}$$

33. Solve the equation

$$0.25x + 0.75(10 - x) = 3.$$

Solution:

$$\begin{aligned} 0.25x + 0.75(10 - x) &= 3 \\ 0.25x + 7.5 - 0.75x &= 3 \\ -0.5x &= -4.5 \\ x &= 9 \end{aligned}$$

35. Solve the equation

$x + 8 = 2(x - 2) - x$, if possible.

Solution:

$$\begin{aligned} x + 8 &= 2(x-2) - x \\ x + 8 &= 2x - 4 - x \\ x + 8 &= x - 4 \\ 8 &= -4 \end{aligned}$$

Not possible.

Thus, the equation has no solution.

39. Solve the equation $\dfrac{5x-4}{5x+4} = \dfrac{2}{3}$.

Solution:

$$\begin{aligned} \frac{5x-4}{5x+4} &= \frac{2}{3} \\ 3(5x-4) &= 2(5x+4) && \text{Cross multiply} \\ 15x - 12 &= 10x + 8 \\ 5x &= 20 \\ x &= 4 \end{aligned}$$

43. Solve the equation

$$\frac{1}{x-3} + \frac{1}{x+3} = \frac{10}{x^2-9}.$$

Solution:

$$\begin{aligned}
\frac{1}{x-3} + \frac{1}{x+3} &= \frac{10}{x^2-9} \\
\frac{1}{x-3}(x+3)(x-3) + \frac{1}{x+3}(x+3)(x-3) &= \frac{10}{x^2-9}(x+3)(x-3) \\
x+3+x-3 &= 10 \\
2x &= 10 \\
x &= 5
\end{aligned}$$

47. Solve the equation

$$\frac{7}{2x+1} - \frac{8x}{2x-1} = -4.$$

Solution:

$$\begin{aligned}
\frac{7}{2x+1} - \frac{8x}{2x-1} &= -4 \\
\frac{7}{2x+1}(2x+1)(2x-1) - \frac{8x}{2x-1}(2x+1)(2x-1) &= -4(2x+1)(2x-1) \\
7(2x-1) - 8x(2x+1) &= -4(4x^2-1) \\
14x - 7 - 16x^2 - 8x &= -16x^2 + 4 \\
-16x^2 + 6x - 7 &= -16x^2 + 4 \\
6x - 7 &= 4 \\
6x &= 11 \\
x &= \frac{11}{6}
\end{aligned}$$

53. Solve the equation

$$(x+2)^2 + 5 = (x+3)^2.$$

Solution:

$$\begin{aligned}
(x+2)^2 + 5 &= (x+3)^2 \\
x^2 + 4x + 4 + 5 &= x^2 + 6x + 9 \\
4x + 9 &= 6x + 9 \\
4x &= 6x \\
-2x &= 0 \\
x &= 0
\end{aligned}$$

57. Solve the equation

$$4(x+1) - ax = x + 5.$$

Solution:

$$\begin{aligned}
4(x+1) - ax &= x + 5 \\
4x + 4 - ax &= x + 5 \\
3x - ax &= 1 \\
x(3-a) &= 1 \\
x &= \frac{1}{3-a}, \quad a \neq 3
\end{aligned}$$

61. Use a calculator to solve $0.275x + 0.725(500 - x) = 300$. Round your answer to three decimal places.

Solution:

$$\begin{aligned} 0.275x + 0.725(500 - x) &= 300 \\ 0.275x + 362.5 - 0.725x &= 300 \\ -0.45x &= -62.5 \\ x &\approx 138.889 \end{aligned}$$

65. Find an equation of the form $ax + b = cx$ that has $x = 2$ as a solution.

Solution:

$$\begin{aligned} ax + b &= cx \\ ax - cx &= -b \\ x(a - c) &= -b \\ x &= \frac{-b}{a - c} \end{aligned}$$

Since $x = 2$, we have $2 = \dfrac{-b}{a - c}$.

Choose any combination of a, b and c that satisfies this equation. Listed below are some possibilities.

$a = 4,$	$b = -2,$	$c = 3$	:	$4x - 2 = 3x$
$a = 6,$	$b = -12,$	$c = 0$	:	$6x - 12 = 0$
$a = 0,$	$b = 3,$	$c = 1.5$	:	$6x - 13 = 1.5x$
$a = -1,$	$b = 5,$	$c = 1.5$	:	$-x + 5 = 1.5x$
$a = 2,$	$b = -2,$	$c = 1$	:	$2x - 2 = x$

69. Calculate

$$\frac{3.33 + \dfrac{1.98}{0.74}}{4 + \dfrac{6.25}{3.15}}$$ in two ways.

(a) Calculate entirely on your calculator by storing intermediate results, and then round to two decimal places.

(b) Round both the numerator and the denominator to two decimal places before dividing, and then round the final answer to two decimal places. Does the second method introduce an additional round-off error?

69. —CONTINUED—

Solution:

(a) $\boxed{(}$ 3.33 $\boxed{+}$ 1.98 $\boxed{\div}$ 0.74 $\boxed{)}$ $\boxed{\div}$ $\boxed{(}$ 4 + 6.25 $\boxed{\div}$ 3.15 $\boxed{)}$ $\boxed{=}$ 1.003600975

≈ 1.00

(b) 3.33 $\boxed{+}$ 1.98 $\boxed{\div}$ 0.74 $\boxed{=}$ 6.005675676 $\approx$ 6.01

4 $\boxed{+}$ 6.25 $\boxed{\div}$ 3.15 $\boxed{=}$ 5.984126984 $\approx$ 5.98

6.01 $\boxed{\div}$ 5.98 $\boxed{=}$ 1.005016722 $\approx$ 1.01

The second method DOES introduce an additional round-off error.

73. Use the following information about a possible negative income tax for a family of two adults and two children. The plan would guarantee the poor a minimum income while encouraging families to increase their private income (see figure).

Family's earned income: $I = x$

Government payment: $G = 8000 - \frac{1}{2}x, \quad 0 \le x \le 16{,}000$

Spendable income: $S = I + G$.

Express the spendable income S in terms of x.

Solution:

For $0 \le x \le 16{,}000$: $S = I + G = x + \left(8000 - \frac{1}{2}x\right)$

$= \frac{1}{2}x + 8000$

For $x > 16{,}000$: $S = I = x$

Putting these together, we have

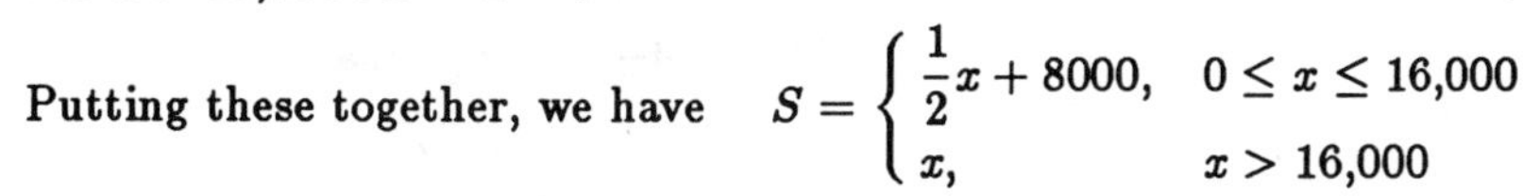

$$S = \begin{cases} \frac{1}{2}x + 8000, & 0 \le x \le 16{,}000 \\ x, & x > 16{,}000 \end{cases}$$

77. The surface area S of the rectangular solid in the figure is

$$S = 2(24) + 2(4x) + 2(6x).$$

Find the length of the box x if the surface area is 248 square inches.

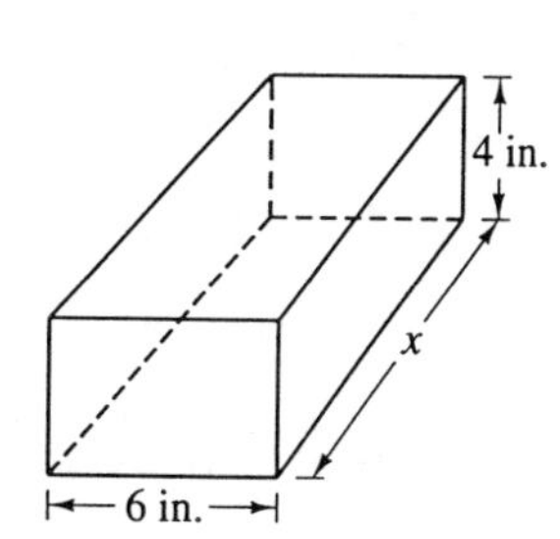

Solution:

$248 = 2(24) + 2(4x) + 2(6x)$

$248 = 48 + 8x + 12x$

$200 = 20x$

$x = 10$ inches

SECTION 2.2

Linear Equations and Modeling

- You should be able to set up mathematical models to solve problems.
- You should be able to translate key words and phrases.

  (a) Consecutive:
  Next, subsequent

  (b) Addition:
  Sum, plus, greater, increased by, more than, exceeds, total of

  (c) Subtraction:
  Difference, minus, less than, decreased by, subtracted from, reduced by, the remainder

  (d) Multiplication:
  Product, multiplied by, twice, times, percent of

  (e) Division:
  Quotient, divided by, ratio, per

  (f) Equality:
  Equals, Equal to, is, are, was, will be, represents

- You should know the following formulas:

  (a) Perimeter
  1. Square: $P = 4s$
  2. Rectangle: $P = 2L + 2W$
  3. Circle: $C = 2\pi r$

  (b) Area
  1. Square: $A = s^2$
  2. Rectangle: $A = LW$
  3. Circle: $A = \pi r^2$
  4. Triangle: $A = (1/2)bh$

  (c) Volume
  1. Cube: $V = s^3$
  2. Rectangular solid: $V = LWH$
  3. Cylinder: $V = \pi r^2 h$
  4. Sphere: $V = (4/3)\pi r^3$

  (d) Simple Interest: $I = Prt$

  (e) Compound Interest: $A = P(1 + \frac{r}{n})^{nt}$

  (f) Distance: $D = r \cdot t$

  (g) Temperature: $F = \frac{9}{5}C + 32$

- You should be able to solve word problems. Study the examples in the text carefully.

Solutions to Selected Exercises

1. Write an algebraic expression for the sum of two consecutive natural numbers.

Solution:

Verbal Model Sum = First number + Second number

Labels Sum = S
First number = n
Second number = $n+1$

Algebraic Equation $S = n + (n+1) = 2n+1$

5. Write an algebraic expression for the amount of acid in x gallons of a 20% acid solution.

Solution:

Verbal Model Amount of acid = 20% • Amount of solution

Labels Amount of acid (in gallons) = A, Amount of solution (in gallons) = x

Algebraic Equation $A = 0.20x$

7. Write an algebraic expression for the perimeter of a rectangle whose width is x and whose length is twice the width.

Solution:

Verbal Model Perimeter = 2 • Width + 2 • Length

Labels Perimeter = P, Width = x, Length = 2 • Width = $2x$

Algebraic Equation $P = 2x + 2(2x) = 6x$

11. The sum of two consecutive natural numbers is 525. Find the two numbers.

Solution:

Verbal Model Sum = First number + Second number

Labels Sum = 525,
First number = n
Second number = $n+1$

Algebraic Equation $525 = n + (n+1)$
$525 = 2n + 1$
$n = 262$

First number = n = 262, Second number = $n + 1$ = 263

13. One number is five times another number. The difference between the two numbers is 148. Find the numbers.

Solution:

Verbal Model	Difference = One number − Another number
Labels	Difference = 148, One number = $5x$, Another number = x
Algebraic Equation	$148 = 5x - x$ $148 = 4x$ $x = 37$

One number = $5x = 185$, Another number = $x = 37$

19. What is 0.045% of 2,650,000?

Solution:

$x =$ Percent • Number

$x = 0.045\% \times 2{,}650{,}000$

$x = (0.00045)(2{,}650{,}000)$

$x = 1192.5$

23. 70 is 40% of what number?

Solution:

$70 = 40\% \cdot \text{Number}$

$$\text{Number} = \frac{70}{40\%} = \frac{70}{0.40} = 175$$

27. A family has annual loan payments equaling 58.6% of their annual income. During the year, their loan payments total $13,077.75. What is their income?

Solution:

Verbal Model	Loan payments = 58.6% • Annual income
Labels	Loan payments = 13,077.75 Annual income = I
Algebraic Equation	$13{,}077.75 = 0.586I$ $I \approx 22{,}316.98$

The family's annual income is $22,316.98.

31. Find the percentage increase.

Item	1940	1990
Worker's annual earnings	$1500	$27,000

31. —CONTINUED—

Solution:

Verbal Model 1990 annual earnings = Percentage increase • 1940 annual earnings + 1940 annual earnings

Labels 1990 annual earnings = \$27,000
Percentage increase = p
1940 annual earnings = 1500

Algebraic Equation

$$27{,}000 = 1500p + 1500$$

$$\frac{25{,}500}{1500} = p$$

$$p = 17 = 1700\%$$

Percentage increase = $p = 1700\%$

35. A room is 1.5 times as long as it is wide, and its perimeter is 75 feet (see figure). Find the dimensions of the room.

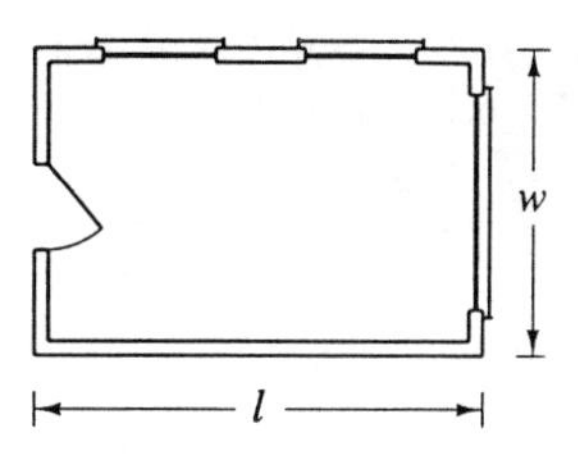

Solution:

Verbal Model Perimeter = 2 • Width + 2 • Length

Labels Perimeter = 75 feet,
Width (in feet) = x,
Length (in feet) = 1.5 • Width = $1.5x$

Algebraic Equation

$$75 = 2x + 2(1.5x)$$

$$75 = 5x$$

$$x = 15$$

Width = x = 15 feet, Length = $1.5x$ = 22.5 feet

37. To get an A in a course a student must have an average of at least 90 on four tests of 100 points each. A student's scores on the first three tests were 87, 92, and 84. What must the student score on the fourth test to get an A for the course?

Solution:

Verbal Model $\dfrac{\text{Sum of scores}}{\text{Number of tests}} = \text{Average}$

Labels Sum of scores $= 87 + 92 + 84 + x$
Number of tests $= 4$
Average $= 90$

Algebraic Equation

$$\frac{87 + 92 + 84 + x}{4} = 90$$

$$\frac{263 + x}{4} = 90$$

$$263 + x = 360$$

$$x = 97 \text{ (or greater)}$$

41. Two cars start at one point and travel in the same direction at average speeds of 40 miles per hour and 55 miles per hour. How much time must elapse before the two cars are 5 miles apart?

Solution:

Distance = Rate • Time

$d_1 = 40 \text{ mph} \cdot t$

$d_2 = 55 \text{ mph} \cdot t$

Distance between cars = Second distance − First distance

$5 = d_2 - d_1$

$5 = 55t - 40t = 15t$

$t = \frac{1}{3}$ hour

43. Two families meet at a park for a picnic. At the end of the day, one family travels east at an average speed of 42 miles per hour and the other travels at an average speed of 50 miles per hour. Both families have approximately 160 miles to travel.

(a) Find the time it takes each family to get home.

(b) Find the time that will have elapsed when they are 100 miles apart.

(c) Find the distance the eastbound family has to travel after the westbound family has arrived home.

43. ——CONTINUED——

Solution:

(a) Time for the first family: $t_1 = \dfrac{d}{r_1} = \dfrac{160}{42} \approx 3.8$ hr

Time for the other family: $t_2 = \dfrac{d}{r_2} = \dfrac{160}{50} = 3.2$ hr

(b) $t = \dfrac{d}{r} = \dfrac{100}{42+50} = \dfrac{100}{92} \approx 1.1$ hr

(c) $d = rt = 42\left(\dfrac{160}{42} - \dfrac{160}{50}\right) = 25.6$ mi

47. Radio waves travel at the speed of light, 3.0×10^8 meters per second. Find the time required for a radio wave to travel from mission control in Houston to NASA astronauts on the surface of the moon, 3.86×10^8 meters away.

Solution:

$$\begin{aligned} d &= rt \\ 3.86 \times 10^8 &= \left(3.0 \times 10^8\right) t \\ t &= \frac{3.86 \times 10^8}{3.0 \times 10^8} \approx 1.29 \text{ sec} \end{aligned}$$

53. Suppose you invest \$25,000 in two funds paying 11% and $12\frac{1}{2}\%$ simple interest. The total annual interest is \$2975. How much is invested in each fund?

Solution:

Interest = Interest rate × Principal

$i_1 = 11\% \times \$x$

$i_2 = 12.5\% \times \$(25{,}000 - x)$

Total interest = Interest in first account + Interest in second account

$$\begin{aligned} 2975 &= i_1 + i_2 \\ 2975 &= 0.11x + 0.125(25{,}000 - x) \\ 2975 &= -0.015x + 3125 \\ x &= \frac{150}{0.015} = 10{,}000 \end{aligned}$$

You have \$10,000 in the 11% fund and \$15,000 in the $12\frac{1}{2}\%$ fund.

57. A 55-gallon barrel contains a mixture with a concentration of 40%. How much of this mixture must be withdrawn and replaced by 100% concentrate to bring the mixture up to 75% concentration?

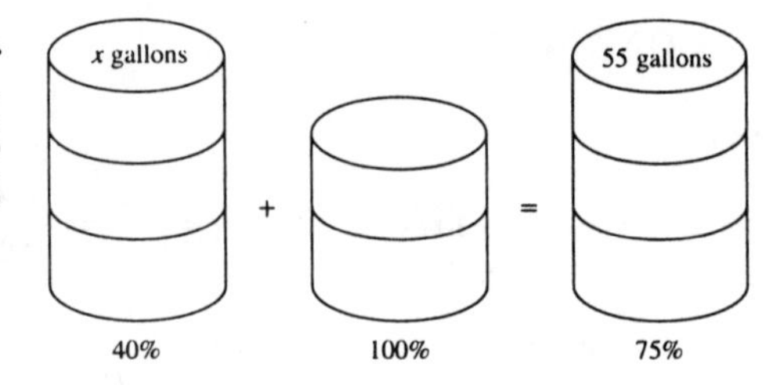

Solution:

$$\begin{aligned}(\text{Final concentration})(\text{Amount}) &= (\text{Solution 1 concentration})(\text{Amount}) + \\ &\quad (\text{Solution 2 concentration})(\text{Amount}) \\ (75\%)(55 \text{ gal}) &= (40\%)(55 - x) + (100\%)x \\ 41.25 &= 0.60x + 22 \\ x &\approx 32.1 \text{ gal}\end{aligned}$$

61. A company has fixed costs of \$10,000 per month and variable costs of \$9.30 per unit manufactured. The company has \$85,000 available to cover the monthly costs. How many units can the company manufacture? (*Fixed costs* occur regardless of the level of production. *Variable costs* depend on the level of production.)

Solution:

Cost = Fixed costs + Variable cost • Number of units

$$\begin{aligned}\$85{,}000 &= \$10{,}000 + \$9.30x \\ x &= \frac{75{,}000}{9.3} \approx 8064.52 \text{ units}\end{aligned}$$

At most the company can manufacture 8064 units.

63. Suppose you have a uniform beam of length L with a fulcrum x feet from one end. If there are objects with weights W_1 and W_2 placed at opposite ends of the beam, then the beam will balance if $W_1x = W_2(L - x)$. Find x so the beam will balance if two children weighing 50 pounds and 75 pounds are going to play on a seesaw that is 10 feet long.

Solution:

$$\begin{aligned}W_1x &= W_2(L - x) \\ 50x &= 75(10 - x) \\ 50x &= 750 - 75x \\ 125x &= 750 \\ x &= 6 \text{ feet from the 50 pound child}\end{aligned}$$

67. Solve for C: $S = C + RC$.

Solution:

$$\begin{aligned} S &= C + RC \\ S &= C(1+R) \\ \frac{S}{1+R} &= C \end{aligned}$$

71. Solve for b: $A = \frac{1}{2}(a+b)h$.

Solution:

$$\begin{aligned} A &= \frac{1}{2}(a+b)h \\ 2A &= (a+b)h \\ 2A &= ah + bh \\ 2A - ah &= bh \\ \frac{2A - ah}{h} &= b \end{aligned}$$

75. Solve for α: $L = L_0[1 + \alpha(\Delta t)]$.

Solution:

$$\begin{aligned} L &= L_0[1 + \alpha(\Delta t)] \\ L &= L_0 + L_0\alpha(\Delta t) \\ L - L_0 &= L_0\alpha(\Delta t) \\ \frac{L - L_0}{L_0(\Delta t)} &= \alpha \end{aligned}$$

79. Solve for R_1: $\frac{1}{f} = (n-1)\left(\frac{1}{R_1} - \frac{1}{R_2}\right)$

Solution:

$$\begin{aligned} \frac{1}{f} &= (n-1)\left(\frac{1}{R_1} - \frac{1}{R_2}\right) \\ \frac{1}{f(n-1)} &= \frac{1}{R_1} - \frac{1}{R_2} \\ \frac{1}{f(n-1)} + \frac{1}{R_2} &= \frac{1}{R_1} \\ \frac{R_2 + f(n-1)}{f(n-1)R_2} &= \frac{1}{R_1} \\ R_1 &= \frac{f(n-1)R_2}{R_2 + f(n-1)} \end{aligned}$$

(R_1 is the reciprocal of $\frac{1}{R_1}$.)

83. Solve for r.

$$S = \frac{rL - a}{r - 1}$$

Solution:

$$\begin{aligned} S &= \frac{rL - a}{r - 1} \\ S(r-1) &= rL - a \\ Sr - S &= rL - a \\ Sr - rL &= S - a \\ r(S - L) &= S - a \\ r &= \frac{S - a}{S - L} \end{aligned}$$

SECTION 2.3

Quadratic Equations

- You should be able to factor a quadratic.
- You should be able to solve a quadratic by extracting square roots.
- You should be able to complete the square on any quadratic.

Solutions to Selected Exercises

5. Write the equation

$$(x-3)^2 = 2$$

in standard quadratic form.

Solution:

$$\begin{aligned}(x-3)^2 &= 2\\ x^2-6x+9 &= 2\\ x^2-6x+7 &= 0\end{aligned}$$

9. Write the equation

$$\frac{3x^2-10}{5} = 12x$$

in standard quadratic form.

Solution:

$$\begin{aligned}\frac{3x^2-10}{5} &= 12x\\ 3x^2-10 &= 60x\\ 3x^2-60x-10 &= 0\end{aligned}$$

13. Solve $x^2-2x-8=0$ by factoring.

Solution:

$$\begin{aligned}x^2-2x-8 &= 0\\ (x+2)(x-4) &= 0\\ x+2=0 \quad &\Rightarrow \quad x=-2\\ x-4=0 \quad &\Rightarrow \quad x=4\end{aligned}$$

17. Solve $3+5x-2x^2=0$ by factoring.

Solution:

$$\begin{aligned}3+5x-2x^2 &= 0\\ (3-x)(1+2x) &= 0\\ 3-x=0 \quad &\Rightarrow \quad x=3\\ 1+2x=0 \quad &\Rightarrow \quad x=-\frac{1}{2}\end{aligned}$$

21. Solve $x^2+2ax+a^2=0$ by factoring.

Solution:

$$\begin{aligned}x^2+2ax+a^2 &= 0\\ (x+a)^2 &= 0\\ x+a &= 0\\ x &= -a\end{aligned}$$

23. Solve $x^2=16$ by extracting square roots.

Solution:

$$\begin{aligned}x^2 &= 16\\ x &= \pm\sqrt{16}\\ x &= \pm 4\end{aligned}$$

27. Solve $3x^2 = 36$ by extracting square roots. List both the exact answer and the decimal answer rounded to two decimal places.

Solution:

$$3x^2 = 36$$
$$x^2 = 12$$
$$x = \pm\sqrt{12}$$
$$x = \pm 2\sqrt{3} \approx \pm 3.46$$

33. Solve $(x-7)^2 = (x+3)^2$ by extracting square roots.

Solution:

$$(x-7)^2 = (x+3)^2$$
$$x - 7 = \pm(x+3)$$

For $x - 7 = +(x+3)$

$$-7 = +3 \qquad \text{No solution}$$

For $x - 7 = -(x+3)$

$$x - 7 = -x - 3$$
$$2x = 4$$
$$x = 2$$

37. Solve $x^2 + 4x - 32 = 0$ by completing the square.

Solution:

$$x^2 + 4x - 32 = 0$$
$$x^2 + 4x = 32$$
$$x^2 + 4x + 2^2 = 32 + 2^2$$
$$(x+2)^2 = 36$$
$$x + 2 = \pm 6$$
$$x = -2 \pm 6$$
$$x = 4 \quad \text{or} \quad x = -8$$

41. Solve $9x^2 - 18x + 3 = 0$ by completing the square.

Solution:

$$9x^2 - 18x + 3 = 0$$
$$x^2 - 2x + \frac{1}{3} = 0$$
$$x^2 - 2x = -\frac{1}{3}$$
$$x^2 - 2x + 1^2 = -\frac{1}{3} + 1^2$$
$$(x-1)^2 = \frac{2}{3}$$
$$x - 1 = \pm\sqrt{\frac{2}{3}}$$
$$x = 1 \pm \sqrt{\frac{2}{3}}$$
$$x = 1 \pm \frac{\sqrt{6}}{3}$$

45. Solve $x^2 = 64$.

Solution:

$$x^2 = 64$$
$$x = \pm\sqrt{64}$$
$$x = \pm 8$$

49. Solve $16x^2 - 9 = 0$.

Solution:

$$16x^2 - 9 = 0$$
$$16x^2 = 9$$
$$x^2 = \frac{9}{16}$$
$$x = \pm\frac{3}{4}$$

53. Solve $(x+3)^2 = 81$.

Solution:

$$(x+3)^2 = 81$$
$$x + 3 = \pm 9$$
$$x = -3 \pm 9$$
$$x = 6 \quad \text{or} \quad x = -12$$

57. Solve $50 + 5x = 3x^2$.

Solution:

$$50 + 5x = 3x^2$$
$$0 = 3x^2 - 5x - 50$$
$$0 = (3x+10)(x-5)$$
$$0 = 3x + 10 \quad \Rightarrow \quad -\frac{10}{3} = x$$
$$0 = x - 5 \quad \Rightarrow \quad 5 = x$$

61. Solve $x^2 - x - \frac{11}{4} = 0$.

Solution:

$$x^2 - x - \frac{11}{4} = 0$$
$$x^2 - x = \frac{11}{4}$$
$$x^2 - x + \left(\frac{1}{2}\right)^2 = \frac{11}{4} + \left(\frac{1}{2}\right)^2$$
$$\left(x - \frac{1}{2}\right)^2 = \frac{12}{4}$$
$$x - \frac{1}{2} = \pm\sqrt{\frac{12}{4}}$$
$$x = \frac{1}{2} \pm \sqrt{3}$$

67. Solve $(x+1)^2 = x^2$.

Solution:

$$(x+1)^2 = x^2$$
$$x^2 = (x+1)^2$$
$$x = \pm(x+1)$$

For $x = +(x+1)$

$0 = 1$ No solution

For $x = -(x+1)$

$$2x = -1$$
$$x = -\frac{1}{2}$$

69. Complete the square for the quadratic portion of

$$\frac{1}{x^2 - 4x - 12}.$$

Solution:

$$\frac{1}{x^2 - 4x - 12} = \frac{1}{x^2 - 4x + 2^2 - 2^2 - 12} = \frac{1}{(x-2)^2 - 16}$$

73. A one-story building is 14 feet longer than it is wide (see figure). It has 1632 square feet of floor space. Find the length and width of the building.

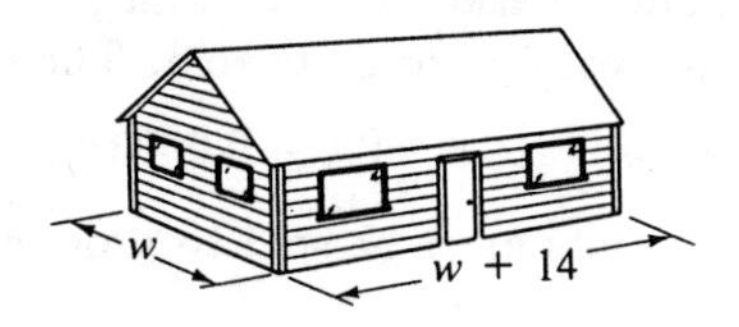

Solution:

$$w(w+14) = 1632$$
$$w^2 + 14w - 1632 = 0$$
$$(w+48)(w-34) = 0$$

Since $w > 0$, we have $w = 34$ feet and the length is $w + 14 = 48$ feet.

77. The hypotenuse of an isosceles right triangle is 5 centimeters long. How long are its sides? (An isosceles triangle has two sides of equal length.)

Solution:

$$x^2 + x^2 = 5^2 \quad \text{Pythagorean Theorem}$$
$$2x^2 = 25$$
$$x^2 = \frac{25}{2}$$
$$x = \sqrt{\frac{25}{2}}$$
$$= \frac{5}{\sqrt{2}}$$
$$= \frac{5\sqrt{2}}{2} \approx 3.54 \text{ centimeters}$$

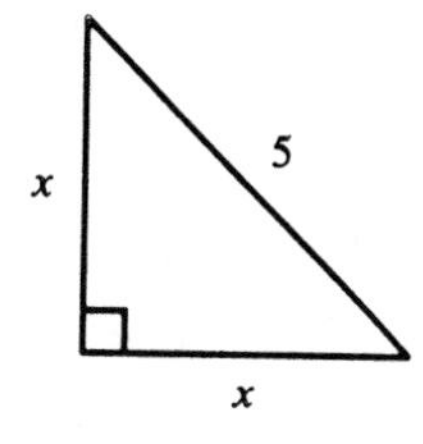

81. The demand equation for a certain product is $p = 20 - 0.0002x$ where p is the price per unit and x is the number of units sold. The total revenue is given by

$$\text{Revenue} = xp = x(20 - 0.0002x).$$

How many units must be sold to produce a revenue of \$500,000?

Solution:

$$\begin{aligned} x(20 - 0.0002x) &= 500{,}000 \\ 0 &= 0.0002x^2 - 20x + 500{,}000 \\ 0 &= x^2 - 100{,}000x + 2{,}500{,}000{,}000 \\ 0 &= (x - 50{,}000)^2 \\ x &= 50{,}000 \text{ units} \end{aligned}$$

85. If a is a nonzero real number, determine the solutions of the equation $ax^2 - ax = 0$.

Solution:

$$\begin{aligned} ax^2 - ax &= 0 \\ ax(x - 1) &= 0 \\ ax &= 0 \quad \Rightarrow \quad x = 0 \\ x - 1 &= 0 \quad \Rightarrow \quad x = 1 \end{aligned}$$

SECTION 2.4

The Quadratic Formula

- You should know the Quadratic Formula: For $ax^2 + bx + c = 0,\ a \neq 0,$

$$x = \frac{-b \pm \sqrt{b^2 - 4ac}}{2a}$$

- You should be able to determine the types of solutions of a quadratic equation by checking the discriminant $b^2 - 4ac$.

  (a) If $b^2 - 4ac > 0$, there are two distinct real solutions.
  (b) If $b^2 - 4ac = 0$, there is one repeating real solution.
  (c) If $b^2 - 4ac < 0$, there are no real solutions.

- You should be able to use your calculator to solve quadratic equations.

- You should be able to solve word problems involving quadratic equations. Study the examples in the text carefully.

Solutions to Selected Exercises

5. Use the discriminant to determine the number of real solutions of $\frac{1}{5}x^2 + \frac{6}{5}x - 8 = 0$.

Solution:

$$\frac{1}{5}x^2 + \frac{6}{5}x - 8 = 0$$

$$x^2 + 6x - 40 = 0; \quad a = 1,\ b = 6,\ c = -40$$

$$b^2 - 4ac = (6)^2 - 4(1)(-40) = 36 + 160 = 196 > 0$$

Two real solutions

9. Use the Quadratic Formula to solve $16x^2 + 8x - 3 = 0$.

Solution:

$16x^2 + 8x - 3 = 0;\ a = 16,\ b = 8,\ c = -3$

$$x = \frac{-8 \pm \sqrt{8^2 - 4(16)(-3)}}{2(16)} = \frac{-8 \pm \sqrt{256}}{32} = \frac{-8 \pm 16}{32}$$

$$x = \frac{-8 + 16}{32} = \frac{1}{4} \quad \text{or} \quad x = \frac{-8 - 16}{32} = -\frac{3}{4}$$

13. Use the Quadratic Formula to solve $x^2 + 14x + 44 = 0$.

Solution:

$x^2 + 14x + 44 = 0$; $a = 1$, $b = 14$, $c = 44$

$$x = \frac{-14 \pm \sqrt{14^2 - 4(1)(44)}}{2(1)} = \frac{-14 \pm \sqrt{20}}{2} = \frac{-14 \pm 2\sqrt{5}}{2}$$

$$x = -7 \pm \sqrt{5}$$

19. Use the Quadratic Formula to solve $36x^2 + 24x - 7 = 0$.

Solution:

$36x^2 + 24x - 7 = 0$; $a = 36$, $b = 24$, $c = -7$

$$x = \frac{-24 \pm \sqrt{24^2 - 4(36)(-7)}}{2(36)} = \frac{-24 \pm \sqrt{1584}}{72} = \frac{-24 \pm 12\sqrt{11}}{72}$$

$$x = \frac{-2 \pm \sqrt{11}}{6} = -\frac{1}{3} \pm \frac{\sqrt{11}}{6}$$

25. Use the Quadratic Formula to solve $25h^2 + 80h + 61 = 0$.

Solution:

$25h^2 + 80h + 61 = 0$; $a = 25$, $b = 80$, $c = 61$

$$h = \frac{-80 \pm \sqrt{(80)^2 - 4(25)(61)}}{2(25)} = \frac{-80 \pm \sqrt{300}}{50} = \frac{-80 \pm 10\sqrt{3}}{50}$$

$$h = \frac{-8 \pm \sqrt{3}}{5} = -\frac{8}{5} \pm \frac{\sqrt{3}}{5}$$

29. Use a calculator to solve $5.1x^2 - 1.7x - 3.2 = 0$. Round your answer to three decimal places.

Solution:

$5.1x^2 - 1.7x - 3.2 = 0$; $a = 5.1$, $b = -1.7$, $c = -3.2$

$$x = \frac{-(-1.7) \pm \sqrt{(-1.7)^2 - 4(5.1)(-3.2)}}{2(5.1)}$$

$$x = \frac{1.7 \pm \sqrt{68.17}}{10.2} \approx \frac{1.7 \pm 8.2565}{10.2}$$

$x \approx 0.976$ or $x \approx -0.643$

31. Use a calculator to solve $422x^2 - 506x - 347 = 0$. Round your answer to three decimal places.

Solution:

$422x^2 - 506x - 347 = 0$; $a = 422$, $b = -506$, $c = -347$

$$x = \frac{-(-506) \pm \sqrt{(-506)^2 - 4(422)(-347)}}{2(422)}$$

$$x = \frac{506 \pm \sqrt{841,772}}{844} \approx \frac{506 \pm 917.481335}{844}$$

$x \approx 1.687$ or $x \approx -0.488$

35. Solve $4x^2 - 15 = 25$.

Solution:

$$4x^2 - 15 = 25$$
$$4x^2 = 40$$
$$x^2 = 10$$
$$x = \pm\sqrt{10}$$

39. Solve $(x - 1)^2 = 9$.

Solution:

$$(x - 1)^2 = 9$$
$$x - 1 = \pm\sqrt{9}$$
$$x = 1 \pm 3$$
$$x = 4 \quad \text{or} \quad x = -2$$

43. Find two numbers whose sum is 100 and whose product is 2500 and make up an applied problem.

Solution:

Verbal model: The sum of Bill and Harry's ages is 100. The product of their ages is 2500. How old is Bill? How old is Harry?

Let x = one number and $100 - x$ = other number.

$$x(100 - x) = 2500$$
$$100x - x^2 = 2500$$
$$0 = x^2 - 100x + 2500$$
$$0 = (x - 50)^2$$
$$x = 50, \quad 100 - x = 50$$

45. Find two consecutive positive integers such that the sum of their squares is 113 and make up an applied problem.

Solution:

Verbal model: One leg of a right triangle is one foot longer than the other leg. The hypotenuse is $\sqrt{113}$ feet. Find the length of each leg.

Let $n =$ first integer and $n+1 =$ next integer.

$$n^2 + (n+1)^2 = 113$$
$$n^2 + n^2 + 2n + 1 = 113$$
$$2n^2 + 2n - 112 = 0$$
$$2(n^2 + n - 56) = 0$$
$$2(n+8)(n-7) = 0$$
$$n = -8 \quad \text{or} \quad n = 7$$

Since we want a positive integer, $n = 7$ and $n + 1 = 8$.

49. Use the cost equation $C = 800 + 0.04x + 0.002x^2$ to find the number of units x that a manufacturer can produce for the given cost $C = \$1680$. Round your answer to the nearest positive integer.

Solution:

$$1680 = 800 + 0.04x + 0.002x^2$$
$$0 = 0.002x^2 + 0.04x - 880$$

Thus, $a = 0.002,\ b = 0.04,\ c = -880.$

$$x = \frac{-0.04 \pm \sqrt{(0.04)^2 - 4(0.002)(-880)}}{2(0.002)} \approx \frac{-0.04 \pm 2.6536}{0.004}$$

$$x \approx \frac{-0.04 + 2.6536}{0.004} \quad \text{or } x \approx \frac{-0.04 - 2.6536}{0.004}$$

$$x \approx 653.4 \qquad \text{or } x \approx -673.4 \qquad \text{Not a valid solution}$$

The manufacturer can produce 653 units for the given cost $C = \$1680$.

53. Two people must mow a rectangular lawn 100 feet by 200 feet. Each wants to mow no more than half of the lawn. The first starts by mowing around the outside of the lawn. How wide a strip must the person mow on each of the four sides? If the mower has a 24-inch cut, approximate the required number of trips around the lawn.

Solution:

$$(200 - 2x)(100 - 2x) = \frac{1}{2}(100)(200)$$

$$20{,}000 - 600x + 4x^2 = 10{,}000$$

$$4x^2 - 600x + 10{,}000 = 0$$

$$4(x^2 - 150x + 2500) = 0$$

Thus, $a = 1,\ b = -150,\ c = 2500.$

$$x = \frac{150 \pm \sqrt{(-150)^2 - 4(1)(2500)}}{2(1)}$$

$$\approx \frac{150 \pm 111.8034}{2}$$

$$x \approx \frac{150 + 111.8034}{2} \approx 130.902 \text{ ft}$$ Not possible since the lot is only 100 ft wide.

$$x \approx \frac{150 - 111.8034}{2} \approx 19.098 \text{ ft}$$

The person must go around the lot

$$\frac{19.098 \text{ ft}}{24 \text{ in}} = \frac{19.098 \text{ ft}}{2 \text{ ft}} = 9.5 \text{ times.}$$

55. An open box with a square base is to be constructed from 108 square inches of material, as shown in the accompanying figure. What should the dimensions of the base be if the height of the box is to be 3 inches? [Hint: The surface area is given by $S = x^2 + 4xh$.]

Solution:

$$S = x^2 + 4xh$$

$$108 = x^2 + 4x(3)$$

$$0 = x^2 + 12x - 108$$

$$0 = (x - 6)(x + 18)$$

$$x = 6 \quad \text{or} \quad x = -18$$

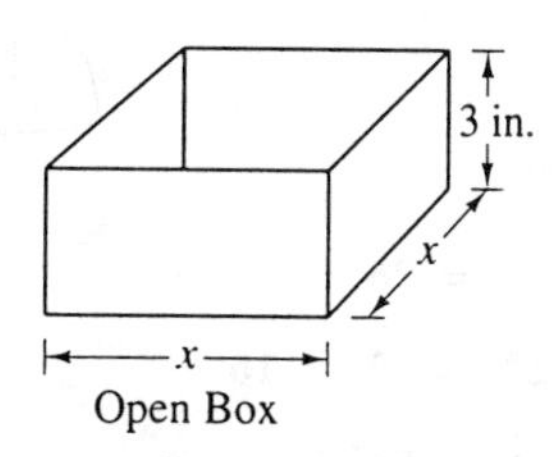

Open Box

Choosing the positive value, we have $x = 6$. The dimensions of the base are 6 in. $\times$ 6 in.

59. Two planes leave simultaneously from the same airport, one flying due east and the other due south, as shown in the accompanying figure. The eastbound plane is flying 50 miles per hour faster than the southbound plane. After three hours the planes are 2440 miles apart. Find the speed of each plane.

Solution:

Let S = rate of the southbound plane and $E = S + 50$ = rate of the eastbound plane.

$$d_S = r_S t = 3S$$

$$d_E = r_E t = 3(S + 50)$$

By the Pythagorean Theorem,

$${d_S}^2 + {d_E}^2 = (2440)^2$$

$$[3S]^2 + [3(S + 50)]^2 = (2440)^2$$

$$9S^2 + 9(S^2 + 100S + 2500) = 5,953,600$$

$$18S^2 + 900S - 5,931,100 = 0$$

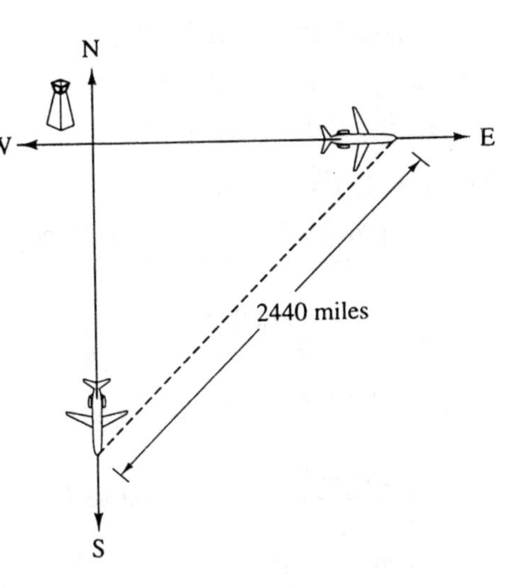

$$S = \frac{-900 \pm \sqrt{(900)^2 - 4(18)(-5931100)}}{2(18)}$$

$$S \approx \frac{-900 \pm 20684.51595}{36}$$

Considering the positive solution,

$$S \approx 550 \text{ mi/hr} \qquad E = S + 50 \approx 600 \text{ mi/hr}$$

61. A windlass is used to tow a boat to the dock. The rope is attached to the boat at a point 15 feet below the level of the windlass (see figure). Find the distance from the boat to the dock when 75 feet of rope has been let out.

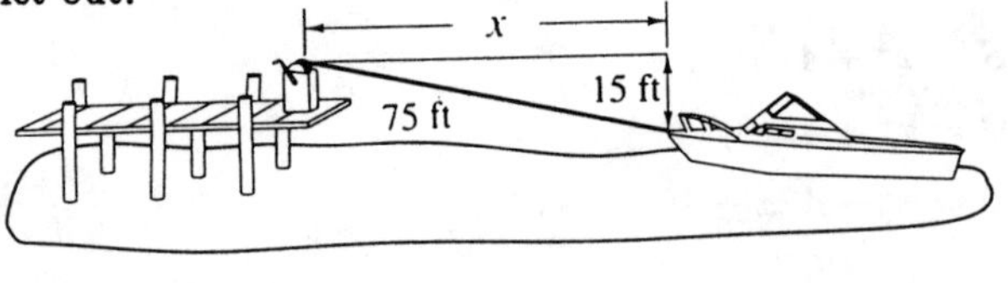

Solution:

$$x^2 + 15^2 = 75^2 \qquad \text{Pythagorean Theorem}$$

$$x^2 = 75^2 - 15^2$$

$$x^2 = 5400$$

$$x = \sqrt{5400} \approx 73.5 \text{ feet}$$

SECTION 2.5

Complex Numbers

- **You should know how to work with complex numbers.**
- **Operations on Complex Numbers**
  (a) Addition: $(a+bi)+(c+di)=(a+c)+(b+d)i$
  (b) Subtraction: $(a+bi)-(c+di)=(a-c)+(b-d)i$
  (c) Multiplication: $(a+bi)(c+di)=(ac-bd)+(ad+bc)i$
  (d) Division: $\dfrac{a+bi}{c+di}=\dfrac{a+bi}{c+di}\cdot\dfrac{c-di}{c-di}=\dfrac{ac+bd}{c^2+d^2}+\dfrac{bc-ad}{c^2+d^2}i$
- **The complex conjugate of $a+bi$ is $a-bi$:**

$$(a+bi)(a-bi)=a^2+b^2$$

- **The additive inverse of $a+bi$ is $-a-bi$.**
- **The multiplicative inverse of $a+bi$ is**

$$\frac{a-bi}{a^2+b^2}.$$

- $\sqrt{-a}=\sqrt{a}\,i$ for $a>0$.

Solutions to Selected Exercises

3. Find real numbers a and b so that the equation $(a-1)+(b+3)i=5+8i$ is true.

Solution:

$$(a-1)+(b+3)i=5+8i$$
$$a-1=5 \quad\Rightarrow\quad a=6$$
$$b+3=8 \quad\Rightarrow\quad b=5$$

7. Write $2-\sqrt{-27}$ in standard form.

Solution:

$$2-\sqrt{-27}=2-\sqrt{27}\,i=2-3\sqrt{3}\,i$$

11. Write $-6i+i^2$ in standard form.

Solution:

$$-6i+i^2=-6i+(-1)=-1-6i$$

15. Write $\sqrt{-0.09}$ in standard form.

Solution:

$$\sqrt{-0.09} = \sqrt{0.09}\,i = 0.3i$$

19. Perform the indicated operation and write the result in standard form.

$$(8 - i) - (4 - i)$$

Solution:

$$(8 - i) - (4 - i) = 8 - i - 4 + i = 4$$

21. Perform the indicated operation and write the result in standard form.

$$(-2 + \sqrt{-8}) + (5 - \sqrt{-50})$$

Solution:

$$(-2 + \sqrt{-8}) + (5 - \sqrt{-50}) = -2 + 2\sqrt{2}i + 5 - 5\sqrt{2}i = 3 - 3\sqrt{2}i$$

25. Perform the indicated operations and write the result in standard form.

$$-\left(\frac{3}{2} + \frac{5}{2}i\right) + \left(\frac{5}{3} + \frac{11}{3}i\right)$$

Solution:

$$\begin{aligned}-\left(\frac{3}{2} + \frac{5}{2}i\right) + \left(\frac{5}{3} + \frac{11}{3}i\right) &= -\frac{3}{2} - \frac{5}{2}i + \frac{5}{3} + \frac{11}{3}i \\ &= -\frac{9}{6} - \frac{15}{6}i + \frac{10}{6} + \frac{22}{6}i \\ &= \frac{1}{6} + \frac{7}{6}i\end{aligned}$$

29. Write the conjugate of $-2 - \sqrt{5}i$ and find the product of the number and its conjugate.

Solution:

The complex conjugate of $-2 - \sqrt{5}i$ is $-2 + \sqrt{5}i$.

$$(-2 - \sqrt{5}i)(-2 + \sqrt{5}i) = 4 - 5i^2 = 4 + 5 = 9.$$

31. Write the conjugate of $20i$ and find the product of the number and its conjugate.

Solution:

The complex conjugate of $20i$ is $-20i$.

$$(20i)(-20i) = -400i^2 = 400.$$

35. Perform the specified operation and write the result in standard form.

$$\sqrt{-6}\sqrt{-2}$$

Solution:

$$\sqrt{-6}\sqrt{-2} = (\sqrt{6}\,i)(\sqrt{2}\,i) = \sqrt{12}\,i^2 = 2\sqrt{3}(-1) = -2\sqrt{3}$$

Note: $\sqrt{-6}\sqrt{-2} \neq \sqrt{12}$

39. Perform the specified operation and write the result in standard form.

$$(1+i)(3-2i)$$

Solution:

$$(1+i)(3-2i) = 3 - 2i + 3i - 2i^2 = 3 + i + 2 = 5 + i$$

41. Perform the specified operation and write the result in standard form.

$$6i(5-2i)$$

Solution:

$$6i(5-2i) = 30i - 12i^2 = 12 + 30i$$

43. Perform the specified operation and write the result in standard form.

$$(\sqrt{14} + \sqrt{10}\,i)(\sqrt{14} - \sqrt{10}\,i)$$

Solution:

$$(\sqrt{14} + \sqrt{10}\,i)(\sqrt{14} - \sqrt{10}\,i) = 14 - 10i^2 = 14 + 10 = 24$$

47. Perform the specified operation and write the result in standard form.

$$(2+3i)^2 + (2-3i)^2$$

Solution:

$$\begin{aligned}(2+3i)^2 + (2-3i)^2 &= (4 + 12i + 9i^2) + (4 - 12i + 9i^2)\\ &= 8 + 18i^2\\ &= 8 - 18\\ &= -10\end{aligned}$$

51. Perform the specified operation and write the result in standard form.

$$\frac{2+i}{2-i}$$

Solution:

$$\frac{2+i}{2-i} = \frac{2+i}{2-i} \cdot \frac{2+i}{2+i} = \frac{4 + 4i + i^2}{4+1} = \frac{3+4i}{5} = \frac{3}{5} + \frac{4}{5}i$$

55. Perform the specified operation and write the result in standard form.

$$\frac{1}{(4-5i)^2}$$

Solution:

$$\frac{1}{(4-5i)^2} = \frac{1}{16-40i+25i^2} = \frac{1}{-9-40i} \cdot \frac{-9+40i}{-9+40i}$$

$$= \frac{-9+40i}{81+1600} = \frac{-9+40i}{1681} = -\frac{9}{1681} + \frac{40}{1681}i$$

59. Use the Quadratic Formula to solve $4x^2 + 16x + 17 = 0$.

Solution:

$4x^2 + 16x + 17 = 0;\ a = 4,\ b = 16,\ c = 17$

$$x = \frac{-16 \pm \sqrt{(16)^2 - 4(4)(17)}}{2(4)}$$

$$= \frac{-16 \pm \sqrt{-16}}{8} = \frac{-16 \pm 4i}{8}$$

$$= -2 \pm \frac{1}{2}i$$

63. Use the Quadratic Formula to solve $16t^2 - 4t + 3 = 0$.

Solution:

$16t^2 - 4t + 3 = 0;\ a = 16,\ b = -4,\ c = 3$

$$t = \frac{-(-4) \pm \sqrt{(-4)^2 - 4(16)(3)}}{2(16)}$$

$$= \frac{4 \pm \sqrt{-176}}{32} = \frac{4 \pm 4\sqrt{11}i}{32}$$

$$= \frac{1}{8} \pm \frac{\sqrt{11}}{8}i$$

65. Write out the first 16 positive powers of i and express each as i, $-i$, 1, or -1.

Solution:

$i = i$	$i^5 = i$	$i^9 = i$	$i^{13} = i$
$i^2 = -1$	$i^6 = -1$	$i^{10} = -1$	$i^{14} = -1$
$i^3 = -i$	$i^7 = -i$	$i^{11} = -i$	$i^{15} = -i$
$i^4 = 1$	$i^8 = 1$	$i^{12} = 1$	$i^{16} = 1$

69. Simplify $-5i^5$ and write it in standard form.

Solution:

$$-5i^5 = -5i$$

73. Simplify $\frac{1}{i^3}$ and write it in standard form.

Solution:

$$\frac{1}{i^3} = \frac{1}{-i} = \frac{1}{-i} \cdot \frac{i}{i} = \frac{i}{-i^2} = \frac{i}{1} = i$$

77. Prove that the sum of a complex number $a + bi$ and its conjugate is a real number.

Solution:

$$\begin{aligned}(a + bi) + (a - bi) &= (a + a) + (b - b)i \\ &= 2a + 0i = 2a \quad \text{which is a real number.}\end{aligned}$$

81. Prove that the conjugate of the sum of two complex numbers $a_1 + b_1 i$ and $a_2 + b_2 i$ is the sum of their conjugates.

Solution:

$$(a_1 + b_1 i) + (a_2 + b_2 i) = (a_1 + a_2) + (b_1 + b_2)i$$

The complex conjugate of the sum is $(a_1 + a_2) - (b_1 + b_2)i$, and the sum of the conjugates is

$$\begin{aligned}(a_1 - b_1 i) + (a_2 - b_2 i) &= (a_1 + a_2) + (-b_1 - b_2)i \\ &= (a_1 + a_2) - (b_1 + b_2)i\end{aligned}$$

Thus, the conjugate of the sum is the sum of the conjugates.

SECTION 2.6

Other Types of Equations

- You should be able to solve certain types of nonlinear or nonquadratic equations.
- For equations involving radicals or fractional powers, raise both sides to the same power.
- For equations that are of the quadratic type, $au^2+bu+c=0,\ a\neq 0$, use either factoring or the quadratic equation.
- Always check for extraneous solutions.

Solutions to Selected Exercises

5. Find all solutions of $x^4 - 81 = 0$.

Solution:

$$\begin{aligned} x^4 - 81 &= 0 \\ (x^2+9)(x^2-9) &= 0 \\ (x^2+9)(x+3)(x-3) &= 0 \\ x^2+9=0 \quad &\Rightarrow \quad x = \pm 3i \\ x+3=0 \quad &\Rightarrow \quad x=-3 \\ x-3=0 \quad &\Rightarrow \quad x=3 \end{aligned}$$

9. Find all solutions of $x^3 - 3x^2 - x + 3 = 0$.

Solution:

$$\begin{aligned} x^3 - 3x^2 - x + 3 &= 0 \\ x^2(x-3) - (x-3) &= 0 \\ (x-3)(x^2-1) &= 0 \\ (x-3)(x+1)(x-1) &= 0 \\ x-3=0 \quad &\Rightarrow \quad x=3 \\ x+1=0 \quad &\Rightarrow \quad x=-1 \\ x-1=0 \quad &\Rightarrow \quad x=1 \end{aligned}$$

15. Find all solutions of $x^4 + 5x^2 - 36 = 0$.

Solution:

$$\begin{aligned} x^4 + 5x^2 - 36 &= 0 \\ (x^2+9)(x^2-4) &= 0 \\ (x^2+9)(x+2)(x-2) &= 0 \\ x^2+9=0 \quad &\Rightarrow \quad x = \pm 3i \\ x+2=0 \quad &\Rightarrow \quad x=-2 \\ x-2=0 \quad &\Rightarrow \quad x=2 \end{aligned}$$

19. Find all solutions of $x^6 + 7x^3 - 8 = 0$.

Solution:

$$x^6 + 7x^3 - 8 = 0$$
$$(x^3 + 8)(x^3 - 1) = 0$$
$$(x + 2)(x^2 - 2x + 4)(x - 1)(x^2 + x + 1) = 0$$
$$x = -2, \quad x = \frac{2 \pm \sqrt{4 - 16}}{2}, \quad x = 1 \quad \text{or} \quad x = \frac{-1 \pm \sqrt{1 - 4}}{2}$$
$$x = -2, \quad x = 1 \pm \sqrt{3}\,i, \quad x = 1 \quad \text{or} \quad x = -\frac{1}{2} \pm \frac{\sqrt{3}}{2}i$$

21. Find all solutions of $\frac{1}{t^2} + \frac{8}{t} + 15 = 0$.

Solution:

$$\frac{1}{t^2} + \frac{8}{t} + 15 = 0$$
$$1 + 8t + 15t^2 = 0$$
$$(1 + 3t)(1 + 5t) = 0$$
$$1 + 3t = 0 \quad \Rightarrow \quad t = -\frac{1}{3}$$
$$1 + 5t = 0 \quad \Rightarrow \quad t = -\frac{1}{5}$$

25. Find all solutions of $5 - 3x^{1/3} - 2x^{2/3} = 0$.

Solution:

$$5 - 3x^{1/3} - 2x^{2/3} = 0$$
$$(5 + 2x^{1/3})(1 - x^{1/3}) = 0$$
$$5 + 2x^{1/3} = 0 \quad \text{or} \quad 1 - x^{1/3} = 0$$
$$x^{1/3} = -\frac{5}{2} \quad \text{or} \quad x^{1/3} = 1$$
$$x = \left(-\frac{5}{2}\right)^3 \quad \text{or} \quad x = (1)^3$$
$$x = -\frac{125}{8} \quad \text{or} \quad x = 1$$

29. Find all solutions of $\sqrt{x-10}-4=0$.

Solution:

$$\begin{aligned}\sqrt{x-10}-4&=0\\ \sqrt{x-10}&=4\\ x-10&=16\\ x&=26\end{aligned}$$

35. Find all solutions of $-\sqrt{26-11x}+4=x$.

Solution:

$$\begin{aligned}-\sqrt{26-11x}+4&=x\\ 4-x&=\sqrt{26-11x}\\ (4-x)^2&=26-11x\\ 16-8x+x^2&=26-11x\\ x^2+3x-10&=0\\ (x+5)(x-2)&=0\\ x+5=0 \quad&\Rightarrow\quad x=-5\\ x-2=0 \quad&\Rightarrow\quad x=2\end{aligned}$$

39. Find all solutions of $\sqrt{x}+\sqrt{x-20}=10$.

Solution:

$$\begin{aligned}\sqrt{x}+\sqrt{x-20}&=10\\ \sqrt{x}&=10-\sqrt{x-20}\\ (\sqrt{x})^2&=(10-\sqrt{x-20})^2\\ x&=100-20\sqrt{x-20}+x-20\\ x&=x+80-20\sqrt{x-20}\\ -80&=-20\sqrt{x-20}\\ 4&=\sqrt{x-20}\\ 16&=x-20\\ 36&=x\end{aligned}$$

43. Find all solutions of $\sqrt{7x+36}-\sqrt{5x+16}=2$.

Solution:

$$\begin{aligned}
\sqrt{7x+36}-\sqrt{5x+16}&=2\\
\sqrt{7x+36}&=2+\sqrt{5x+16}\\
\left(\sqrt{7x+36}\right)^2&=\left(2+\sqrt{5x+16}\right)^2\\
7x+36&=4+4\sqrt{5x+16}+5x+16\\
7x+36&=5x+20+4\sqrt{5x+16}\\
2x+16&=4\sqrt{5x+16}\\
x+8&=2\sqrt{5x+16}\\
(x+8)^2&=\left(2\sqrt{5x+16}\right)^2\\
x^2+16x+64&=4(5x+16)\\
x^2+16x+64&=20x+64\\
x^2-4x&=0\\
x(x-4)&=0\\
x&=0\\
x-4=0 \;\Rightarrow\; x&=4
\end{aligned}$$

47. Find all solutions of $(x+3)^{3/4}=27$.

Solution:

$$\begin{aligned}
(x+3)^{3/4}&=27\\
x+3&=27^{4/3}\\
x+3&=81\\
x&=78
\end{aligned}$$

49. Find all solutions of $(x^2-5)^{2/3}=16$.

Solution:

$$\begin{aligned}
(x^2-5)^{2/3}&=16\\
x^2-5&=\pm 16^{3/2}\\
x^2-5&=\left(\pm\sqrt{16}\right)^3\\
x^2-5&=\pm 4^3\\
x^2&=5\pm 64\\
x^2&=\quad 69 \quad\text{or}\quad x^2=-59\\
x&=\pm\sqrt{69}\quad\text{or}\quad x=\pm\sqrt{59}\,i
\end{aligned}$$

55. Find all solutions of $\frac{1}{x} - \frac{1}{x+1} = 3.$

Solution:

$$\begin{aligned}
\frac{1}{x} - \frac{1}{x+1} &= 3 \\
x(x+1)\frac{1}{x} - x(x+1)\frac{1}{x+1} &= x(x+1)(3) \\
x + 1 - x &= 3x(x+1) \\
1 &= 3x^2 + 3x \\
0 &= 3x^2 + 3x - 1; \quad a = 3,\ b = 3,\ c = -1 \\
x &= \frac{-3 \pm \sqrt{(3)^2 - 4(3)(-1)}}{2(3)} = \frac{-3 \pm \sqrt{21}}{6} = -\frac{1}{2} \pm \frac{\sqrt{21}}{6}
\end{aligned}$$

57. Find all solutions of $x = \frac{3}{x} + \frac{1}{2}.$

Solution:

$$\begin{aligned}
x &= \frac{3}{x} + \frac{1}{2} \\
(2x)(x) &= (2x)\left(\frac{3}{x}\right) + (2x)\left(\frac{1}{2}\right) \\
2x^2 &= 6 + x \\
2x^2 - x - 6 &= 0 \\
(2x+3)(x-2) &= 0
\end{aligned}$$

$$2x + 3 = 0 \quad \Rightarrow \quad x = -\frac{3}{2}$$

$$x - 2 = 0 \quad \Rightarrow \quad x = 2$$

61. Find all solutions of $\frac{4}{x+1} - \frac{3}{x+2} = 1.$

Solution:

$$\begin{aligned}
\frac{4}{x+1} - \frac{3}{x+2} &= 1 \\
(x+1)(x+2)\frac{4}{x+1} - (x+1)(x+2)\frac{3}{x+2} &= (x+1)(x+2)(1) \\
4(x+2) - 3(x+1) &= (x+1)(x+2) \\
4x + 8 - 3x - 3 &= x^2 + 3x + 2 \\
x + 5 &= x^2 + 3x + 2 \\
0 &= x^2 + 2x - 3 \\
0 &= (x+3)(x-1)
\end{aligned}$$

$$x + 3 = 0 \quad \Rightarrow \quad x = -3$$

$$x - 1 = 0 \quad \Rightarrow \quad x = 1$$

65. Find all solutions of $|2x-1|=5$.

Solution:

$$\begin{aligned} 2x-1&=-5 &\quad\text{or}\quad 2x-1&=5\\ 2x&=-4 & 2x&=6\\ x&=-2 & x&=3 \end{aligned}$$

69. Find all solutions of $|x-10|=x^2-10x$.

Solution:

$$\begin{aligned} x-10 &= x^2-10x & \quad\text{or}\quad -(x-10) &= x^2-10x\\ 0 &= x^2-11x+10 & 0 &= x^2-9x-10\\ 0 &= (x-1)(x-10) & 0 &= (x+1)(x-10)\\ x-1 &= 0 \Rightarrow x=1 & x+1 &= 0 \Rightarrow x=-1\\ x-10 &= 0 \Rightarrow x=10 & x-10 &= 0 \Rightarrow x=10 \end{aligned}$$

By checking, we see that $x=10$ and $x=-1$ are the only real solutions. $x=1$ is an extraneous solution.

73. Use a calculator to find the real solutions of $1.8x-6\sqrt{x}-5.6=0$. Round your answer to three decimal places.

Solution:

$$\begin{aligned} 1.8x-6\sqrt{x}-5.6 &= 0 && \text{Given equation}\\ 1.8(\sqrt{x})^2-6\sqrt{x}-5.6 &= 0 && \text{Quadratic form with } u=\sqrt{x} \end{aligned}$$

Use the Quadratic Formula with $a=1.8$, $b=-6$, and $c=-5.6$.

$$\sqrt{x}=\frac{6\pm\sqrt{36-4(1.8)(-5.6)}}{2(1.8)}\approx\frac{6\pm 8.7361}{3.6}$$

Considering only the positive value for $\sqrt{x}$, we have

$$\begin{aligned} \sqrt{x}&\approx 4.0934\\ x&\approx 16.756 \end{aligned}$$

77. An airline runs a commuter flight between two cities that are 720 miles apart. If the average speed of the plane is increased by 40 miles per hour, the travel time is decreased by 12 minutes. What airspeed is required to obtain this decrease in travel time?

Solution:

Formula: $\text{Time} = \dfrac{\text{Distance}}{\text{Rate}}$

Let x = average speed of the plane. Then we have a travel time of $t = \dfrac{720}{x}$.

If the average speed is increased by 40 mph, then

$$t - \frac{12}{60} = \frac{720}{x+40}$$

$$t = \frac{720}{x+40} + \frac{1}{5}$$

Now, we equate these two equations and solve for x.

$$\begin{aligned}
\frac{720}{x} &= \frac{720}{x+40} + \frac{1}{5} \\
720(5)(x+40) &= 720(5)x + x(x+40) \\
3600x + 144{,}000 &= 3600x + x^2 + 40x \\
0 &= x^2 + 40x - 144{,}000 \\
0 &= (x+400)(x-360)
\end{aligned}$$

Using the positive value for x we have $x = 360$ mph and $x + 40 = 400$ mph.

The airspeed required to obtain the decrease in travel time is 400 miles per hour.

83. The demand equation for a certain product is

$$p = 40 - \sqrt{0.01x + 1}$$

where x is the number of units demanded per day and p is the price per unit. Find the demand if the price is set at \$37.55.

Solution:

$$\begin{aligned}
37.55 &= 40 - \sqrt{0.01x + 1} \\
\sqrt{0.01x + 1} &= 2.45 \\
0.01x + 1 &= 6.0025 \\
0.01x &= 5.0025 \\
x &= 500.25
\end{aligned}$$

Rounding x to the nearest whole unit yields $x \approx 500$ units.

85. A power station is on one side of a river that is 1/2–mile wide. A factory is 6 miles downstream on the other side of the river. It costs \$18 per foot to run power lines overland and \$24 per foot to run them underwater. The total cost of the project is \$616,877.27. Find the length x as labeled in the figure.

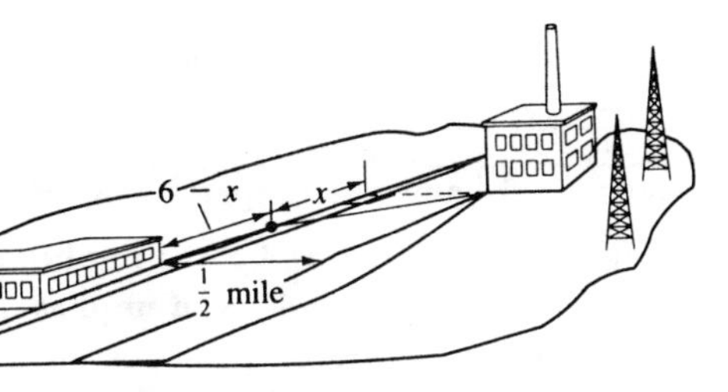

Solution:

Verbal Model Cost = Cost underwater • Distance underwater + Cost overland • Distance overland

Labels
Cost = \$616,877.27
Cost underwater = \$24 (per foot)
Distance underwater (in feet) $= 5280\sqrt{x^2 + (1/4)} = 2640\sqrt{4x^2 + 1}$
Cost overland = \$18 (per foot)
Distance overland (in feet) $= 5280(6 - x)$

Algebraic Equation

$$616{,}877.27 = 24(2640)\sqrt{4x^2+1} + 18(5280)(6-x)$$
$$616{,}877.27 = 12(2640)[2\sqrt{4x^2+1} + 3(6-x)]$$
$$19.472136 = 2\sqrt{4x^2+1} + 18 - 3x$$
$$3x + 1.472136 = 2\sqrt{4x^2+1}$$
$$9x^2 + 8.8328163x + 2.16718455 = 4(4x^2+1)$$
$$7x^2 - 8.8328163x + 1.83281545 = 0$$
$$x = \frac{8.8328163 \pm \sqrt{26.69981119}}{14} \approx \frac{8.8328 \pm 5.1672}{14}$$
$$x = 1 \text{ mi} \quad \text{or} \quad x = 0.262 \text{ mi}$$

87. The surface area of a cone is given by $S = \pi r\sqrt{r^2 + h^2}$. Solve this equation for h.

Solution:

$$S = \pi r\sqrt{r^2 + h^2}$$
$$S^2 = \pi^2 r^2 (r^2 + h^2)$$
$$S^2 = \pi^2 r^4 + \pi^2 r^2 h^2$$
$$\frac{S^2 - \pi^2 r^4}{\pi^2 r^2} = h^2$$
$$h = \frac{\sqrt{S^2 - \pi^2 r^4}}{\pi r}$$

SECTION 2.7

Linear Inequalities

- You should know the properties of inequalities.
  (a) Transitive: $a < b$ and $b < c$ implies $a < c$.
  (b) Addition: $a < b$ and $c < d$ implies $a + c < b + d$.
  (c) Adding or Subtracting a Constant: $a \pm c < b \pm c$ if $a < b$.
  (d) Multiplying or Dividing a Constant: For $a < b$,
    1. If $c > 0$, then $ac < bc$ and $\frac{a}{c} < \frac{b}{c}$.
    2. If $c < 0$, then $ac > bc$ and $\frac{a}{c} > \frac{b}{c}$.
- You should know that

$$|x| = \begin{cases} x & \text{if } x \geq 0 \\ -x & \text{if } x < 0 \end{cases}.$$

- You should be able to solve absolute value inequalities.
  (a) $|x| < a$ if and only if $-a < x < a$.
  (b) $|x| > a$ if and only if $x < -a$ or $x > a$.

Solutions to Selected Exercises

1. Write an inequality to represent the interval $[-1, 3]$ and state whether the interval is bounded or unbounded.

Solution:

$[-1, 3]$ corresponds to $-1 \leq x \leq 3$. Bounded

5. Determine whether the values of x are solutions of the inequality $5x - 12 > 0$.

(a) $x = 3$ (b) $x = -3$ (c) $x = \frac{5}{2}$ (d) $x = \frac{3}{2}$

5. —CONTINUED—

Solution:

(a) $x = 3$

$5(3) - 12 \overset{?}{>} 0$

$3 > 0$

$x = 3$ is a solution.

(b) $x = -3$

$5(-3) - 12 \overset{?}{>} 0$

$-27 \not> 0$

$x = -3$ is not a solution.

(c) $x = \frac{5}{2}$

$5\left(\frac{5}{2}\right) - 12 \overset{?}{>} 0$

$\frac{1}{2} > 0$

$x = \frac{5}{2}$ is a solution.

(d) $x = \frac{3}{2}$

$5\left(\frac{3}{2}\right) - 12 \overset{?}{>} 0$

$-\frac{9}{2} \not> 0$

$x = \frac{3}{2}$ is not a solution.

7. Determine whether the given values of x are solutions of the inequality.

$$0 < \frac{x-2}{4} < 2$$

(a) $x = 4$ (b) $x = 10$ (c) $x = 0$ (d) $x = \frac{7}{2}$

Solution:

(a) $x = 4$

$0 \overset{?}{<} \frac{4-2}{4} \overset{?}{<} 2$

$0 < \frac{1}{2} < 2$

$x = 4$ is a solution.

(b) $x = 10$

$0 \overset{?}{<} \frac{10-2}{4} \overset{?}{<} 2$

$0 < 2 \not< 2$

$x = 10$ is not a solution.

(c) $x = 0$

$0 \overset{?}{<} \frac{0-2}{4} \overset{?}{<} 2$

$0 \not< -\frac{1}{2} < 2$

$x = 0$ is not a solution.

(d) $x = \frac{7}{2}$

$0 \overset{?}{<} \frac{(7/2)-2}{4} \overset{?}{<} 2$

$0 < \frac{3}{8} < 2$

$x = \frac{7}{2}$ is a solution.

13. Match $|x| < 4$ with its graph.

Solution:

$|x| < 4$

$-4 < x < 4$

Matches graph (g)

−5 −4 −3 −2 −1 0 1 2 3 4 5 x

17. Solve $4x < 12$ and sketch the solution on the real number line.

Solution:

$$4x < 12$$
$$x < 3$$

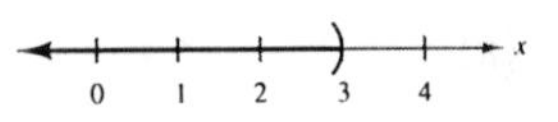

23. Solve $4(x + 1) < 2x + 3$ and sketch the solution on the real number line.

Solution:

$$4(x + 1) < 2x + 3$$
$$4x + 4 < 2x + 3$$
$$2x < -1$$
$$x < -\frac{1}{2}$$

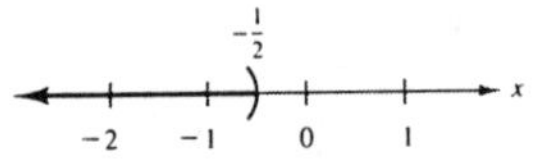

27. Solve $4 - 2x < 3$ and sketch the solution on the real number line.

Solution:

$$4 - 2x < 3$$
$$-2x < -1$$
$$x > \frac{1}{2}$$

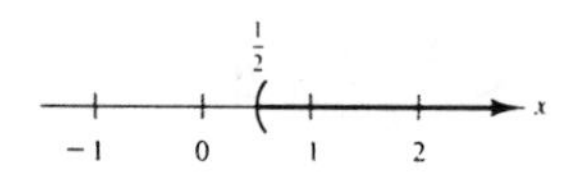

31. Solve

$$-4 < \frac{2x - 3}{3} < 4$$

and sketch the solution on the real number line.

Solution:

$$-4 < \frac{2x - 3}{3} < 4$$
$$-12 < 2x - 3 < 12$$
$$-9 < 2x < 15$$
$$-\frac{9}{2} < x < \frac{15}{2}$$

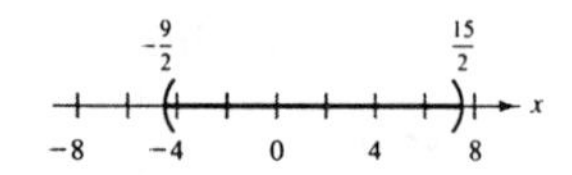

37. Solve $|x/2| > 3$ and sketch the solution on the real number line.

Solution:

$$\left|\frac{x}{2}\right| > 3$$
$$\frac{x}{2} < -3 \quad \text{or} \quad \frac{x}{2} > 3$$
$$x < -6 \quad \text{or} \quad x > 6$$

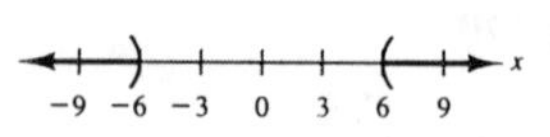

39. Solve $|x - 20| \leq 4$ and sketch the solution on the real number line.

Solution:

$$|x - 20| \leq 4$$
$$-4 \leq x - 20 \leq 4$$
$$16 \leq x \leq 24$$

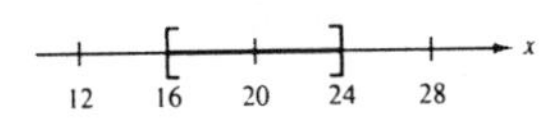

43. Solve

$$\left|\frac{x-3}{2}\right| \geq 5$$

and sketch the solution on the real number line.

Solution:

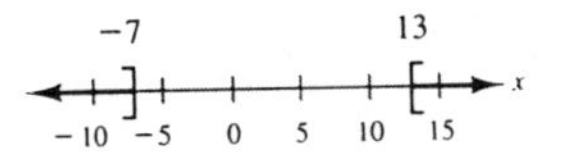

$$\left|\frac{x-3}{2}\right| \geq 5$$

$$\frac{x-3}{2} \leq -5 \quad \text{or} \quad \frac{x-3}{2} \geq 5$$
$$x - 3 \leq -10 \quad \text{or} \quad x - 3 \geq 10$$
$$x \leq -7 \quad \text{or} \quad x \geq 13$$

47. Solve $2|x + 10| \geq 9$ and sketch the solution on the real number line.

Solution:

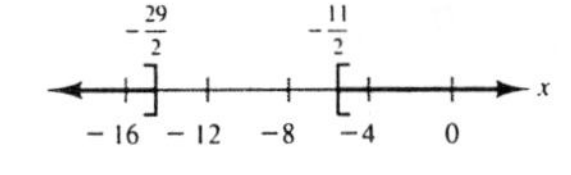

$$2|x + 10| \geq 9$$
$$|x + 10| \geq \frac{9}{2}$$
$$x + 10 \leq -\frac{9}{2} \quad \text{or} \quad x + 10 \geq \frac{9}{2}$$
$$x \leq -\frac{29}{2} \quad \text{or} \quad x \geq -\frac{11}{2}$$

51. Find the interval on the real number line for which the radicand in $\sqrt{x-5}$ is nonnegative.

Solution:

The radicand of $\sqrt{x-5}$ is $x - 5$.

$$x - 5 \geq 0$$
$$x \geq 5$$

Therefore, the interval is $[5, \infty)$.

55. Find the interval on the real number line for which the radicand in $\sqrt[4]{7-2x}$ is nonnegative.

Solution:

The radicand in $\sqrt[4]{7-2x}$ is $7-2x$.

$$7-2x \geq 0$$
$$-2x \geq -7$$
$$x \leq \frac{7}{2}$$

Therefore, the interval is $\left(-\infty, \frac{7}{2}\right]$.

59. Use absolute value notation to define the pair of intervals on the real line.

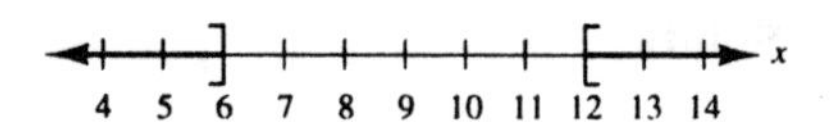

Solution:

$$|x-9| \geq 3$$

63. Use absolute value notation to define following the interval on the real line:

All real numbers whose distances from -3 are more than 5.

Solution:

$$|x-(-3)| > 5$$
$$|x+3| > 5$$

65. Suppose you can rent a midsize car from Company A for \$250 per week with no extra charge for mileage. A similar car can be rented from Company B for \$150 per week plus \$0.25 cents for each mile driven. How many miles must you drive in a week to make the rental fee for Company B *greater than* that for Company A?

Solution:

$$150 + 0.25x > 250$$
$$0.25x > 100$$
$$x > 400$$

If you drive more than 400 miles in a week, the rental fee for Company B is greater than the rental fee for Company A.

69. The revenue for selling x units of a product is $R = 115.95x$. The cost of producing x units is $C = 95x + 750$. In order to obtain a profit, the revenue must be greater than the cost. For what values of x will this product return a profit?

Solution:

$$R > C$$
$$115.95x > 95x + 750$$
$$20.95x > 750$$
$$x > 35.7995$$
$$x \geq 36 \text{ units}$$

75. The heights, h, of two–thirds of the members of a certain population satisfy the inequality

$$\left|\frac{h - 68.5}{2.7}\right| \leq 1$$

where h is measured in inches. Determine the interval on the real line in which these heights lie.

Solution:

$$\left|\frac{h - 68.5}{2.7}\right| \leq 1$$
$$-1 \leq \frac{h - 68.5}{2.7} \leq 1$$
$$-2.7 \leq h - 68.5 \leq 2.7$$
$$65.8 \text{ inches } \leq h \leq 71.2 \text{ inches}$$

65.8 71.2
h
64 66 68 70 72

77. Given two real numbers a and b, such that $a > b > 0$, prove that $1/a < 1/b$.

Solution:

$$a > b > 0$$
$$b < a \quad \text{where } a \text{ and } b \text{ are both positive}$$
$$1 < \frac{a}{b}$$
$$1 \cdot \frac{1}{a} < \frac{a}{b} \cdot \frac{1}{a}$$
$$\frac{1}{a} < \frac{1}{b}$$

SECTION 2.8

Other Types of Inequalities

- You should be able to solve inequalities.
  (a) Find the critical numbers.
    1. Values that make the expression zero
    2. Values that make the expression undefined
  (b) Test one value in each interval on the real number line resulting from the critical numbers.
  (c) Determine the solution intervals.

Solutions to Selected Exercises

3. Solve the inequality $x^2 > 4$ and give the answer in interval notation.

Solution:

$$x^2 > 4$$
$$x^2 - 4 > 0$$
$$(x+2)(x-2) > 0$$

Critical numbers:

$x = \pm 2$

Test intervals:

$(-\infty, -2)$, $(-2, 2)$, $(2, \infty)$

Solution intervals:

$(-\infty, -2) \cup (2, \infty)$

7. Solve the inequality $x^2+4x+4 \geq 9$ and give the answer in interval notation.

Solution:

$$x^2 + 4x + 4 \geq 9$$
$$x^2 + 4x - 5 \geq 0$$
$$(x+5)(x-1) \geq 0$$

Critical numbers:

$x = -5$, $x = 1$

Test intervals:

$(-\infty, -5)$, $(-5, 1)$, $(1, \infty)$

Solution intervals:

$(-\infty, -5] \cup [1, \infty)$

11. Solve the inequality

$$3(x-1)(x+1) > 0$$

and give the answer in interval notation.

Solution:

$$3(x-1)(x+1) > 0$$

Critical numbers:

$x = -1,\ x = 1$

Test intervals:

$(-\infty,\ -1),\ (-1,\ 1),\ (1,\ \infty)$

Solution intervals:

$(-\infty,\ -1) \cup (1,\ \infty)$

15. Solve the inequality

$$4x^3 - 6x^2 < 0$$

and give the answer in interval notation.

Solution:

$$4x^3 - 6x^2 < 0$$

$$2x^2(2x-3) < 0$$

Critical numbers:

$x = 0,\ x = \frac{3}{2}$

Test intervals:

$(-\infty,\ 0),\ (0,\ \frac{3}{2}),\ (\frac{3}{2},\ \infty)$

Solution intervals:

$(-\infty,\ 0) \cup (0,\ \frac{3}{2})$

Note: $x = 0$ is *not* a solution.

19. Solve the inequality

$$(x-1)^2(x+2)^3 \geq 0$$

and give the answer in interval notation.

Solution:

$$(x-1)^2(x+2)^3 \geq 0$$

Critical numbers:

$x = 1,\ x = -2$

Test intervals:

$(-\infty,\ -2),\ (-2,\ 1),\ (1,\ \infty)$

Solution intervals:

$[-2,\ 1] \cup [1,\ \infty)$ or $[-2,\ \infty)$

21. Solve the inequality

$$\frac{1}{x} - x > 0$$

and give the answer in interval notation.

Solution:

$$\frac{1}{x} - x > 0$$

$$\frac{1-x^2}{x} > 0$$

$$\frac{(1+x)(1-x)}{x} > 0$$

Critical numbers:

$x = -1,\ x = 0,\ x = 1$

Test intervals:

$(-\infty,\ -1),\ (-1,\ 0),\ (0,\ 1),\ (1,\ \infty)$

Solution intervals:

$(-\infty,\ -1) \cup (0,\ 1)$

25. Solve the inequality $\frac{3x-5}{x-5} > 4$ and give the answer in interval notation.

Solution:

$$\frac{3x-5}{x-5} > 4$$

$$\frac{3x-5-4(x-5)}{x-5} > 0$$

$$\frac{15-x}{x-5} > 0$$

Critical numbers: $x = 5,\ x = 15$

Test intervals: $(-\infty,\ 5),\ (5,\ 15),\ (15,\ \infty)$

Solution interval: $(5,\ 15)$

29. Solve the inequality $\frac{1}{x-3} \le \frac{9}{4x+3}$ and give the answer in interval notation.

Solution:

$$\frac{1}{x-3} \le \frac{9}{4x+3}$$

$$\frac{1}{x-3} - \frac{9}{4x+3} \le 0$$

$$\frac{4x+3-9(x-3)}{(x-3)(4x+3)} \le 0$$

$$\frac{30-5x}{(x-3)(4x+3)} \le 0$$

$$\frac{-5(x-6)}{(x-3)(4x+3)} \le 0$$

Critical numbers: $x = -\frac{3}{4},\ x = 3,\ x = 6$

Test intervals: $\left(-\infty,\ -\frac{3}{4}\right),\ \left(-\frac{3}{4},\ 3\right),\ (3,\ 6),\ (6,\ \infty)$

Solution intervals: $\left(-\frac{3}{4},\ 3\right) \cup [6,\ \infty)$

Note: We have $x = 6$ in the solution intervals, but not $x = -\frac{3}{4}$ and $x = 3$ since they yield a zero in the denominator.

33. Solve the inequality $\frac{3}{x-1} - \frac{2}{x+1} < 1$ and give the answer in interval notation.

Solution:

$$\frac{3}{x-1} - \frac{2}{x+1} < 1$$

$$\frac{3}{x-1} - \frac{2}{x+1} - 1 < 0$$

$$\frac{3(x+1) - 2(x-1) - (x-1)(x+1)}{(x-1)(x+1)} < 0$$

$$\frac{-x^2 + x + 6}{(x-1)(x+1)} < 0$$

$$\frac{-(x-3)(x+2)}{(x-1)(x+1)} < 0$$

Critical numbers: $x = 3,\ x = -2,\ x = \pm 1$

Test intervals: $(-\infty,\ -2),\ (-2,\ -1),\ (-1,\ 1),\ (1,\ 3),\ (3,\ \infty)$

Solution intervals: $(-\infty,\ -2) \cup (-1,\ 1) \cup (3,\ \infty)$

37. Find the domain of x in the expression $\sqrt{x^2 - 7x + 12}$.

Solution:

The radicand is $x^2 - 7x + 12$.

$$x^2 - 7x + 12 \geq 0$$

$$(x-3)(x-4) \geq 0$$

Critical numbers: $x = 3,\ x = 4$

Test intervals: $(-\infty,\ 3),\ (3,\ 4),\ (4,\ \infty)$

Domain: $(-\infty,\ 3] \cup [4,\ \infty)$

41. Use a calculator to solve the inequality $-0.5x^2 + 12.5x + 1.6 > 0$.

Solution:

$$-0.5x^2 + 12.5x + 1.6 > 0$$

Critical numbers: $x = \dfrac{-12.5 \pm \sqrt{(12.5)^2 - 4(-0.5)(1.6)}}{2(-0.5)}$

$$x \approx -0.13, \quad x \approx 25.13$$

Test intervals: $(-\infty,\ -0.13),\ (-0.13,\ 25.13),\ (25.13,\ \infty)$

Solution interval: $(-0.13,\ 25.13)$

45. A rectangular playing field with a perimeter of 100 meters is to have an area of at least 500 square meters. Within what bounds must the length of the rectangle lie?

Solution:

$$2L + 2W = 100$$
$$W = \frac{100 - 2L}{2} = 50 - L$$
$$LW \geq 500$$
$$L(50 - L) \geq 500$$
$$50L - L^2 \geq 500$$
$$0 \geq L^2 - 50L + 500$$

By the quadratic formula the critical numbers are:

$$L = \frac{50 \pm \sqrt{(50)^2 - 4(500)}}{2} = \frac{50 \pm \sqrt{500}}{2} = \frac{50 \pm 10\sqrt{5}}{2} = 25 \pm 5\sqrt{5}$$

Solution interval: $[25 - 5\sqrt{5},\ 25 + 5\sqrt{5}]$, or $13.8197 \text{ meters} \leq L \leq 36.1803 \text{ meters}$

49. The percentage of the American population that graduated from college between 1940 and 1987 is approximated by the model

$$\text{Percent of graduates} = 5.136 + 0.0069t^2$$

where the time t represents the calendar year with $t = 0$ corresponding to 1940. According to this model, when will the percentage of college graduates exceed 25% of the population? (*Source: U.S. Bureau of Census*)

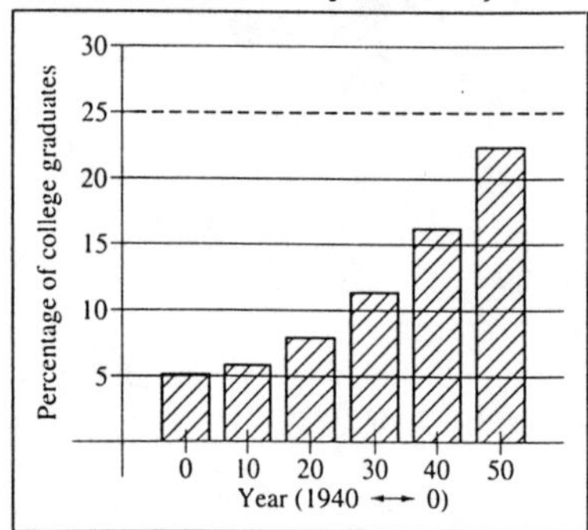

Solution:

$$5.136 + 0.0069t^2 > 25$$
$$0.0069t^2 > 19.864$$
$$t^2 > 2878.84058$$
$$t > 53.65$$

The percentage of college graduates will exceed 25% of the population in $1940 + 53.65 = 1993.65$ or in the year 1993.

REVIEW EXERCISES FOR CHAPTER 2

Solutions to Selected Exercises

1. Determine whether $6 - (x-2)^2 = 2 + 4x - x^2$ is an identity or conditional equation.

Solution:

$$\begin{aligned}
6 - (x-2)^2 &= 2 + 4x - x^2 \\
6 - (x^2 - 4x + 4) &= 2 + 4x - x^2 \\
2 + 4x - x^2 &= 2 + 4x - x^2 \\
0 &= 0 \qquad \text{Identity}
\end{aligned}$$

All real numbers are solutions.

5. Solve the equation $3x - 2(x+5) = 10$.

Solution:

$$\begin{aligned}
3x - 2(x+5) &= 10 \\
3x - 2x - 10 &= 10 \\
x &= 20
\end{aligned}$$

9. Solve the equation

$$3\left(1 - \frac{1}{5t}\right) = 0.$$

Solution:

$$\begin{aligned}
3\left(1 - \frac{1}{5t}\right) &= 0 \\
1 - \frac{1}{5t} &= 0 \\
1 &= \frac{1}{5t} \\
5t &= 1 \\
t &= \frac{1}{5}
\end{aligned}$$

13. Solve the equation $6x^2 = 5x + 4$.

Solution:

$$\begin{aligned}
6x^2 &= 5x + 4 \\
6x^2 - 5x - 4 &= 0 \\
(3x - 4)(2x + 1) &= 0 \\
x = \frac{4}{3} \quad \text{or} \quad x &= -\frac{1}{2}
\end{aligned}$$

19. Solve the equation $5x^4 - 12x^3 = 0$.

Solution:

$$\begin{aligned}
5x^4 - 12x^3 &= 0 \\
x^3(5x - 12) &= 0 \\
x^3 = 0 \quad \text{or} \quad 5x - 12 &= 0 \\
x = 0 \quad \text{or} \quad x &= \frac{12}{5}
\end{aligned}$$

21. Solve the equation $4t^3 - 12t^2 + 8t = 0$.

Solution:

$$4t^3 - 12t^2 + 8t = 0$$
$$4t(t^2 - 3t + 2) = 0$$
$$4t(t-1)(t-2) = 0$$
$$t = 0, \ t = 1, \ t = 2$$

25. Solve the equation

$$\frac{4}{(x-4)^2} = 1.$$

Solution:

$$\frac{4}{(x-4)^2} = 1$$
$$4 = (x-4)^2$$
$$\pm 2 = x - 4$$
$$4 \pm 2 = x$$
$$x = 6 \quad \text{or} \quad x = 2$$

29. Solve the equation $\sqrt{x+4} = 3$.

Solution:

$$\sqrt{x+4} = 3$$
$$(\sqrt{x+4})^2 = (3)^2$$
$$x + 4 = 9$$
$$x = 5$$

33. Solve the equation $\sqrt{2x+3} + \sqrt{x-2} = 2$.

Solution:

$$\sqrt{2x+3} + \sqrt{x-2} = 2$$
$$(\sqrt{2x+3})^2 = (2 - \sqrt{x-2})^2$$
$$2x + 3 = 4 - 4\sqrt{x-2} + x - 2$$
$$x + 1 = -4\sqrt{x-2}$$
$$(x+1)^2 = (-4\sqrt{x-2})^2$$
$$x^2 + 2x + 1 = 16(x-2)$$
$$x^2 - 14x + 33 = 0$$
$$(x-3)(x-11) = 0$$
$$x = 3, \text{ extraneous} \quad \text{or} \quad x = 11, \text{ extraneous}$$

No solution

37. Solve the equation $(x+4)^{1/2}+5x(x+4)^{3/2}=0$.

Solution:

$$(x+4)^{1/2}+5x(x+4)^{3/2}=0$$
$$(x+4)^{1/2}[1+5x(x+4)]=0$$
$$(x+4)^{1/2}(5x^2+20x+1)=0$$
$$(x+4)^{1/2}=0$$
$$x=-4 \quad \text{OR}$$

$$5x^2+20x+1=0$$
$$x=\frac{-20\pm\sqrt{400-20}}{10}$$
$$x=\frac{-20\pm 2\sqrt{95}}{10}$$
$$x=-2\pm\frac{\sqrt{95}}{5}$$

39. Solve the equation $|x-5|=10$.

Solution:

$$|x-5|=10$$
$$x-5=-10 \quad \text{or} \quad x-5=10$$
$$x=-5 \qquad\qquad x=15$$

43. Solve for $r: V=\frac{1}{3}\pi r^2 h$.

Solution:

$$V=\frac{1}{3}\pi r^2 h$$
$$3V=\pi r^2 h$$
$$\frac{3V}{\pi h}=r^2$$
$$r=\sqrt{\frac{3V}{\pi h}}$$

Since r represents the radius of a cone, r is positive only.

45. Solve for $p: L=\frac{k}{3\pi r^2 p}$.

Solution:

$$L=\frac{k}{3\pi r^2 p}$$
$$3\pi r^2 pL=k$$
$$p=\frac{k}{3\pi r^2 L}$$

49. Solve the inequality $x^2-4\le 0$.

Solution:

$$x^2-4\le 0$$
$$(x+2)(x-2)\le 0$$

Critical numbers:

$x=2,\ x=-2$

Test intervals:

$(-\infty,\ -2),\ (-2,\ 2),\ (2,\ \infty)$

Solution interval:

$[-2,\ 2]$

53. Solve the inequality $|x - 2| < 1$.

Solution:

$$|x-2| < 1$$
$$-1 < x - 2 < 1$$
$$1 < x < 3$$

55. Solve the inequality.

$$\left|x - \frac{3}{2}\right| \geq \frac{3}{2}$$

Solution:

$$\left|x - \frac{3}{2}\right| \geq \frac{3}{2}$$

$$x - \frac{3}{2} \leq -\frac{3}{2} \quad \text{or} \quad x - \frac{3}{2} \geq \frac{3}{2}$$
$$x \leq 0 \quad \text{or} \quad x \geq 3$$

61. Perform the indicated operations and write the result in standard form.

$$\left(\frac{\sqrt{2}}{2} - \frac{\sqrt{2}}{2}i\right) - \left(\frac{\sqrt{2}}{2} + \frac{\sqrt{2}}{2}i\right)$$

Solution:

$$\left(\frac{\sqrt{2}}{2} - \frac{\sqrt{2}}{2}i\right) - \left(\frac{\sqrt{2}}{2} + \frac{\sqrt{2}}{2}i\right) = \left(\frac{\sqrt{2}}{2} - \frac{\sqrt{2}}{2}\right) + \left(-\frac{\sqrt{2}}{2} - \frac{\sqrt{2}}{2}\right)i$$
$$= 0 - \sqrt{2}i$$
$$= -\sqrt{2}i$$

65. Perform the indicated operations and write the result in standard form.

$$(10 - 8i)(2 - 3i)$$

Solution:

$$(10 - 8i)(2 - 3i) = 20 - 30i - 16i + 24i^2 = -4 - 46i$$

69. Perform the indicated operations and write the result in standard form.

$$\frac{4}{-3i}$$

Solution:

$$\frac{4}{-3i} = \frac{4}{-3i} \cdot \frac{3i}{3i} = \frac{12i}{9} = \frac{4i}{3} = \frac{4}{3}i$$

73. A car radiator contains 10 quarts of a 30% antifreeze solution. How many quarts will have to be replaced with pure antifreeze if the resulting solution is to be 50% antifreeze?

Solution:

Let $x =$ the number of quarts of pure antifreeze.

$$\begin{aligned} 30\% \text{ of } (10 - x) + 100\% \text{ of } x &= 50\% \text{ of } 10 \\ 0.30(10 - x) + 1.00x &= 0.50(10) \\ 3 - 0.30x + 1.00x &= 5 \\ 0.70x &= 2 \\ x &= \frac{2}{0.70} \approx 2.857 \text{ quarts} \end{aligned}$$

77. A group of farmers agree to share equally in the cost of a $48,000 piece of machinery. If they could find two more farmers to join the group, each person's share of the cost would decrease by $4000. How many farmers are presently in the group?

Solution:

Let $x =$ number of farmers in the group

Cost per farmer $= \dfrac{48,000}{x}$

If two more farmers join the group, the cost per farmer will be $\dfrac{48,000}{x+2}$.

Since this new cost is $4000 less than the original cost,

$$\begin{aligned} \frac{48,000}{x} - 4000 &= \frac{48,000}{x+2} \\ 48,000(x+2) - 4000x(x+2) &= 48,000x \\ 12(x+2) - x(x+2) &= 12x && \text{Divide both sides by 4000} \\ 12x + 24 - x^2 - 2x &= 12x \\ 0 &= x^2 + 2x - 24 \\ 0 &= (x+6)(x-4) \\ x &= -6, \text{ extraneous} \quad \text{or} \quad x = 4 \\ x &= 4 \text{ farmers} \end{aligned}$$

81. The revenue for selling x units of a product is

$$R = 125.95x.$$

The cost of producing x units is

$$C = 92x + 1200.$$

In order to obtain a profit, the revenue must be greater than the cost. For what values of x will this product return a profit?

Solution:

$$R > C$$
$$125.95x > 92x + 1200$$
$$33.95 > 1200$$
$$x > 35.346$$

When $x \geq 36$ units, this product will return a profit.

Practice Test for Chapter 2

1. Solve $5x + 4 = 7x - 8$.

2. Solve $\frac{x}{3} - 5 = \frac{x}{5} + 1$.

3. Solve $\frac{3x+1}{6x-7} = \frac{2}{5}$.

4. Solve $(x-3)^2 + 4 = (x+1)^2$.

5. Solve $A = \frac{1}{2}(a+b)h$ for a.

6. Find three consecutive natural numbers whose sum is 132.

7. 301 is what percent of 4300?

8. Cindy has \$6.05 in quarters and nickles. How many of each coin does she have if there are 53 coins in all?

9. Ed has \$15,000 invested in two funds paying $9\frac{1}{2}\%$ and 11% simple interest, respectively. How much is invested in each if the yearly interest is \$1582.50?

10. Solve $28 + 5x - 3x^2 = 0$ by factoring.

11. Solve $(x-2)^2 = 24$ by taking the square root of both sides.

12. Solve $x^2 - 4x - 9 = 0$ by completing the square.

13. Complete the square on the denominator of $\frac{1}{x^2 - 6x + 1}$.

14. Solve $x^2 + 5x - 1 = 0$ by the Quadratic Formula.

15. Solve $3x^2 - 2x + 4 = 0$ by the Quadratic Formula.

16. The perimeter of a rectangle is 1100 feet. Find the dimensions so that the enclosed area will be 60,000 square feet.

17. Find two consecutive even positive integers whose product is 624.

18. Solve $x^3 - 10x^2 + 24x = 0$ by factoring.

19. Solve $\sqrt[3]{6-x} = 4$.

20. Solve $(x^2 - 8)^{2/5} = 4$.

21. Solve $x^4 - x^2 - 12 = 0$.

22. Solve $4 - 3x > 16$.

23. Solve $\left|\frac{x-3}{2}\right| < 5$.

24. Solve $\frac{x+1}{x-3} < 2$.

25. Solve $|3x - 4| \geq 9$.

CHAPTER 3

Functions and Graphs

SECTION 3.1

The Cartesian Plane

- You should be able to plot points.
- You should know that the distance between (x_1, y_1) and (x_2, y_2) in the plane is

  $$d = \sqrt{(x_2 - x_1)^2 + (y_2 - y_1)^2}.$$

- You should know that the midpoint of the line segment joining (x_1, y_1) and (x_2, y_2) is

  $$\left(\frac{x_1 + x_2}{2}, \frac{y_1 + y_2}{2}\right).$$

Solutions to Selected Exercises

3. Sketch the square with vertices $(2, 4)$, $(5, 1)$, $(2, -2)$, and $(-1, 1)$.

Solution:

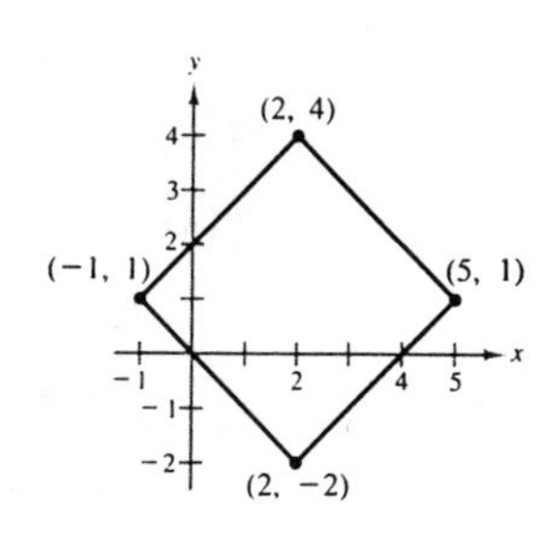

5. Find the coordinates of the vertices of the figure in its new position.

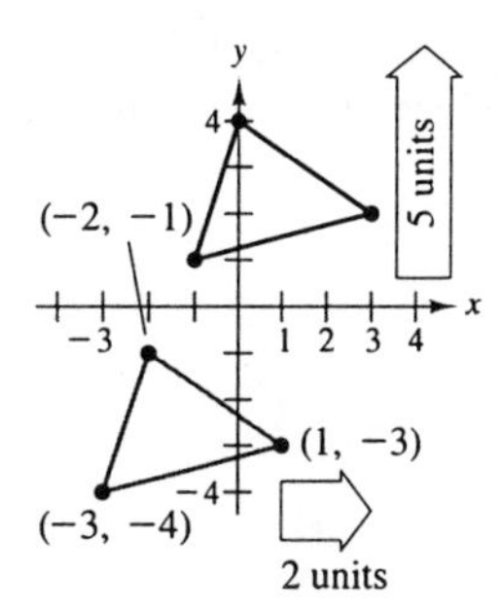

Solution:

$$(-2 + 2, -1 + 5) = (0, 4)$$
$$(-3 + 2, -4 + 5) = (-1, 1)$$
$$(1 + 2, -3 + 5) = (3, 2)$$

9. Find the distance between the points $(-3, -1)$ and $(2, -1)$.

Solution:

Since the points $(-3, -1)$ and $(2, -1)$ lie on the same vertical line, the distance between the points is given by the absolute value of the difference of their x–coordinates.

$$d = |-3-2| = 5$$

13. For the indicated triangle (a) find the length of the two sides of the right triangle and use the Pythagorean Theorem to find the length of the hypotenuse, and (b) use the Distance Formula to find the length of the hypotenuse of the triangle.

Solution:

(a) $a = |-3-7| = 10$

$b = |4-1| = 3$

$c = \sqrt{10^2+3^2} = \sqrt{109}$

(b) $c = \sqrt{(7-(-3))^2+(4-1)^2}$

$= \sqrt{10^2+3^2}$

$= \sqrt{109}$

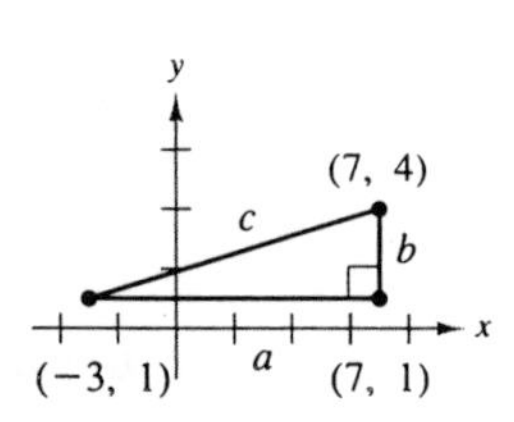

17. (a) Plot the points $(-4, 10)$ and $(4, -5)$, (b) find the distance between the points, and (c) find the midpoint of the line segment joining the points.

Solution:

(a)

(b) $d = \sqrt{(-4-4)^2+(10-(-5))^2}$

$= \sqrt{(-8)^2+(15)^2}$

$= \sqrt{289} = 17$

(c) $m = \left(\frac{-4+4}{2}, \frac{10+(-5)}{2}\right)$

$= \left(0, \frac{5}{2}\right)$

23. (a) Plot the points (6.2, 5.4), and (−3.7, 1.8), (b) find the distance between the points, and (c) find the midpoint of the line segment joining the points.

Solution:

(a)

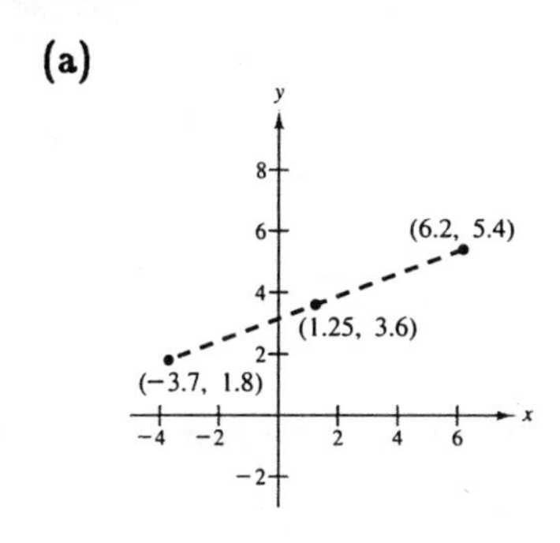

(b) $d = \sqrt{(6.2 - (-3.7))^2 + (5.4 - 1.8)^2}$

$= \sqrt{(9.9)^2 + (3.6)^2}$

$= \sqrt{110.97} \approx 10.5342$

(c) $m = \left(\dfrac{6.2 + (-3.7)}{2}, \dfrac{5.4 + 1.8}{2}\right)$

$= (1.25, 3.6)$

27. Use the Midpoint Formula to estimate the sales of a company for 1991. Assume the annual sales followed a linear pattern.

Year	1989	1993
Sales	$520,000	$740,000

Solution:

$$\frac{520{,}000 + 740{,}000}{2} = 630{,}000$$

The estimated sales for 1991 is $630,000.

29. Show that the points (4, 0), (2, 1), and (−1, −5) form the vertices of a right triangle.

Solution:

$d_1 = \sqrt{(-1 - 2)^2 + (-5 - 1)^2} = \sqrt{45}$

$d_2 = \sqrt{(2 - 4)^2 + (1 - 0)^2} = \sqrt{5}$

$d_3 = \sqrt{(4 - (-1))^2 + (0 - (-5))^2} = \sqrt{50}$

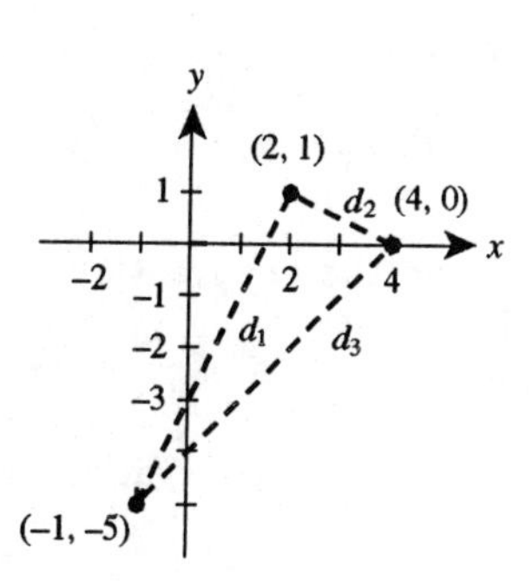

Since $d_1{}^2 + d_2{}^2 = d_3{}^2$, we can conclude by the Pythagorean Theorem that the triangle is a right triangle.

33. Find x so that the distance between $(1, 2)$ and $(x, -10)$ is 13.

Solution:

$$\sqrt{(x-1)^2 + (-10-2)^2} = 13$$
$$\sqrt{x^2 - 2x + 1 + 144} = 13$$
$$x^2 - 2x + 145 = 169$$
$$x^2 - 2x - 24 = 0$$
$$(x+4)(x-6) = 0$$
$$x = -4 \quad \text{or} \quad x = 6$$

37. Find a relationship between x and y so that (x, y) is equidistant from the points $(4, -1)$ and $(-2, 3)$.

Solution:

The distance between $(4, -1)$ and (x, y) is equal to the distance between $(-2, 3)$ and (x, y).

$$\sqrt{(x-4)^2 + (y+1)^2} = \sqrt{(x+2)^2 + (y-3)^2}$$
$$(x-4)^2 + (y+1)^2 = (x+2)^2 + (y-3)^2$$
$$x^2 - 8x + 16 + y^2 + 2y + 1 = x^2 + 4x + 4 + y^2 - 6y + 9$$
$$-12x + 8y + 4 = 0$$
$$3x - 2y - 1 = 0 \qquad \text{Divide both sides by } -4$$

41. Determine the quadrant(s) in which (x, y) is located so that the conditions $x > 0$ and $y > 0$ are satisfied.

Solution:

$x > 0 \rightarrow x$ lies in Quadrant I or in Quadrant IV

$y > 0 \rightarrow y$ lies in Quadrant I or in Quadrant II

$x > 0$ and $y > 0 \rightarrow (x, y)$ lies in Quadrant I

45. Determine the quadrant(s) in which (x, y) is located so that the condition $y < -5$ is satisfied.

Solution:

$y < -5 \rightarrow y$ is negative $\rightarrow y$ lies in either Quadrant III or Quadrant IV.

51. Use the Midpoint Formula twice to find the three points that divide the line segment joining (x_1, y_1) and (x_2, y_2) into four parts.

Solution:

The midpoint of the given line segment is $\left(\dfrac{x_1+x_2}{2}, \dfrac{y_1+y_2}{2}\right)$.

The midpoint between (x_1, y_1) and $\left(\dfrac{x_1+x_2}{2}, \dfrac{y_1+y_2}{2}\right)$ is

$$\left(\frac{x_1+\dfrac{x_1+x_2}{2}}{2}, \frac{y_1+\dfrac{y_1+y_2}{2}}{2}\right) = \left(\frac{3x_1+x_2}{4}, \frac{3y_1+y_2}{4}\right).$$

The midpoint between $\left(\dfrac{x_1+x_2}{2}, \dfrac{y_1+y_2}{2}\right)$ and (x_2, y_2) is

$$\left(\frac{\dfrac{x_1+x_2}{2}+x_2}{2}, \frac{\dfrac{y_1+y_2}{2}+y_2}{2}\right) = \left(\frac{x_1+3x_2}{4}, \frac{y_1+3y_2}{4}\right).$$

Thus, the three points are

$$\left(\frac{3x_1+x_2}{4}, \frac{3y_1+y_2}{4}\right), \left(\frac{x_1+x_2}{2}, \frac{y_1+y_2}{2}\right), \quad \text{and} \quad \left(\frac{x_1+3x_2}{4}, \frac{y_1+3y_2}{4}\right).$$

55. The normal temperature y (Fahrenheit) for Duluth, Minnesota for each month of the year is given in the table. The months are numbered 1 through 12, with 1 corresponding to January. *(Source: NOAA)* Plot the points.

x	1	2	3	4	5	6	7	8	9	10	11	12
y	6	12	23	38	50	59	65	63	54	44	28	14

Solution:

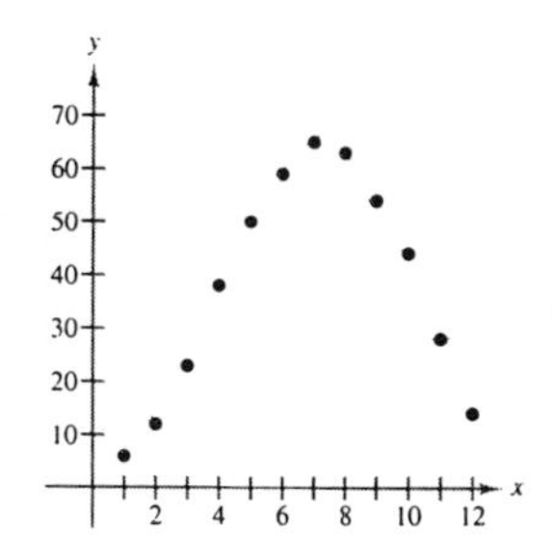

59. Find the percentage increase in the cost of a 30–second spot from Super Bowl I to Super Bowl XXV.

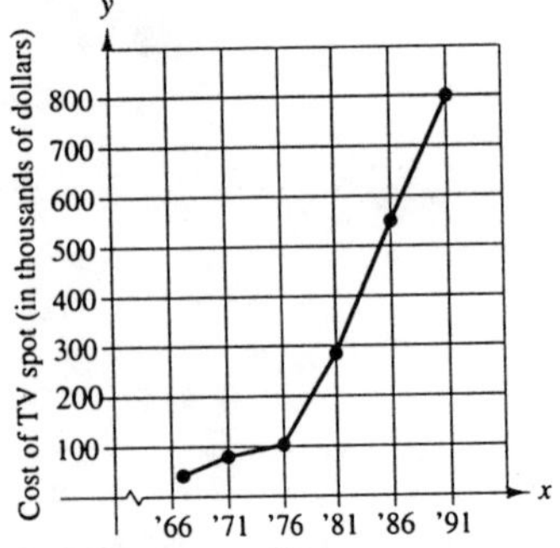

Solution:

$$\frac{800{,}000 - 42{,}500}{42{,}500} \approx 17.8235 = 1782.35\%$$

63. Prove that the diagonals of the parallelogram in the figure bisect each other.

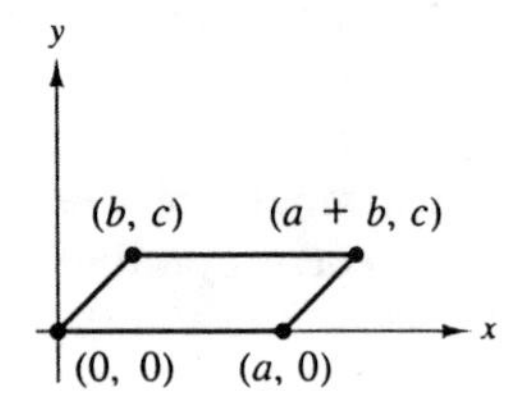

Solution:

The midpoint of the diagonal connecting $(0, 0)$ and $(a + b, c)$ is

$$\left(\frac{a+b}{2}, \frac{c}{2}\right).$$

The midpoint of the diagonal connecting $(a, 0)$ and (b, c) is

$$\left(\frac{a+b}{2}, \frac{c}{2}\right).$$

Thus, the diagonals bisect each other.

SECTION 3.2

Graphs of Equations

- You should be able to use the point-plotting method of graphing.
- You should be able to find x- and y-intercepts.
  (a) To find the x-intercepts, let $y = 0$ and solve for x.
  (b) To find the y-intercepts, let $x = 0$ and solve for y.
- You should be able to test for symmetry.
  (a) To test for x-axis symmetry, replace y with $-y$.
  (b) To test for y-axis symmetry, replace x with $-x$.
  (c) To test for origin symmetry, replace x with $-x$ and y with $-y$.
- You should know the standard equation of a circle with center (h, k) and radius r:

  $$(x-h)^2 + (y-k)^2 = r^2$$

Solutions to Selected Exercises

5. Determine whether the points (a) $(1, \frac{1}{5})$, and (b) $(2, \frac{1}{2})$ lie on the graph of the equation $x^2y - x^2 + 4y = 0$.

Solution:

(a) $\left(1, \dfrac{1}{5}\right)$ lies on the graph since $(1)^2\left(\dfrac{1}{5}\right) - (1)^2 + 4\left(\dfrac{1}{5}\right) = \dfrac{1}{5} - 1 + \dfrac{4}{5} = 0$.

(b) $\left(2, \dfrac{1}{2}\right)$ lies on the graph since $(2)^2\left(\dfrac{1}{2}\right) - (2)^2 + 4\left(\dfrac{1}{2}\right) = 2 - 4 + 2 = 0$.

7. Find the constant C so that the ordered pair $(2, 6)$ is a solution point of the equation $y = x^2 + C$.

Solution:

$$y = x^2 + C$$
$$6 = (2)^2 + C$$
$$6 = 4 + C$$
$$C = 2$$

11. Complete the table (shown in the textbook) and use the solution points to sketch the graph of the equation $2x + y = 3$.

Solution:

$$2x + y = 3$$
$$y = -2x + 3$$

x	-4	-2	0	2	4
y	11	7	3	-1	-5
(x, y)	$(-4, 11)$	$(-2, 7)$	$(0, 3)$	$(2, -1)$	$(4, -5)$

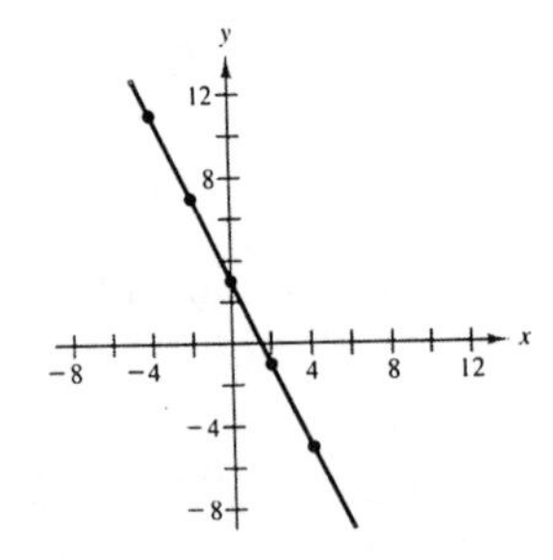

15. Find the x- and y-intercepts of the graph of the equation $y = x^2 + x - 2$.

Solution:

Let $y = 0$. Then $0 = x^2 + x - 2 = (x + 2)(x - 1)$ and $x = -2$ or $x = 1$.

x-intercepts: $(-2, 0)$ and $(1, 0)$

Let $x = 0$. Then $y = -2$.

y-intercept: $(0, -2)$

19. Find the x- and y-intercepts of the graph of the equation $xy - 2y - x + 1 = 0$.

Solution:

Let $y = 0$. Then $-x + 1 = 0$ and $x = 1$.

x-intercept: $(1, 0)$

Let $x = 0$. Then $-2y + 1 = 0$ and $y = \frac{1}{2}$.

y-intercept: $\left(0, \frac{1}{2}\right)$

23. Check for symmetry with respect to both axes and the origin for $x - y^2 = 0$.

Solution:

By replacing y with $-y$, we have

$$x - (-y)^2 = 0$$
$$x - y^2 = 0$$

which is the original equation. Replacing x with $-x$ or replacing both x and y with $-x$ and$-y$ does not yield equivalent equations. Thus, $x - y^2 = 0$ is symmetric with respect to the x-axis.

27. Check for symmetry with respect to both axes and the origin for

$$y = \frac{x}{x^2 + 1}.$$

Solution:

Replacing x with $-x$ or y with $-y$ does not yield equivalent equations. Replacing x with $-x$ and y with $-y$ yields the following.

$$-y = \frac{-x}{(-x)^2 + 1}$$

$$-y = \frac{-x}{x^2 + 1} \qquad \text{Multiply both sides by } -1$$

$$y = \frac{x}{x^2 + 1}$$

Thus, $y = \dfrac{x}{x^2 + 1}$ is symmetric with respect to the origin.

Note: An equation is symmetric with respect to the origin if it is symmetric with respect to both the x-axis and the y-axis. Also, if an equation is symmetric with respect to the origin, then one of the following is true:

1. The equation has both x-axis and y-axis symmetry or
2. The equation has neither x-axis nor y-axis symmetry.

31. Use symmetry to sketch the complete graph.

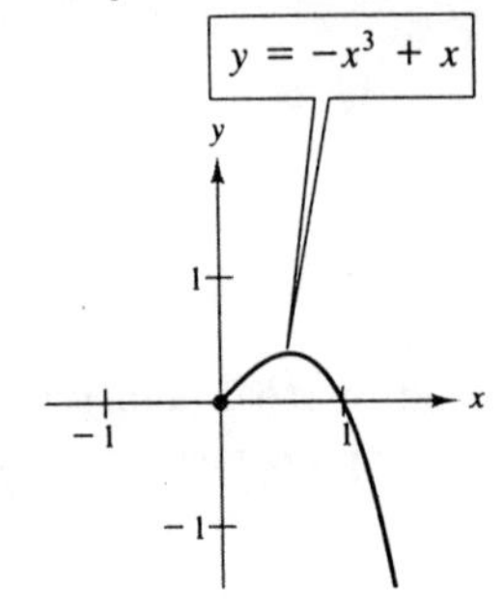

Solution: Origin Symmetry

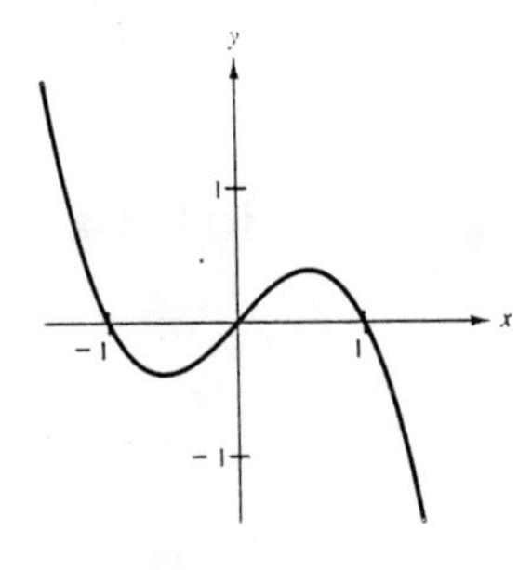

$y = -x^3 + x$

37. Match $y = x^3 - x$ with its graph.

Solution:

$y = x^3 - x$

x-intercepts: $(-1, 0)$, $(0, 0)$, $(1, 0)$

y-intercept: $(0, 0)$

Symmetry: Origin

Matches graph (e)

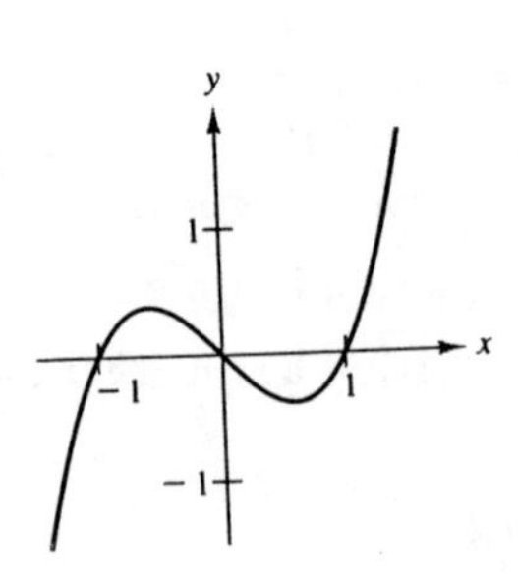

39. Sketch the graph of $y = -3x + 2$. Identify any intercepts and test for symmetry.

Solution:

$y = -3x + 2$

x-intercept: $\left(\frac{2}{3}, 0\right)$

y-intercept: $(0, 2)$

No symmetry

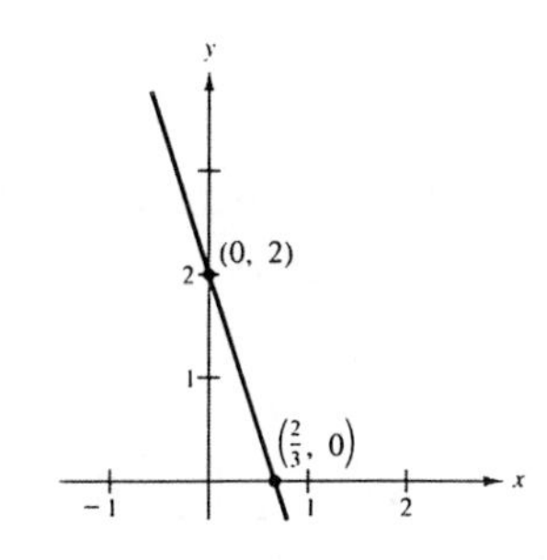

43. Sketch the graph of $y = x^2 - 4x + 3$. Identify any intercepts and test for symmetry.

Solution:

$y = x^2 - 4x + 3 = (x - 1)(x - 3)$

x-intercepts: $(1, 0)$, $(3, 0)$

y-intercept: $(0, 3)$

No symmetry

x	-1	0	1	2	3	4
y	8	3	0	-1	0	3

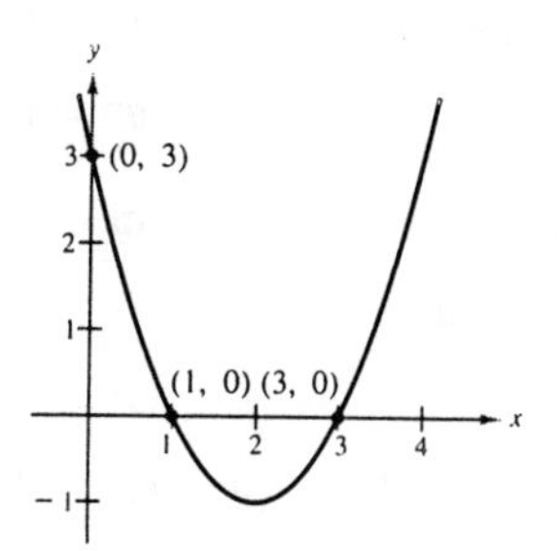

49. Sketch the graph of $y = \sqrt{x - 3}$. Identify any intercepts and test for symmetry.

Solution:

$y = \sqrt{x - 3}$

x-intercept: $(3, 0)$

No y-intercept

No symmetry

x	3	4	7	12
y	0	1	2	3

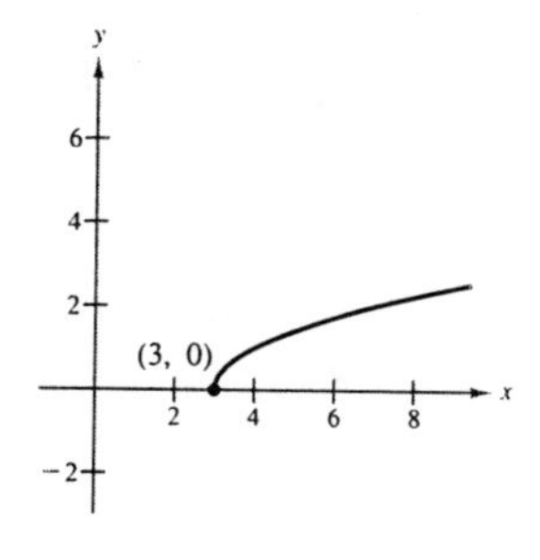

Note: The domain is $[3, \infty)$ and the range is $[0, \infty)$.

55. Sketch the graph of $x = y^2 - 1$. Identify any intercepts and test for symmetry.

Solution:

$x = y^2 - 1$

x-intercept: $(-1,\ 0)$

y-intercepts: $(0,\ -1)$, $(0,\ 1)$

x-axis symmetry

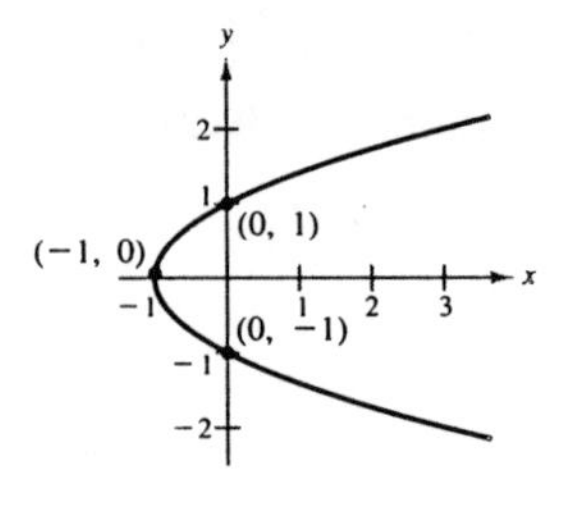

x	-1	0	3
y	0	± 1	± 2

59. Find the standard form of the equation of the circle with center $(0,\ 0)$ and radius 3.

Solution:

$$(x - 0)^2 + (y - 0)^2 = 3^2$$
$$x^2 + y^2 = 9$$

61. Find the standard form of the equation of the circle with center $(2,\ -1)$ and radius 4.

Solution:

$$(x - 2)^2 + (y + 1)^2 = 4^2$$
$$(x - 2)^2 + (y + 1)^2 = 16$$

63. Find the standard form of the equation of the circle with center $(-1,\ 2)$ and passing through $(0,\ 0)$.

Solution:

$$(x + 1)^2 + (y - 2)^2 = r^2$$
$$(0 + 1)^2 + (0 - 2)^2 = r^2 \quad \rightarrow \quad r^2 = 5$$
$$(x + 1)^2 + (y - 2)^2 = 5$$

67. Find the center and radius, and sketch the graph of the equation.

$x^2 + y^2 - 2x + 6y + 6 = 0$

Solution:

$$x^2 + y^2 - 2x + 6y + 6 = 0$$
$$(x^2 - 2x + 1^2) + (y^2 + 6y + 3^2) = -6 + 1 + 9$$
$$(x - 1)^2 + (y + 3)^2 = 4$$

Center: $(1, -3)$

Radius: 2

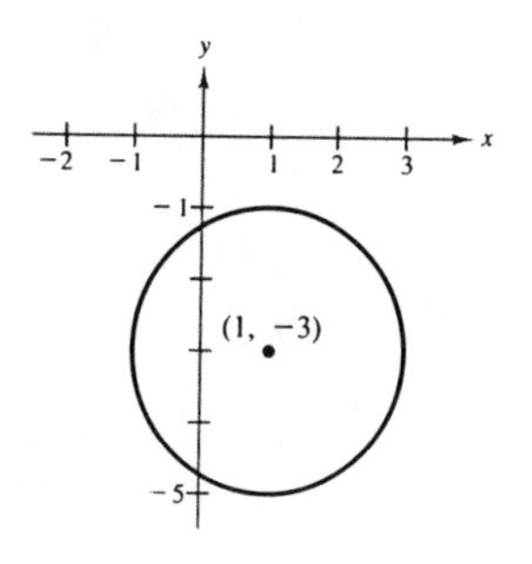

69. Find the center and radius, and sketch the graph of the equation.

$x^2 + y^2 - 2x + 6y + 10 = 0$

Solution:

$$x^2 + y^2 - 2x + 6y + 10 = 0$$
$$(x^2 - 2x + 1^2) + (y^2 + 6y + 3^2) = -10 + 1 + 9$$
$$(x - 1)^2 + (y + 3)^2 = 0$$

Center: $(1, -3)$

Radius: 0

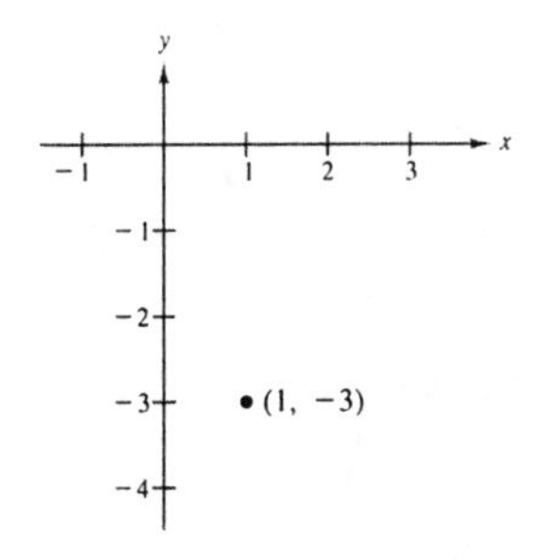

The graph is a single point, the center $(1, -3)$.

73. Find the center and radius, and sketch the graph of the equation.

$$16x^2 + 16y^2 + 16x + 40y - 7 = 0$$

Solution:

$$16x^2 + 16y^2 + 16x + 40y - 7 = 0$$

$$x^2 + y^2 + x + \frac{5}{2}y - \frac{7}{16} = 0 \qquad \text{Divide both sides by 16}$$

$$\left(x^2 + x + \left(\frac{1}{2}\right)^2\right) + \left(y^2 + \frac{5}{2}y + \left(\frac{5}{4}\right)^2\right) = \frac{7}{16} + \frac{1}{4} + \frac{25}{16}$$

$$\left(x + \frac{1}{2}\right)^2 + \left(y + \frac{5}{4}\right)^2 = \frac{36}{16}$$

$$\left(x + \frac{1}{2}\right)^2 + \left(y + \frac{5}{4}\right)^2 = \frac{9}{4}$$

Center: $\left(-\frac{1}{2}, -\frac{5}{4}\right)$ Radius: $\frac{3}{2}$

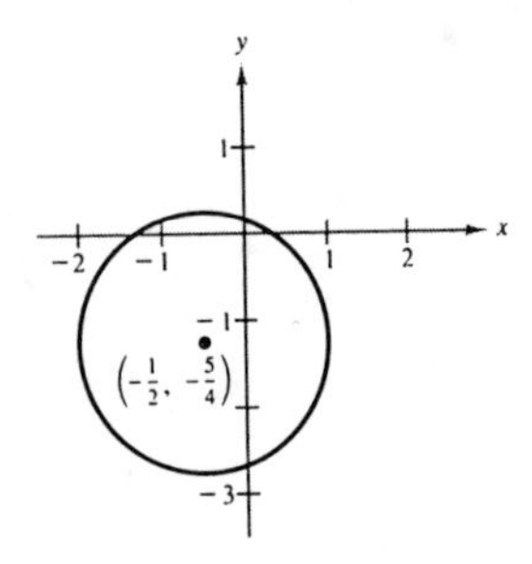

77. The following table gives the per capita federal debt for the United States for selected years from 1950 to 1990. *(Source: U.S. Treasury Department)*

Year	1950	1960	1970	1980	1985	1990
Per capita debt	\$1688	\$1572	\$1807	\$3981	\$7614	\$12,848

A mathematical model for the per capita debt during this period is

$$y = 0.40t^3 - 9.42t^2 + 1053.24$$

where y represents per capita debt and t is the time in years with $t = 0$ corresponding to 1950.

(a) Sketch a graph to compare the given data and the model for that data; (b) use the model to predict y for the year 1994, and (c) for the year 2000.

Solution:

$$y = 0.40t^3 - 9.42t^2 + 1053.24$$

(a)

t	y	Per capita debt
0	\$1,053	\$1,688
10	\$511	\$1,572
20	\$485	\$1,807
30	\$3,375	\$3,981
35	\$6,664	\$7,614
40	\$11,581	\$12,848

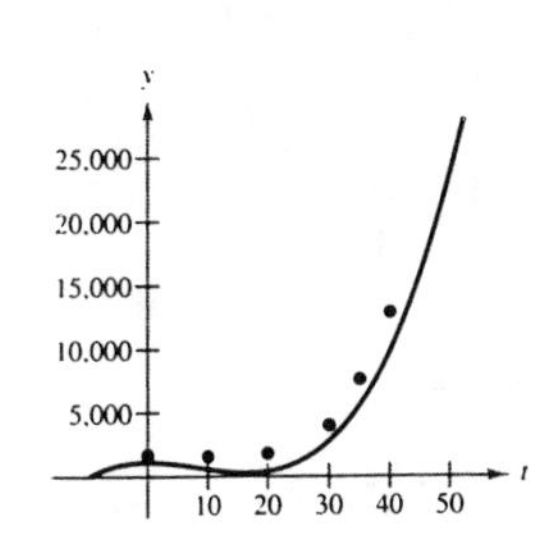

(b) When $t = 44$, $y \approx \$16,890$

(c) When $t = 50$, $y \approx \$27,503$

SECTION 3.3

Lines in the Plane

You should know the following important facts about lines.

- The slope of the line through $(x_1,\ y_1)$ and $(x_2,\ y_2)$ is

$$m = \frac{y_2 - y_1}{x_2 - x_1}.$$

- (a) If $m > 0$, the line rises from left to right.
  (b) If $m = 0$, the line is horizontal.
  (c) If $m < 0$, the line falls from left to right.
  (d) If m is undefined, the line is vertical.

- Equations of Lines
  (a) Point-Slope: $y - y_1 = m(x - x_1)$
  (b) Two-Point: $y - y_1 = \dfrac{y_2 - y_1}{x_2 - x_1}(x - x_1)$
  (c) Slope-Intercept: $y = mx + b$
  (d) General: $Ax + By + C = 0$
  (d) Vertical: $x = a$
  (e) Horizontal: $y = b$

- Given two distinct nonvertical lines

$$L_1 : y = m_1x + b_1 \quad \text{and} \quad L_2 : y = m_2x + b_2$$

  (a) L_1 is parallel to L_2 if and only if $m_1 = m_2$ and $b_1 \neq b_2$.
  (b) L_1 is perpendicular to L_2 if and only if $m_1 = -1/m_2$.

Solutions to Selected Exercises

3. Estimate the slope from its graph.

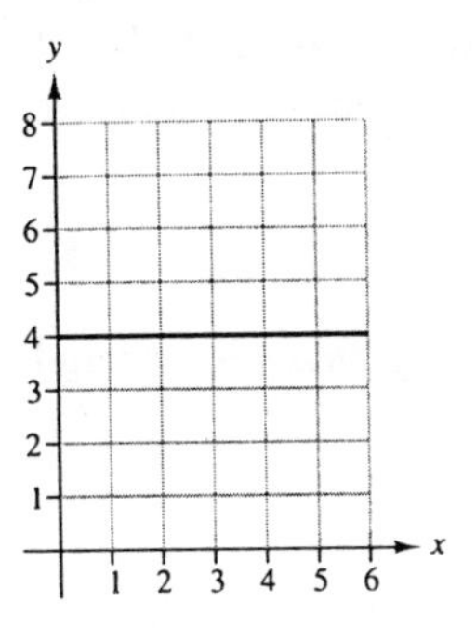

Solution:

Since the line is horizontal, it has a slope of zero.

9. Plot the points $(-3, -2)$ and $(1, 6)$ and find the slope of the line passing through the points.

Solution:

$$m = \frac{6-(-2)}{1-(-3)} = \frac{8}{4} = 2$$

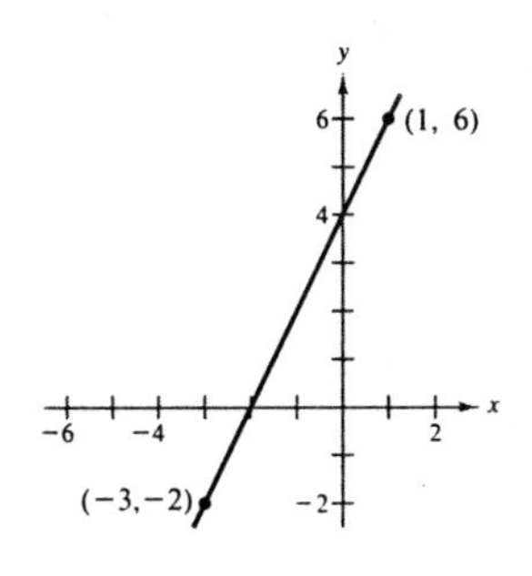

11. Plot the points $(-6, -1)$ and $(-6, 4)$ and find the slope of the line passing through the points.

Solution:

$$m = \frac{4-(-1)}{-6-(-6)} = \frac{5}{0}$$

The slope is undefined.

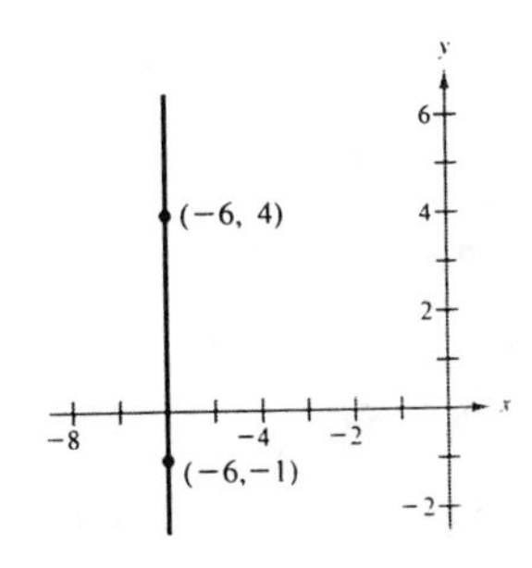

15. Use the point $(2, 1)$ on the line and the slope $m = 0$ of the line to find three additional points that the line passes through. (The solution is not unique.)

Solution:

Since $m = 0$, the line is horizontal, and since the line passes through $(2, 1)$, all other points on the line will be of the form $(x, 1)$. Three additional points are: $(0, 1)$, $(1, 1)$, $(3, 1)$.

19. Use the point $(-8,\ 1)$ on the line and the undefined slope of the line to find three additional points that the line passes through. (The solution is not unique.)

Solution:

Since m is undefined, the line is vertical, and since the line passes through $(-8,\ 1)$, all other points on the line will be of the form $(-8,\ y)$. Three additional points are: $(-8,\ -1)$, $(-8,\ 0)$, $(-8,\ 2)$.

21. Determine if the lines L_1 and L_2 passing through the given pairs of points are parallel, perpendicular, or neither.

$L_1 : (0,\ -1),\ (5,\ 9)$

$L_2 : (0,\ 3),\ (4,\ 1)$

Solution:

The slope of L_1 is $m_1 = \dfrac{9-(-1)}{5-0} = \dfrac{10}{5} = 2.$

The slope of L_2 is $m_2 = \dfrac{1-3}{4-0} = -\dfrac{2}{4} = -\dfrac{1}{2}.$

Since m_1 and m_2 are negative reciprocals of each other, the lines are perpendicular.

25. When driving down a mountain road, you notice signs warning of a "12% grade". This means that the slope of the road is $-12/100$. Determine the amount of horizontal change in your position if you note from elevation markers that you have descended 2000 feet vertically.

Solution:

$$\text{Slope} = \frac{\text{Rise}}{\text{Run}}$$

$$-\frac{12}{100} = -\frac{2000}{y}$$

$$-12y = -200{,}000$$

$$y = 16{,}666\tfrac{2}{3} \text{ feet } \approx 3.16 \text{ miles}$$

27. Find the slope and y-intercept, if possible, of the line specified by $5x - y + 3 = 0$. Sketch a graph of the line.

Solution:

$$5x - y + 3 = 0$$
$$-y = -5x - 3$$
$$y = 5x + 3$$

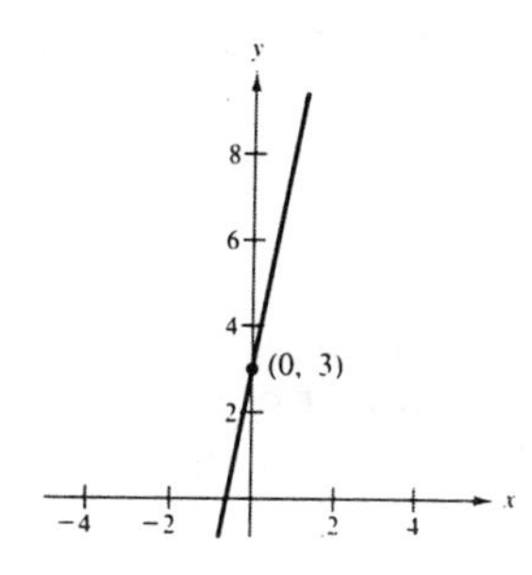

Slope: $m = 5$

y-intercept: $(0, 3)$

29. Find the slope and y-intercept, if possible, of the line specified by $5x - 2 = 0$. Sketch a graph of the line.

Solution:

$$5x - 2 = 0$$
$$5x = 2$$
$$x = \frac{2}{5} \qquad \text{Vertical line}$$

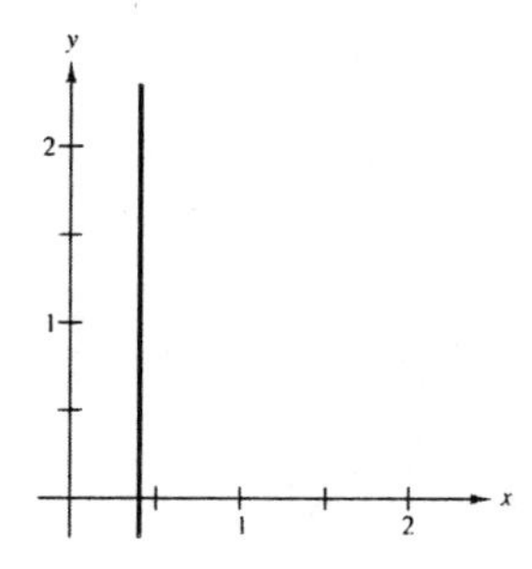

Slope: Undefined

y-intercept: None

35. Find an equation for the line passing through the points $(2, \frac{1}{2})$, $(\frac{1}{2}, \frac{5}{4})$.

Solution:

$$m = \frac{\frac{5}{4} - \frac{1}{2}}{\frac{1}{2} - 2} = \frac{\frac{3}{4}}{-\frac{3}{2}} = -\frac{1}{2}$$

$$y - \frac{1}{2} = -\frac{1}{2}(x - 2)$$
$$2y - 1 = -(x - 2)$$
$$2y - 1 = -x + 2$$
$$x + 2y - 3 = 0$$

39. Find an equation for the line passing through the points (1, 0.6), and (−2, −0.6).

Solution:

$$m = \frac{-0.6 - 0.6}{-2 - 1} = 0.4$$

$$y - 0.6 = 0.4(x - 1)$$

$$y - 0.6 = 0.4x - 0.4$$

$$y = 0.4x + 0.2 \quad \text{or} \quad 2x - 5y + 1 = 0$$

43. Find an equation of the line that passes through the point (−3, 6) and has a slope of $m = -2$. Sketch a graph of the line.

Solution:

$$y - 6 = -2[x - (-3)]$$

$$y - 6 = -2x - 6$$

$$y = -2x$$

$$2x + y = 0$$

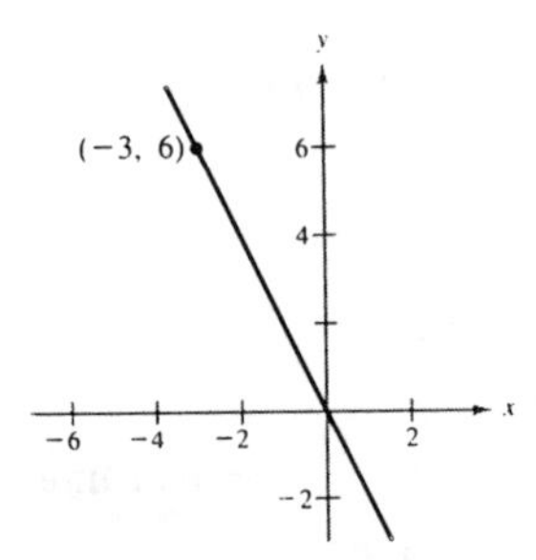

47. Find an equation of the line that passes through the point (6, −1) and has an undefined slope. Sketch a graph of the line.

Solution:

Since the slope is undefined, the line is vertical and since the line passes through (6, −1), its equation is $x = 6$ or $x - 6 = 0$.

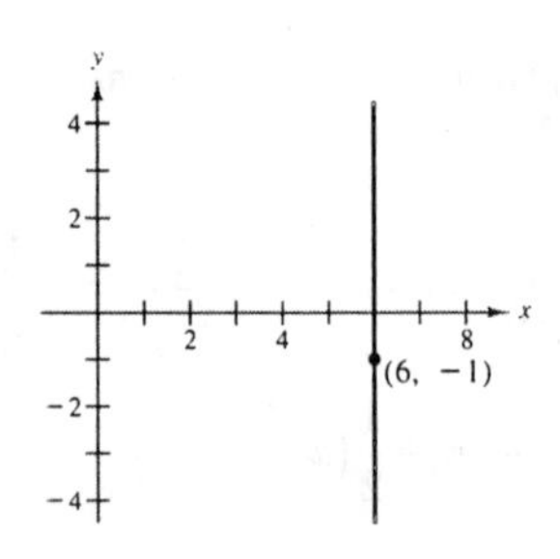

53. Use the Intercept Form to write an equation of the line with x-intercept $(-\frac{1}{6},\ 0)$ and y-intercept $(0,\ -\frac{2}{3})$.

Solution:

x-intercept: $\left(-\frac{1}{6},\ 0\right)$

y-intercept: $\left(0,\ -\frac{2}{3}\right)$

$$\frac{x}{-1/6} + \frac{y}{-2/3} = 1$$

$$-6x - \frac{3}{2}y = 1$$

$$-12x - 3y = 2$$

$$12x + 3y = -2$$

$$12x + 3y + 2 = 0$$

59. Write the equation of the line through the point $(-6,\ 4)$, (a) parallel to the line $3x+4y = 7$, and (b) perpendicular to the line $3x + 4y = 7$.

Solution:

$$3x + 4y = 7$$

$$4y = -3x + 7$$

$$y = -\frac{3}{4}x + \frac{7}{4}$$

The slope of the given line is $m_1 = -3/4$.

(a) The slope of the parallel line is $m_2 = m_1 = -3/4$.

$$y - 4 = -\frac{3}{4}[x - (-6)]$$

$$y - 4 = -\frac{3}{4}x - \frac{9}{2}$$

$$y = -\frac{3}{4}x - \frac{1}{2}$$

$$4y = -3x - 2$$

$$3x + 4y + 2 = 0$$

(b) The slope of the perpendicular line is $m_2 = -1/m_1 = 4/3$.

$$y - 4 = \frac{4}{3}[x - (-6)]$$

$$y - 4 = \frac{4}{3}x + 8$$

$$y = \frac{4}{3}x + 12$$

$$3y = 4x + 36$$

$$4x - 3y + 36 = 0$$

63. The 1990 value of a product is \$2540 and the rate at which the value is expected to change during the next five years is a \$125 increase per year. Write a linear equation for the dollar value V of the product in terms of the year t. (Let $t = 0$ represent 1990.)

Solution:

Value = \$2540 + (\$125)(the number of years t after 1990)

$$V = 125t + 2540, \quad 0 \le t \le 5$$

67. Match the description "A person is paying \$10 per week to a friend to repay a \$100 loan" with a graph. Determine the slope and how it is interpreted in the situation.

Solution:

A person is paying \$10 per week to a friend to repay a \$100 loan. Matches graph (b).

The slope is $m = -10$. This represents the decrease in the amount of the loan each week.

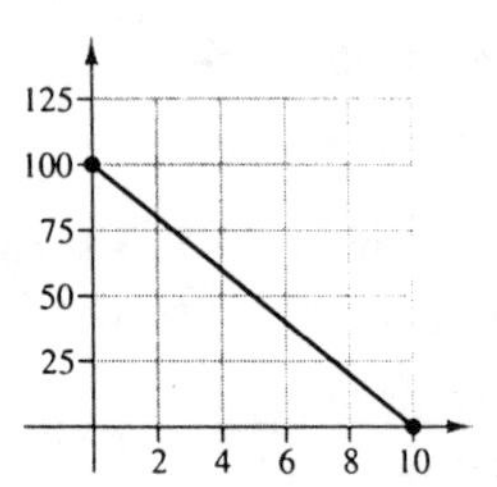

71. Find the equation of the line giving the relationship between the temperature in degrees Celsius, C, and degrees Fahrenheit, F. Use the fact that water freezes at 0° Celsius (32° Fahrenheit) and boils at 100° Celsius (212° Fahrenheit).

Solution:

Using the points (0, 32) and (100, 212), we have

$$m = \frac{212 - 32}{100 - 0} = \frac{180}{100} = \frac{9}{5}$$

$$\text{F} - 32 = \frac{9}{5}(\text{C} - 0)$$

$$\text{F} = \frac{9}{5}\text{C} + 32.$$

75. A small business purchases a piece of equipment for \$875. After five years the equipment will be outdated and have no value. Write a linear equation giving the value V of the equipment during the five years it will be used.

Solution:

Using the points (0, 875) and (5, 0), where the first coordinate represents the year t and the second coordinate represents the value V, we have

$$m = \frac{0 - 875}{5 - 0} = -175$$

$$V = -175t + 875, \quad 0 \le t \le 5.$$

77. A store is offering a 15% discount on all items in its inventory. Write a linear equation giving the sale price S for an item with a list price L.

Solution:

Sale Price = List Price − 15% of the List Price

$$S = L - 0.15L$$

$$S = 0.85L$$

81. A contractor purchases a piece of equipment for \$36,500. The equipment requires an average expenditure of \$5.25 per hour for fuel and maintenance, and the operator is paid \$11.50 per hour.

(a) Write a linear equation giving the total cost C of operating this equipment for t hours. (Include the purchase cost for the equipment.)

(b) If customers are charged \$27 per hour of machine use, write an equation for the revenue R derived from t hours of use.

(c) Use the formula for profit ($P = R - C$) to write an equation for the profit derived from t hours of use.

(d) *Break-Even Point* Use the result of part (c) to find the number of hours this equipment must be used to yield a profit of 0 dollars.

Solution:

(a) $C = 36{,}500 + 5.25t + 11.50t$
$\quad = 16.75t + 36{,}500$

(b) $R = 27t$

(c) $P = R - C$
$\quad = 27t - (16.75t + 36{,}500)$
$\quad = 10.25t - 36{,}500$

(d) $0 = 10.25t - 36{,}500$
$36{,}500 = 10.25t$
$t \approx 3561$ hours

SECTION 3.4

Functions

- Given a set or an equation, you should be able to determine if it represents a function.
- Given a function, you should be able to do the following.
  (a) Find the domain.
  (b) Evaluate it at specific values.

Solutions to Selected Exercises

1. Determine which of the sets of ordered pairs represents a function from A to B. Give reasons for your answers.

$A = \{0, 1, 2, 3\}$ and $B = \{-2, -1, 0, 1, 2\}$

(a) $\{(0, 1), (1, -2), (2, 0), (3, 2)\}$

(b) $\{(0, -1), (2, 2), (1, -2), (3, 0), (1, 1)\}$

(c) $\{(0, 0), (1, 0), (2, 0), (3, 0)\}$

(d) $\{(0, 2), (3, 0), (1, 1)\}$

Solution:

(a) Each element of A is matched with exactly one element of B, so it does represent a function.

(b) The element 1 in A is matched with two elements, -2 and 1 of B, so it does not represent a function.

(c) Each element of A is matched with exactly one element of B, so it does represent a function.

(d) The element 2 in A is not matched with an element of B, so it does not represent a function.

3. Determine if y is a function of x for $x^2 + y^2 = 4$.

Solution:

y is not a function of x since some values of x give two values for y. For example, if $x = 0$, then $y = \pm 2$.

7. Determine if y is a function of x for $2x + 3y = 4$.

Solution:

$$2x + 3y = 4$$
$$y = \frac{1}{3}(4 - 2x)$$

y is a function of x. No value of x yields more than one value of y.

13. Evaluate the function at the specified value of the independent variable and simplify.

$$f(x) = 2x - 3$$

(a) $f(1)$ (b) $f(-3)$ (c) $f(x-1)$

Solution:

(a) $f(1) = 2(1) - 3 = -1$

(b) $f(-3) = 2(-3) - 3 = -9$

(c) $f(x-1) = 2(x-1) - 3 = 2x - 5$

17. Evaluate the function at the specified value of the independent variable and simplify.

$$f(y) = 3 - \sqrt{y}$$

(a) $f(4)$ (b) $f(0.25)$ (c) $f(4x^2)$

Solution:

(a) $f(4) = 3 - \sqrt{4} = 1$

(b) $f(0.25) = 3 - \sqrt{0.25} = 2.5$

(c) $f(4x^2) = 3 - \sqrt{4x^2} = 3 - 2|x|$

21. Evaluate the function at the specified value of the independent variable and simplify.

$$f(x) = \frac{|x|}{x}$$

(a) $f(2)$ (b) $f(-2)$ (c) $f(x-1)$

Solution:

(a) $f(2) = \frac{|2|}{2} = 1$

(b) $f(-2) = \frac{|-2|}{-2} = -1$

(c) $f(x-1) = \frac{|x-1|}{x-1}$

23. Evaluate the function at the specified value of the independent variable and simplify.

$$f(x) = \begin{cases} 2x+1, & x < 0 \\ 2x+2, & x \geq 0 \end{cases}$$

(a) $f(-1)$ (b) $f(0)$ (c) $f(2)$

Solution:

(a) $f(-1) = 2(-1) + 1 = -1$

(b) $f(0) = 2(0) + 2 = 2$

(c) $f(2) = 2(2) + 2 = 6$

27. Find all real values x such that $f(x) = 0$ for $f(x) = x^2 - 9$.

Solution:

$$\begin{aligned} x^2 - 9 &= 0 \\ x^2 &= 9 \\ x &= \pm 3 \end{aligned}$$

31. Find the domain of $h(t) = 4/t$.

Solution:

The domain includes all real numbers except 0, i.e. $t \neq 0$.

35. Find the domain of $f(x) = \sqrt[4]{1 - x^2}$.

Solution:

Choose x-values for which $1 - x^2 \geq 0$. Using methods of Section 2.8, we find that the domain is $-1 \leq x \leq 1$.

39. Assume that the domain of $f(x) = x^2$ is the set $A = \{-2, -1, 0, 1, 2\}$. Determine the set of ordered pairs representing the function f.

Solution:

$$\begin{aligned} &\{(-2,\ f(-2)),\ (-1,\ f(-1)),\ (0,\ f(0)),\ (1,\ f(1)),\ 2,\ f(2))\} \\ = &\{(-2,\ 4),\ (-1,\ 1),\ (0,\ 0),\ (1,\ 1),\ (2,\ 4)\} \end{aligned}$$

45. Find the value(s) of x for which $f(x) = g(x)$ where $f(x) = \sqrt{3x} + 1$ and $g(x) = x + 1$.

Solution:

$$\begin{aligned} f(x) &= g(x) \\ \sqrt{3x} + 1 &= x + 1 \\ \sqrt{3x} &= x \\ 3x &= x^2 \\ 0 &= x^2 - 3x \\ 0 &= x(x - 3) \\ x = 0 \quad &\text{or} \quad x = 3 \end{aligned}$$

47. For $f(x) = x^2 - x + 1$, find $\dfrac{f(2+h) - f(2)}{h}$ and simplify your answer.

Solution:

$$\begin{aligned} f(x) &= x^2 - x + 1 \\ f(2+h) &= (2+h)^2 - (2+h) + 1 \\ &= 4 + 4h + h^2 - 2 - h + 1 \\ &= h^2 + 3h + 3 \\ f(2) &= (2)^2 - 2 + 1 = 3 \\ f(2+h) - f(2) &= h^2 + 3h \\ \frac{f(2+h) - f(2)}{h} &= h + 3 \end{aligned}$$

53. Express the area A of a circle as a function of its circumference C.

Solution:

$$A = \pi r^2, \quad C = 2\pi r$$

$$r = \frac{C}{2\pi}$$

$$\begin{aligned} A &= \pi\left(\frac{C}{2\pi}\right)^2 \\ &= \frac{C^2}{4\pi} \end{aligned}$$

55. A right triangle is formed in the first quadrant by the x- and y-axes and a line through the point (1, 2), as shown in the figure. Write the area of the triangle as a function of x, and determine the domain of the function.

Solution:

$$A = \frac{1}{2}bh = \frac{1}{2}xy$$

Since (0, y), (1, 2) and (x, 0) all lie on the same line, the slopes between any pair are equal.

$$\frac{2-y}{1-0} = \frac{0-2}{x-1}$$

$$2 - y = -\frac{2}{x-1}$$

$$y = \frac{2}{x-1} + 2$$

$$y = \frac{2x}{x-1}$$

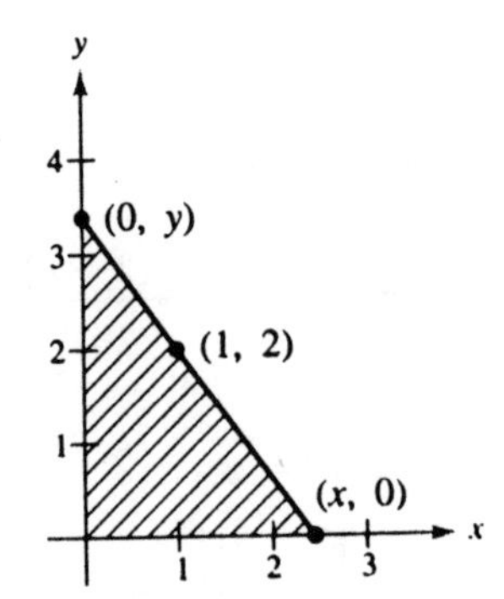

Therefore,

$$A = \frac{1}{2}x\left(\frac{2x}{x-1}\right) = \frac{x^2}{x-1}$$

The domain of A includes x-values such that $x^2/(x-1) > 0$. Using methods of Section 2.8, we find that the domain is $x > 1$.

59. A balloon carrying a transmitter ascends vertically from a point 2000 feet from the receiving station (see figure). Let d be the distance between the balloon and the receiving station. Express the height of the balloon as a function of d. What is the domain of the function?

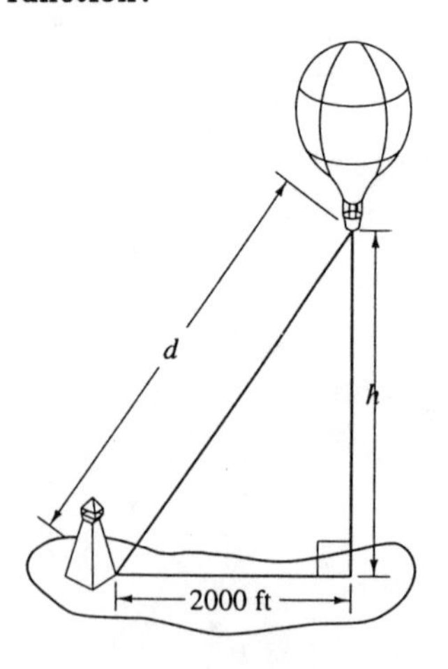

59. —-CONTINUED—-

Solution:

By the Pythagorean Theorem we have

$$h^2 + 2000^2 = d^2$$
$$h^2 = d^2 - 2000^2$$
$$h = \sqrt{d^2 - 2000^2}.$$

Since $d^2 - 2000^2 \geq 0$ and $d \geq 0$, we have a domain of $d \geq 2000$ feet.

61. A company produces a product for which the variable cost is $12.30 per unit and the fixed costs are $98,000. The product sells for $17.98. Let x be the number of units produced and sold.

(a) Write the total cost C as a function of the number of units produced.

(b) Write the revenue R as a function of the number of units sold.

(c) Write the profit P as a function of the number of units sold.

(**Note:** $P = R - C$.)

Solution:

(a) Cost = variable costs + fixed costs

$$C = 12.30x + 98{,}000$$

(b) Revenue = price per unit × number of units

$$R = 17.98x$$

(c) Profit = Revenue − Cost

$$P = 17.98x - (12.30x + 98{,}000)$$
$$P = 5.68x - 98{,}000$$

SECTION 3.5

Graphs of Functions

- You should be able to determine the domain and range of a function from its graph.
- You should be able to use the vertical line test for functions.
- You should know that the graph of $f(x) = c$ is a horizontal line through $(0, c)$.
- You should be able to determine when a function is constant, increasing, or decreasing.
- You should know that f is
  (a) Odd if $f(-x) = -f(x)$.
  (b) Even if $f(-x) = f(x)$.
- You should know the basic types of transformations.
  (a) Vertical and horizontal shifts
  (b) Reflections in the coordinate axes

Solutions to Selected Exercises

5. Determine the domain and range of the function $f(x) = \sqrt{25 - x^2}$.

Solution:

From the graph we see that the x-values do not extend beyond $x = -5$ (on the left) and $x = 5$ (on the right). The domain is $[-5, 5]$. Similarly, the y-values do not extend beyond $y = 0$ and $y = 5$. The range is $[0, 5]$.

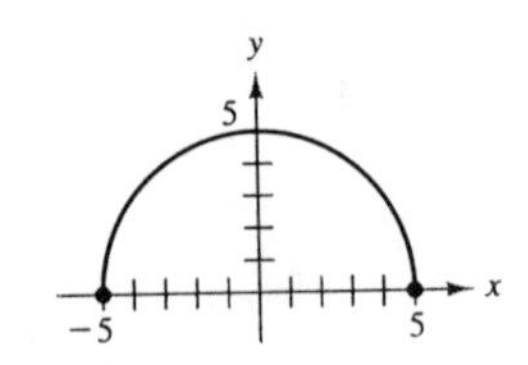

7. Use the vertical line test to determine if y is a function of x where $y = x^2$.

Solution:

Since no vertical line would ever cross the graph more than one time, y *is* a function of x.

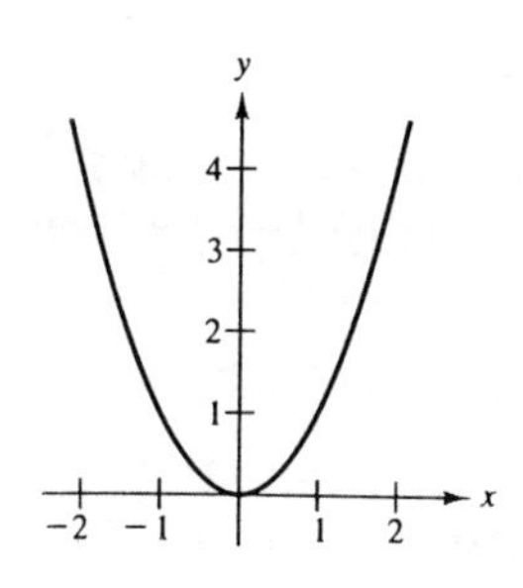

11. Use the vertical line test to determine if y is a function of x where $x^2 = xy - 1$.

Solution:

Since no vertical line would ever cross the graph more than one time, y *is* a function of x.

$$y = \frac{x^2 + 1}{x}$$

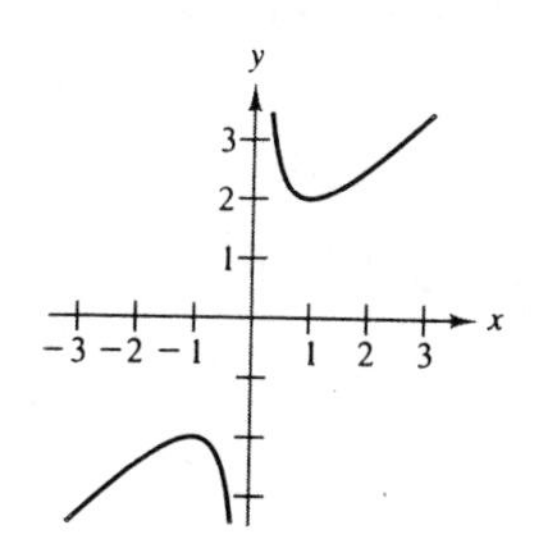

15. (a) Determine the intervals over which the function is increasing, decreasing,or constant, and (b) determine if the function is even, odd, or neither for $f(x) = x^3 - 3x^2$.

Solution:

(a) By its graph we see that f is increasing on $(-\infty, 0)$ and $(2, \infty)$ and is decreasing on $(0, 2)$.

(b) $f(-x) = (-x)^3 - 3(-x)^2$
$= -x^3 - 3x^2$

$f(-x) \neq f(x)$ and $f(-x) \neq -f(x)$, so the function is neither odd nor even.

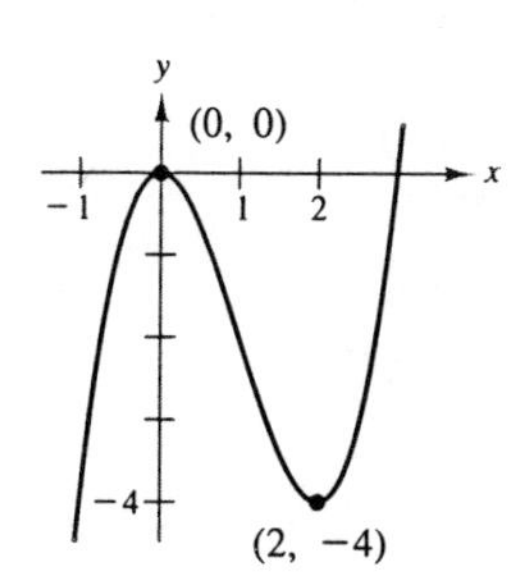

19. (a) Determine the intervals over which the function is increasing, decreasing, or constant, and

(b) determine if the function is even, odd, or neither for $f(x) = x\sqrt{x+3}$.

Solution:

(a) By its graph we see that f is increasing on $(-2, \infty)$ and decreasing on $(-3, -2)$.

(b) $f(-x) = -x\sqrt{-x+3}$

$f(-x) \neq f(x)$ and $f(-x) \neq -f(x)$, so the function is neither odd nor even.

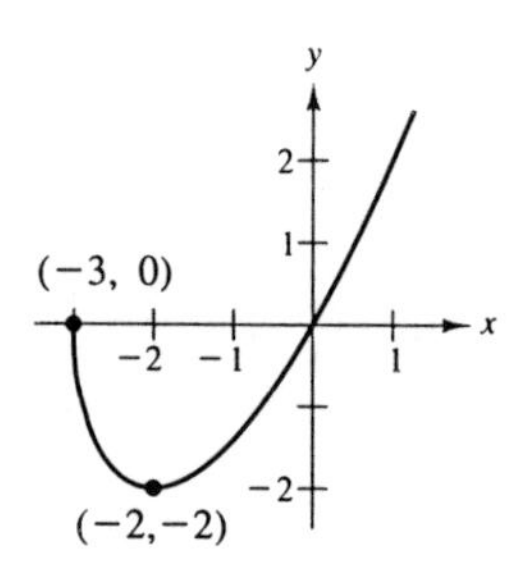

23. Determine whether $g(x) = x^3 - 5x$ is even, odd, or neither.

Solution:

$$\begin{aligned} g(x) &= x^3 - 5x \\ g(-x) &= (-x)^3 - 5(-x) \\ &= -x^3 + 5x \\ &= -(x^3 - 5x) \\ &= -g(x) \end{aligned}$$

Therefore, g is odd.

27. Sketch the graph of $f(x) = 3$ and determine whether the function is odd, even, or neither.

Solution:

$f(x) = 3$

Domain: $(-\infty, \infty)$

Range: $\{3\}$

y-intercept: $(0, 3)$

y-axis symmetry

$f(-x) = 3 = f(x)$

Therefore, f is even.

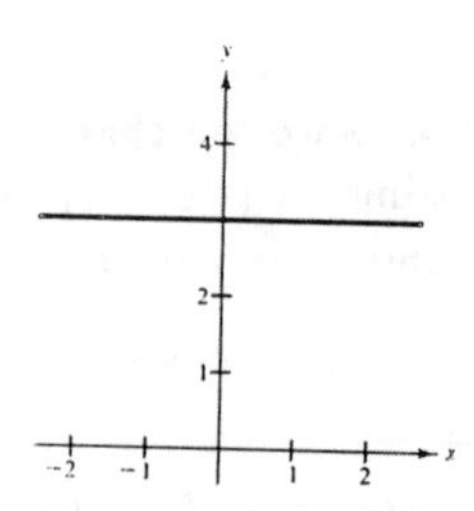

31. Sketch the graph of $g(s) = s^3/4$ and determine whether the function is odd, even, or neither.

Solution:

$g(s) = s^3/4$
Intercept: $(0, 0)$
Origin symmetry
Domain: $(-\infty, \infty)$
Range: $(-\infty, \infty)$
$g(-s) = -g(s)$
Therefore, g is odd.

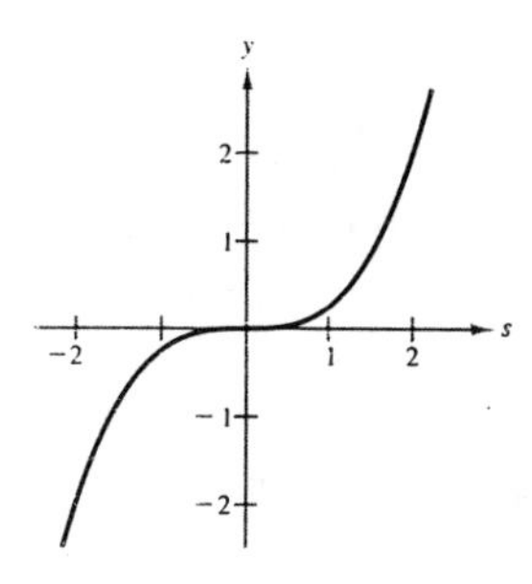

35. Sketch the graph of $g(t) = (t-1)^2 + 2$ and determine whether the function is odd, even, or neither.

Solution:

$g(t) = (t-1)^2 + 2$
Intercept: $(0, 3)$
No symmetries
Domain: $(-\infty, \infty)$
Range: $[2, \infty)$
Neither odd nor even

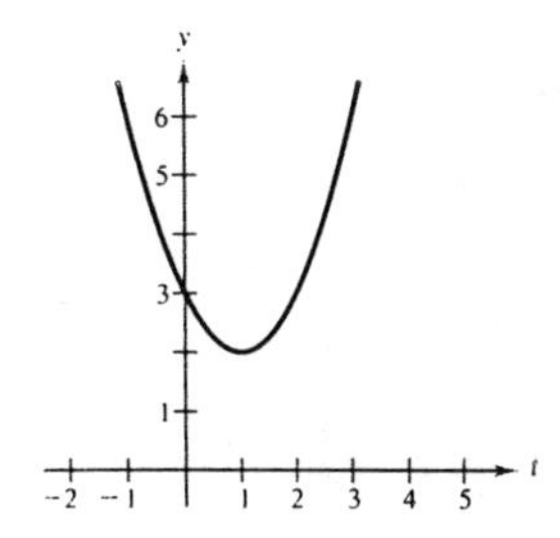

37. Sketch the graph of $f(x) = \begin{cases} x+3, & \text{if } x \le 0 \\ 3, & \text{if } 0 < x \le 2 \\ 2x-1, & \text{if } x > 2 \end{cases}$ and determine whether the function is odd, even, or neither.

Solution:

For $x \le 0$, $f(x) = x + 3$.
For $0 < x \le 2$, $f(x) = 3$.
For $x > 2$, $f(x) = 2x - 1$.
Thus, the graph of f is as shown.

$$f(-x) = \begin{cases} -x+3, & \text{if } x \le 0 \\ 3, & \text{if } 0 < x \le 2 \\ -2x-1, & \text{if } x > 2 \end{cases}$$

So, f is neither odd nor even.

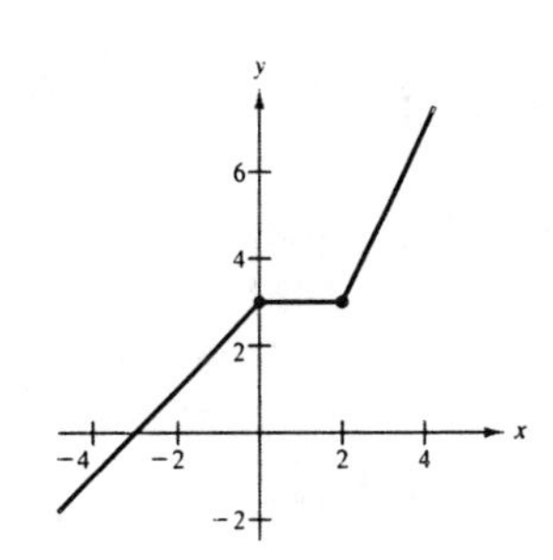

43. Sketch the graph of $f(x) = x^2 - 9$ and determine the interval(s), if any, on the real axis for which $f(x) \geq 0$.

Solution:

$f(x) = x^2 - 9$
x-intercepts: $(-3, 0)$, $(0, 3)$
y-intercept: $(0, -9)$
y-axis symmetry
Domain: $(-\infty, \infty)$
Range: $[-9, \infty)$
$f(x) \geq 0$ on the intervals $(-\infty, -3]$ and $[3, \infty)$.

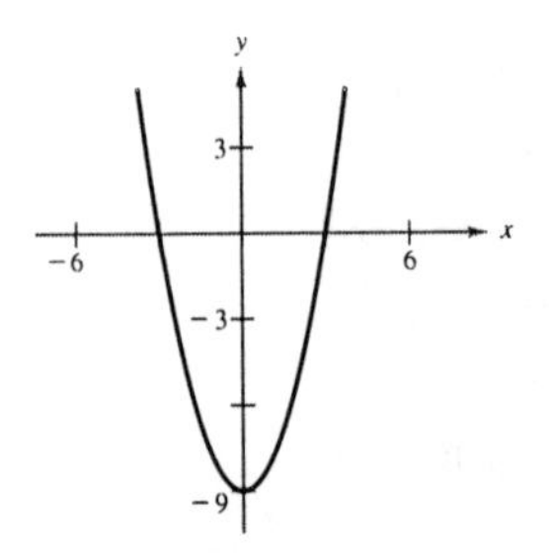

47. Sketch the graph of $f(x) = x^2 + 1$ and determine the interval(s), if any, on the real axis for which $f(x) \geq 0$.

Solution:

$f(x) = x^2 + 1$
x-intercept: None
y-intercept: $(0, 1)$
y-axis symmetry
Domain: $(-\infty, \infty)$
Range: $[1, \infty)$
$f(x) \geq 0$ for all real numbers.

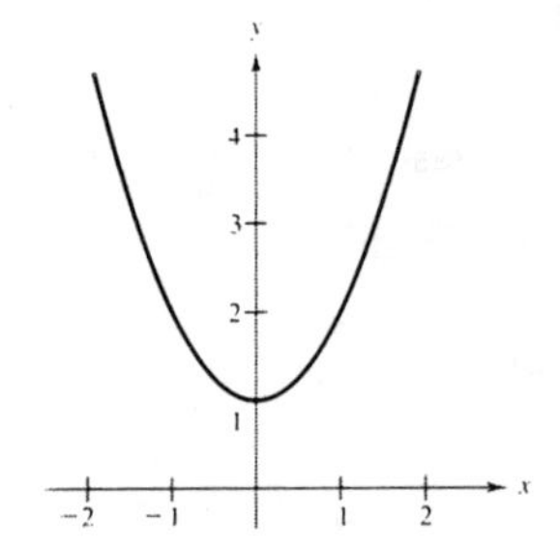

51. Sketch (on the same set of coordinate axes) a graph of f for $c = -2$, 0 and 2.

(a) $f(x) = \frac{1}{2}x + c$ (b) $f(x) = \frac{1}{2}(x - c)$ (c) $f(x) = \frac{1}{2}(cx)$

Solution:

(a) $f(x) = \frac{1}{2}x + c$

$c = -2 : f(x) = \frac{1}{2}x - 2$

$c = 0 : f(x) = \frac{1}{2}x$

$c = 2 : f(x) = \frac{1}{2}x + 2$

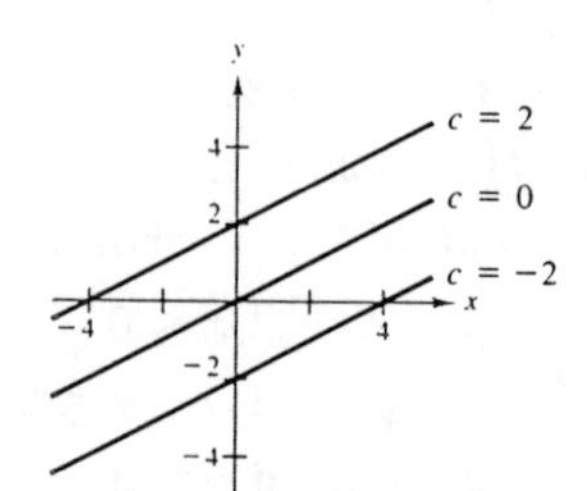

–CONTINUED ON NEXT PAGE–

51. –CONTINUED–

(b) $f(x) = \frac{1}{2}(x - c)$

$c = -2 : f(x) = \frac{1}{2}(x + 2) = \frac{1}{2}x + 1$

$c = \;\; 0 : f(x) = \frac{1}{2}x$

$c = \;\; 2 : f(x) = \frac{1}{2}(x - 2) = \frac{1}{2}x - 1$

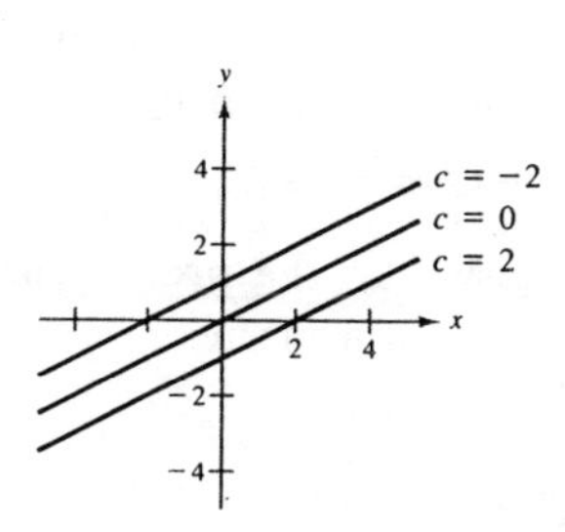

(c) $f(x) = \frac{1}{2}(cx)$

$c = -2 : f(x) = \frac{1}{2}(-2x) = -x$

$c = \;\; 0 : f(x) = \frac{1}{2}(0x) = 0$

$c = \;\; 2 : f(x) = \frac{1}{2}(2x) = x$

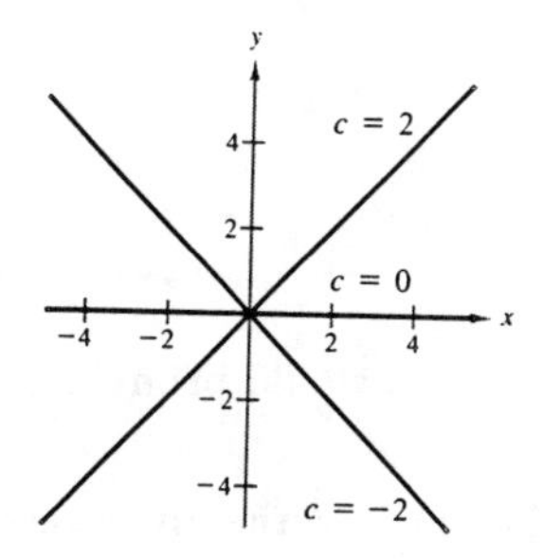

55. Use the graph of $f(x) = x^2$ to write formulas for the functions whose graphs are shown.

(a)

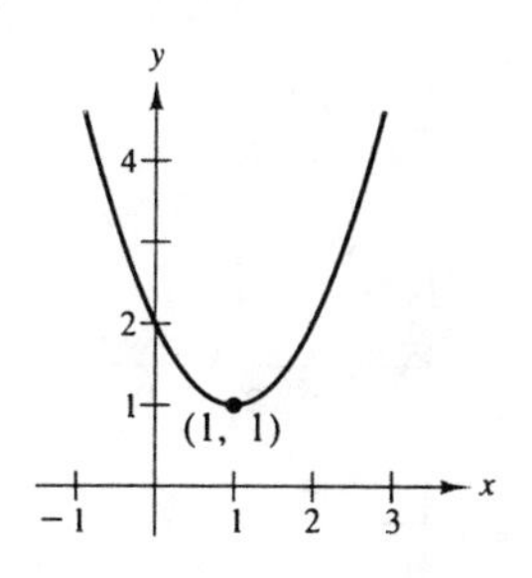

(b)

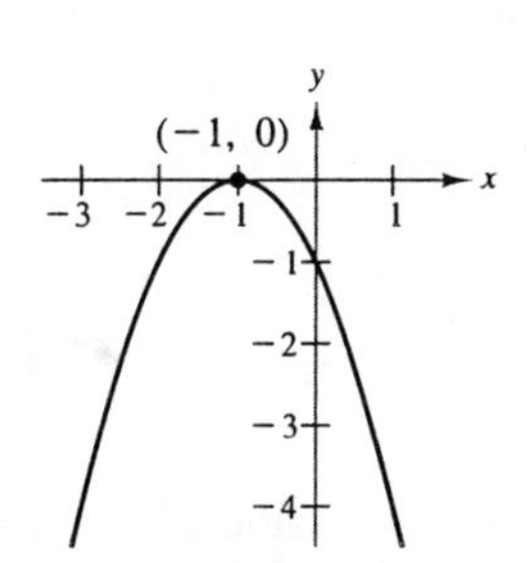

Solution:

(a) Horizontal shift one unit to the right and vertical shift one unit upward:

$h(x) = f(x - 1) + 1 = (x - 1)^2 + 1$

(b) Horizontal shift one unit to the left and reflection in the x-axis:

$h(x) = -f(x + 1) = -(x + 1)^2$

59. The marketing department for a company estimates that the demand for a product is given by $p = 100 - 0.0001x$ where p is the price per unit and x is the number of units. The cost of producing x units is given by $C = 350{,}000 + 30x$, and the profit for producing and selling x units is given by

$$P = R - C = xp - C.$$

Sketch the graph of the profit function and estimate the number of units that would produce a maximum profit.

Solution:

$$\begin{aligned} P &= R - C = xp - C \\ &= x(100 - 0.0001x) - (350{,}000 + 30x) \\ &= -0.0001x^2 + 70x - 350{,}000 \end{aligned}$$

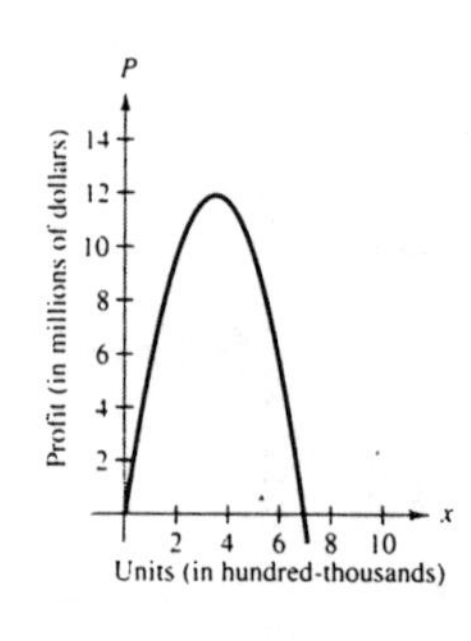

$x = 350{,}000$ units would produce a maximum profit.

63. Write the height h of the given rectangle as a function of x.

Solution:

$$\begin{aligned} h &= \text{top} - \text{bottom} \\ &= (4x - x^2) - x^2 \\ &= 4x - 2x^2 \end{aligned}$$

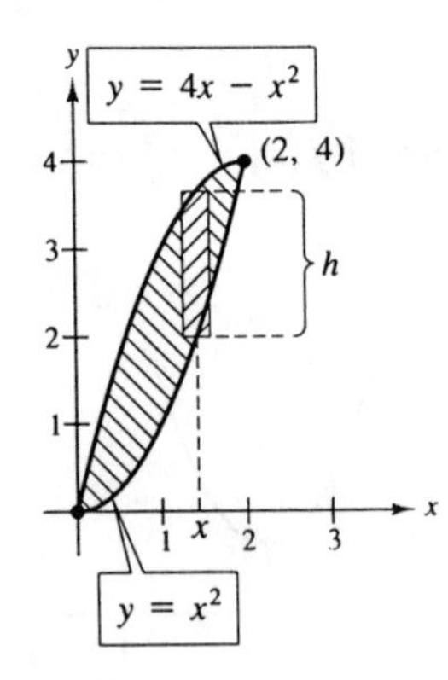

67. Prove that a function of the following form is odd.

$$f(x) = a_{2n+1}x^{2n+1} + a_{2n-1}x^{2n-1} + \ldots + a_3x^3 + a_1x$$

Solution:

$$\begin{aligned} f(x) &= a_{2n+1}x^{2n+1} + a_{2n-1}x^{2n-1} + \ldots + a_3x^3 + a_1x \\ f(-x) &= a_{2n+1}(-x)^{2n+1} + a_{2n-1}(-x)^{2n-1} + \ldots + a_3(-x)^3 + a_1(-x) \\ &= -a_{2n+1}x^{2n+1} - a_{2n-1}x^{2n-1} - \ldots - a_3x^3 - a_1x = -f(x) \end{aligned}$$

Therefore, $f(x)$ is odd.

SECTION 3.6

Combinations of Functions

- Given two functions, f and g, you should be able to form the following functions (if defined):
  1. Sum: $(f+g)(x) = f(x) + g(x)$
  2. Difference: $(f-g)(x) = f(x) - g(x)$
  3. Product: $(fg)(x) = f(x)g(x)$
  4. Quotient: $(f/g)(x) = f(x)/g(x),\ g(x) \neq 0$
  5. Composition of f with g: $(f \circ g)(x) = f(g(x))$
  6. Composition of g with f: $(g \circ f)(x) = g(f(x))$

Solutions to Selected Exercises

5. Find (a) $(f+g)(x)$, (b) $(f-g)(x)$, (c) $(fg)(x)$, and (d) $(f/g)(x)$. What is the domain of f/g?

Solution:

$$f(x) = x^2 + 5, \quad g(x) = \sqrt{1-x}$$

(a) $(f+g)(x) = f(x) + g(x) = x^2 + 5 + \sqrt{1-x}$

(b) $(f-g)(x) = f(x) - g(x) = x^2 + 5 - \sqrt{1-x}$

(c) $(fg)(x) = f(x)g(x) = (x^2+5)\sqrt{1-x}$

(d) $\left(\dfrac{f}{g}\right)(x) = \dfrac{f(x)}{g(x)} = \dfrac{x^2+5}{\sqrt{1-x}},\quad x < 1$

The domain of f/g is $(-\infty,\ 1)$.

9. Evaluate $(f+g)(3)$ for $f(x) = x^2 + 1$ and $g(x) = x - 4$.

Solution:

$$\begin{aligned}(f+g)(3) &= f(3) + g(3) \\ &= [(3)^2 + 1] + (3-4) \\ &= 10 - 1 \\ &= 9\end{aligned}$$

13. Evaluate $(f-g)(2t)$ for $f(x)=x^2+1$ and $g(x)=x-4$.

Solution:

$$\begin{aligned}(f-g)(2t) &= f(2t)-g(2t)\\ &= [(2t)^2+1]-[(2t)-4]\\ &= 4t^2+1-2t+4\\ &= 4t^2-2t+5\end{aligned}$$

17. Evaluate $(f/g)(5)$ for $f(x)=x^2+1$ and $g(x)=x-4$.

Solution:

$$\left(\frac{f}{g}\right)(5)=\frac{f(5)}{g(5)}=\frac{(5)^2+1}{5-4}=26$$

19. Evaluate $(f/g)(-1)-g(3)$ for $f(x)=x^2+1$ and $g(x)=x-4$.

Solution:

$$\left(\frac{f}{g}\right)(-1)-g(3)=\frac{f(-1)}{g(-1)}-g(3)=\frac{(-1)^2+1}{-1-4}-(3-4)=-\frac{2}{5}+1=\frac{3}{5}$$

23. Find (a) $f\circ g$, (b) $g\circ f$, and (c) $f\circ f$ for $f(x)=3x+5$ and $g(x)=5-x$.

Solution:

(a) $$\begin{aligned}f\circ g &= f(g(x))\\ &= f(5-x)\\ &= 3(5-x)+5\\ &= 20-3x\end{aligned}$$

(b) $$\begin{aligned}g\circ f &= g(f(x))\\ &= g(3x+5)\\ &= 5-(3x+5)\\ &= -3x\end{aligned}$$

(c) $$\begin{aligned}f\circ f &= f(f(x))\\ &= f(3x+5)\\ &= 3(3x+5)+5\\ &= 9x+20\end{aligned}$$

25. Find (a) $f\circ g$ and (b) $g\circ f$ for $f(x)=\sqrt{x+4}$ and $g(x)=x^2$.

Solution:

(a) $$\begin{aligned}f\circ g &= f(g(x))\\ &= f(x^2)\\ &= \sqrt{x^2+4}\end{aligned}$$

(b) $$\begin{aligned}g\circ f &= g(f(x))\\ &= g(\sqrt{x+4})\\ &= (\sqrt{x+4})^2\\ &= x+4\end{aligned}$$

29. Find (a) $f \circ g$ and (b) $g \circ f$ for $f(x) = \sqrt{x}$ and $g(x) = \sqrt{x}$.

Solution:

(a) $f \circ g = f(g(x)) = f(\sqrt{x}) = \sqrt{\sqrt{x}} = \sqrt[4]{x}$

(b) Same as (a)

31. Find (a) $f \circ g$ and (b) $g \circ f$ for $f(x) = |x|$ and $g(x) = x + 6$.

Solution:

(a) $f \circ g = f(g(x)) = f(x+6) = |x+6|$

(b) $g \circ f = g(f(x)) = g(|x|) = |x| + 6$

33. Use the graphs of f and g to evaluate (a) $(f+g)(3)$ and (b) $(f/g)(2)$.

Solution:

(a) $(f+g)(3) = f(3) + g(3) = 2 + 1 = 3$

(b) $\left(\frac{f}{g}\right)(2) = \frac{f(2)}{g(2)} = \frac{0}{2} = 0$

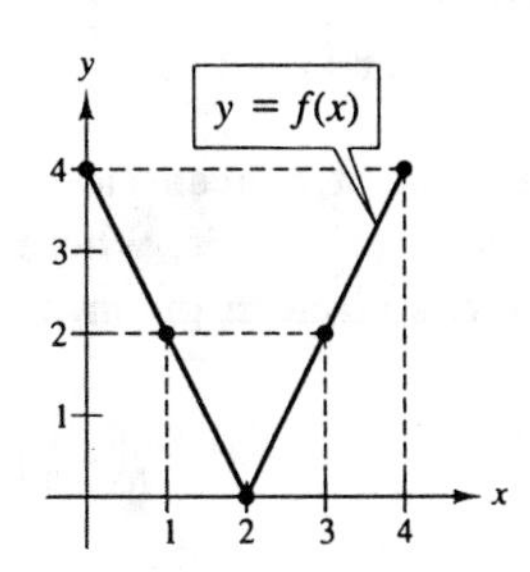

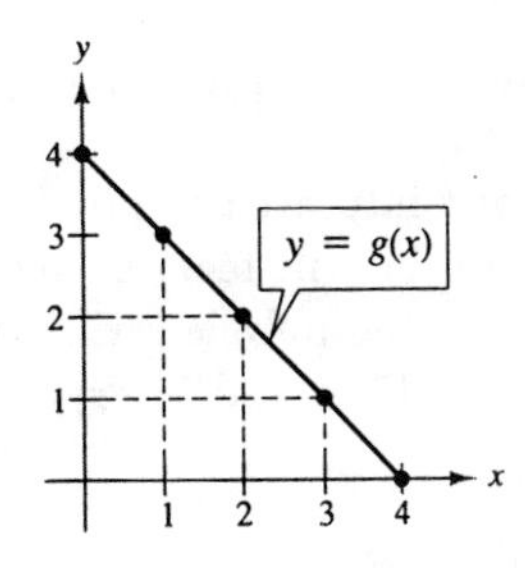

37. Find two functions f and g such that $(f \circ g)(x) = h(x)$ for $h(x) = (2x+1)^2$.

Solution:

Let $f(x) = x^2$ and $g(x) = 2x + 1$, then $(f \circ g)(x) = h(x)$.

This is not a unique solution. For example, if $f(x) = (x+1)^2$ and $g(x) = 2x$, then $(f \circ g)(x) = h(x)$ as well.

41. Find two functions f and g such that $(f \circ g)(x) = h(x)$ for $h(x) = 1/(x+2)$.

Solution:

Let $f(x) = 1/x$ and $g(x) = x + 2$, then $(f \circ g)(x) = h(x)$. Again, this is not a unique solution. Other possibilities are:

$$f(x) = \frac{1}{x+2} \quad \text{and} \quad g(x) = x$$

OR

$$f(x) = \frac{1}{x+1} \quad \text{and} \quad g(x) = x + 1$$

OR

$$f(x) = \frac{1}{x^2+2} \quad \text{and} \quad g(x) = \sqrt{x}$$

47. Determine the domain of (a) f, (b) g, and (c) $f \circ g$ for $f(x) = 3/(x^2-1)$ and $g(x) = x+1$.

Solution:

(a) The domain of $f(x) = 3/(x^2 - 1)$ includes all real numbers except $x = \pm 1$.

(b) The domain of $g(x) = x + 1$ includes all real numbers.

(c) $f \circ g = f(g(x)) = f(x+1) = \dfrac{3}{(x+1)^2 - 1} = \dfrac{3}{x^2 + 2x} = \dfrac{3}{x(x+2)}$

The domain of $f \circ g$ includes all real numbers except $x = 0$ and $x = -2$.

51. A pebble is dropped into a calm pond, causing ripples in the form of concentric circles (see figure). The radius (in feet) of the outer ripple is given by $r(t) = 0.6t$, where t is time in seconds after the pebble strikes the water. The area of the circle is given by the function $A(r) = \pi r^2$. Find and interpret $(A \circ r)(t)$.

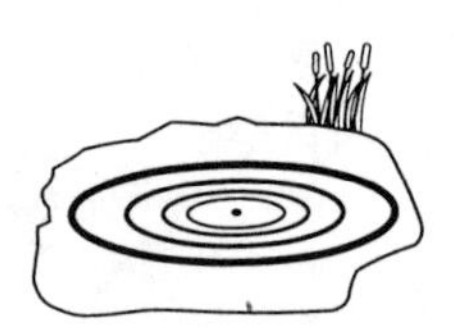

Solution:

$(A \circ r)(t) = A(r(t)) = A(0.6t) = \pi(0.6t)^2 = 0.36\pi t^2$

This equation gives the area as a function of time.

55. Prove that the product of two odd functions is an even function.

Solution:

Let $f(x)$ and $g(x)$ be two odd functions and define $h(x) = f(x)g(x)$.

Then

$$\begin{aligned} h(-x) &= f(-x)g(-x) \\ &= [-f(x)][-g(x)] \quad \text{Since } f(x) \text{ and } g(x) \text{ are odd} \\ &= f(x)g(x) \\ &= h(x) \end{aligned}$$

Thus, h is even.

SECTION 3.7

Inverse Functions

- Two functions f and g are inverses of each other if $f(g(x)) = x$ for every x in the domain of g and $g(f(x)) = x$ for every x in the domain of f.
- A function f has an inverse if and only if f is one-to-one.
- Be able to find the inverse of a function, if it exists.

Solutions to Selected Exercises

1. Find f^{-1} informally for $f(x) = 8x$. Verify that if $f(f^{-1}(x))$ and $f^{-1}(f(x))$ are equal to the identity function.

Solution:

To "undo" 8 times x, we let f^{-1} divide x by 8.

$$f^{-1}(x) = \frac{x}{8}$$

Now, we have $f(f^{-1}(x)) = f\left(\frac{x}{8}\right) = 8\left(\frac{x}{8}\right) = x$ and $f^{-1}(f(x)) = f^{-1}(8x) = \frac{8x}{8} = x$.

7. (a) Show that $f(x) = 2x$ and $g(x) = \frac{x}{2}$ are inverse functions by showing that $f(g(x)) = x$ and $g(f(x)) = x$, and (b) graph f and g on the same set of coordinate axes.

Solution:

$f(x) = 2x,\ g(x) = \frac{x}{2}$

(a) $f(g(x)) = f\left(\frac{x}{2}\right) = 2\left(\frac{x}{2}\right) = x$

$g(f(x)) = g(2x) = \frac{(2x)}{2} = x$

(b)

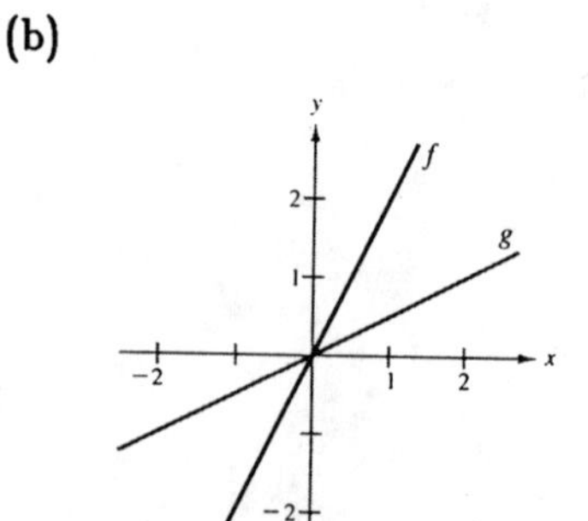

11. (a) Show that $f(x) = x^3$ and $g(x) = \sqrt[3]{x}$ are inverse functions by showing that $f(g(x)) = x$ and $g(f(x)) = x$, and (b) graph f and g on the same set of coordinate axes.

Solution:

$f(x) = x^3, \quad g(x) = \sqrt[3]{x}$

(a) $f(g(x)) = f(\sqrt[3]{x}) = (\sqrt[3]{x})^3 = x$

$g(f(x)) = g(x^3) = \sqrt[3]{x^3} = x$

(b)

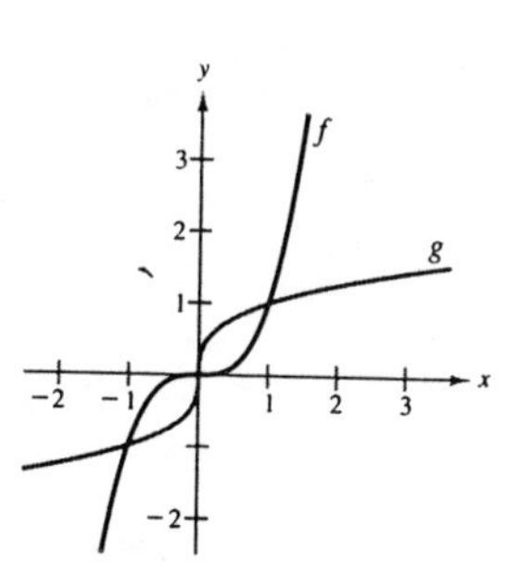

13. (a) Show that $f(x) = \sqrt{x-4}$ and $g(x) = x^2 + 4, \; x \geq 0$ are inverse functions by showing that $f(g(x)) = x$ and $g(f(x)) = x$, and (b) graph f and g on the same set of coordinate axes.

Solution:

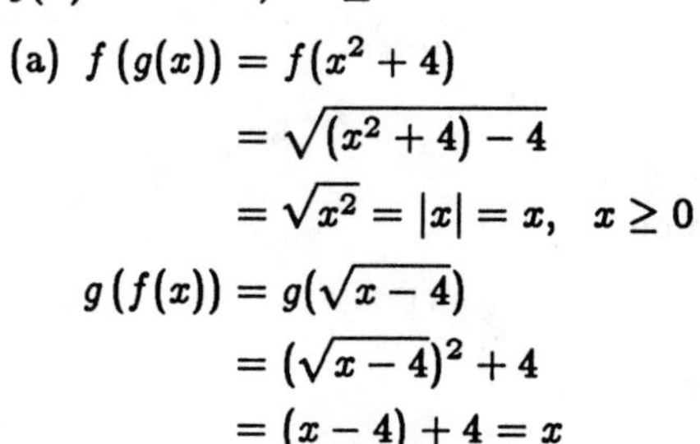

$f(x) = \sqrt{x-4},$

$g(x) = x^2 + 4, \quad x \geq 0$

(a) $f(g(x)) = f(x^2 + 4)$

$= \sqrt{(x^2+4) - 4}$

$= \sqrt{x^2} = |x| = x, \quad x \geq 0$

$g(f(x)) = g(\sqrt{x-4})$

$= (\sqrt{x-4})^2 + 4$

$= (x-4) + 4 = x$

(b)

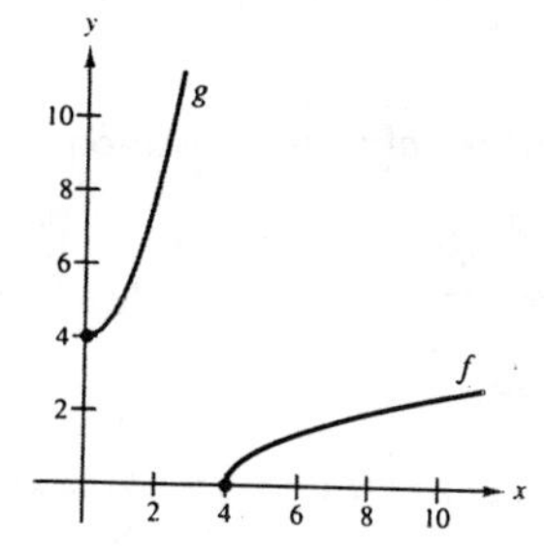

17. Determine whether the function shown is one-to-one.

Solution:

Since the function is decreasing on its entire domain, it is one-to-one.

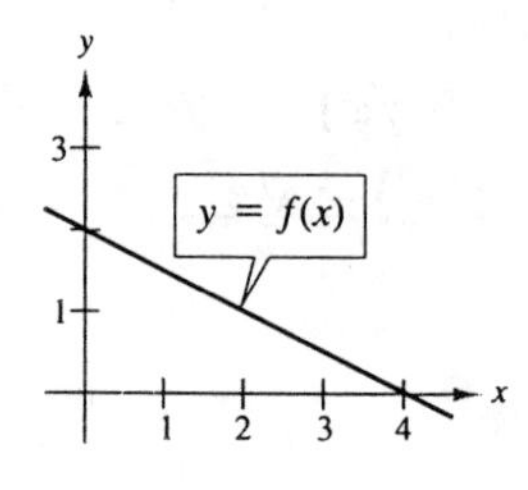

21. Determine whether the function $g(x) = (4 - x)/6$ is one-to-one.

Solution:

Let a and b be real numbers with $g(a) = g(b)$. Then we have

$$\frac{4-a}{6} = \frac{4-b}{6}$$

$$4 - a = 4 - b$$

$$-a = -b$$

$$a = b$$

Therefore, $g(x)$ is one-to-one.

25. Determine whether the function $f(x) = -\sqrt{16 - x^2}$ is one-to-one.

Solution:

Since $f(4) = 0$ and $f(-4) = 0$, the function is not one-to-one.

29. Find the inverse of the one-to-one function $f(x) = x^5$. Then graph both f and f^{-1} on the same coordinate plane.

Solution:

$$f(x) = x^5$$
$$y = x^5$$
$$x = y^5$$
$$\sqrt[5]{x} = y$$
$$f^{-1}(x) = \sqrt[5]{x}$$

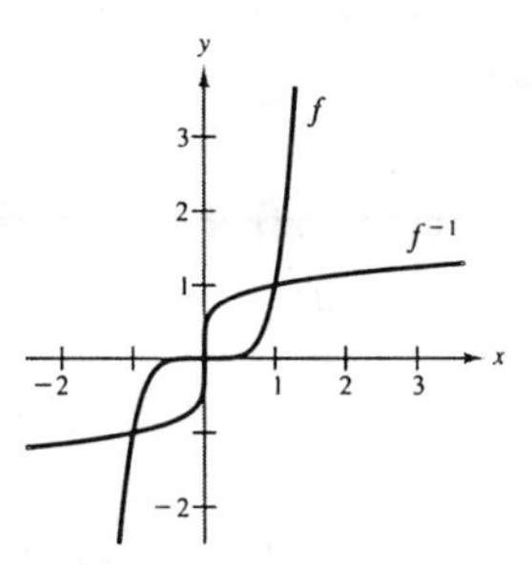

35. Find the inverse of the one-to-one function $f(x) = \sqrt[3]{x-1}$. Then graph both f and f^{-1} on the same coordinate plane.

Solution:

$$f(x) = \sqrt[3]{x-1}$$
$$y = \sqrt[3]{x-1}$$
$$x = \sqrt[3]{y-1}$$
$$x^3 = y - 1$$
$$x^3 + 1 = y$$
$$f^{-1}(x) = x^3 + 1$$

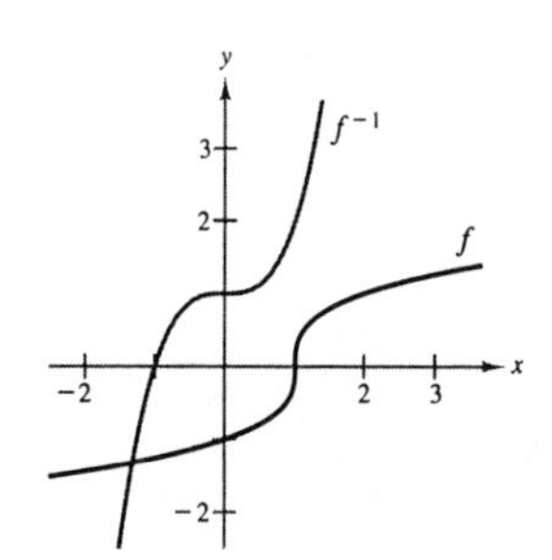

39. Determine whether the function $g(x) = x/8$ is one-to-one. If it is, find its inverse.

Solution:

$$g(a) = g(b)$$
$$\frac{a}{8} = \frac{b}{8}$$
$$a = b$$

Therefore, g is one-to-one.

$$g(x) = \frac{x}{8}$$
$$y = \frac{x}{8}$$
$$x = \frac{y}{8}$$
$$8x = y$$
$$g^{-1}(x) = 8x$$

41. Determine whether the function $p(x) = -4$ is one-to-one. If it is, find its inverse.

Solution:

$p(x) = -4$ for all real numbers x. Therefore, $p(x)$ is not one-to-one and does not have an inverse.

45. Determine whether the function $h(x) = 1/x$ is one-to-one. If it is, find its inverse.

Solution:

$$\begin{aligned} h(a) &= h(b) \\ \frac{1}{a} &= \frac{1}{b} \\ a &= b \end{aligned}$$

Therefore, h is one-to-one.

$$\begin{aligned} h(x) &= \frac{1}{x} \\ y &= \frac{1}{x} \\ x &= \frac{1}{y} \\ xy &= 1 \\ y &= \frac{1}{x} \\ h^{-1}(x) &= \frac{1}{x} \end{aligned}$$

49. Determine whether the function $g(x) = x^2 - x^4$ is one-to-one. If it is, find its inverse.

Solution:

Since $g(0) = 0$ and $g(1) = 0$, g is not one-to-one and does not have an inverse.

51. Determine whether the function $f(x) = 25 - x^2$, $x \leq 0$ is one-to-one. If it is, find its inverse.

Solution:

$$f(a) = f(b)$$
$$25 - a^2 = 25 - b^2$$
$$a^2 = b^2 \qquad \text{Since } a, b \leq 0, \text{ we have } a = b \text{ and } f \text{ is one-to-one.}$$
$$f(x) = 25 - x^2, \quad x \leq 0$$
$$y = 25 - x^2$$
$$x = 25 - y^2$$
$$y^2 = 25 - x$$
$$y = -\sqrt{25 - x} \qquad \text{Since } y \leq 0$$
$$f^{-1}(x) = -\sqrt{25 - x}$$

55. Delete part of the graph of the function so that the part that remains is one-to-one. Find the inverse of the remaining part and give the domain of the inverse. [**Note:** The answer is not unique.]

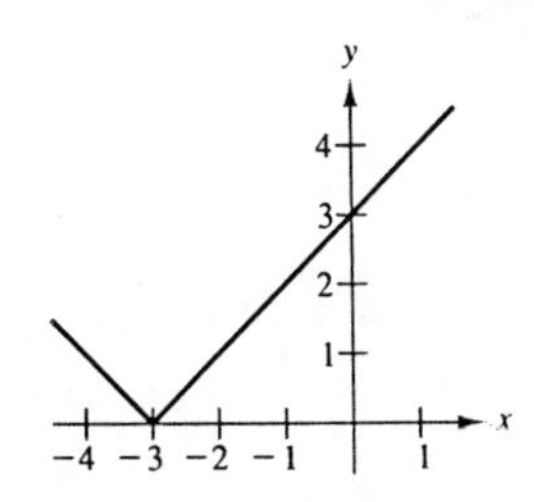

$f(x) = |x + 3|$

55. —CONTINUED—

Solution:

Delete either the part corresponding to $x < -3$ or the part corresponding to $x > -3$.

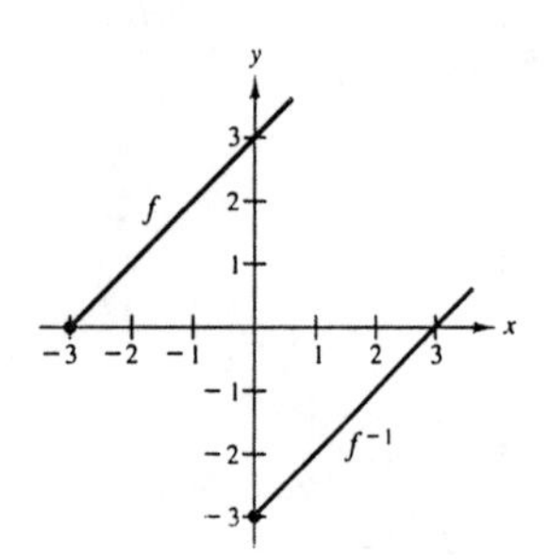

$f(x) = x + 3$

Domain: $[-3, \infty)$

Range: $[0, \infty)$

$f^{-1}(x) = x - 3$

Domain: $[0, \infty)$

Range: $[-3, \infty)$

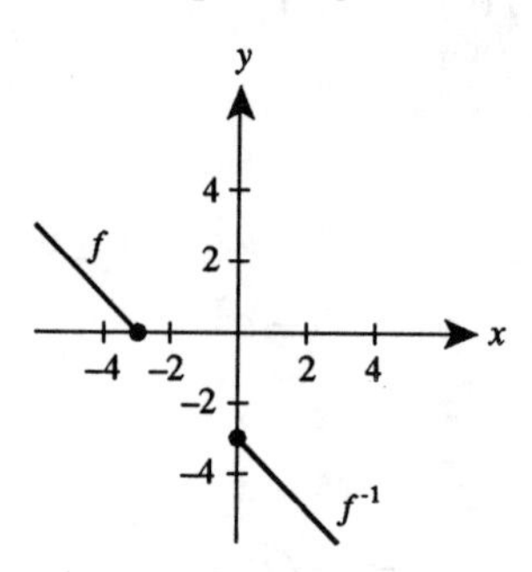

$f(x) = -x - 3$

Domain: $(-\infty, -3]$

Range: $[0, \infty)$

$f^{-1}(x) = -x - 3$

Domain: $[0, \infty)$

Range: $(-\infty, -3]$

59. Use the functions $f(x) = (1/8)x - 3$ and $g(x) = x^3$ to find $(f^{-1} \circ g^{-1})(1)$.

Solution:

$$f(x) = (1/8)x - 3 \implies f^{-1}(x) = 8(x + 3)$$

$$g(x) = x^3 \implies g^{-1}(x) = \sqrt[3]{x}$$

$$(f^{-1} \circ g^{-1})(1) = f^{-1}(g^{-1}(1))$$

$$= f^{-1}(\sqrt[3]{1}) = f^{-1}(1) = 8(1 + 3) = 32$$

63. Use the functions $f(x) = \frac{1}{8}x - 3$ and $g(x) = x^3$ to find $(f \circ g)^{-1}(x)$.

Solution:

$$f(x) = \frac{1}{8}x - 3, \quad g(x) = x^3$$

$$(f \circ g)(x) = f(g(x)) = f(x^3) = \frac{1}{8}x^3 - 3$$

$$(f \circ g)^{-1}(x) = \sqrt[3]{8(x + 3)} = 2\sqrt[3]{x + 3}$$

67. True or False: If f is an even function, then f^{-1} exists.

Solution:

If f is an even function, then $f(-x) = f(x)$.

This implies that f is not one-to-one.

Thus f^{-1} does not exist.

The statement is false.

71. Prove that if f is a one-to-one odd function, then f^{-1} is an odd function.

Solution:

Suppose f^{-1} is not an odd function. Then there exists a value $x = a$ such that

$$f^{-1}(-a) \neq -f^{-1}(a).$$

Since f is one-to-one, we have $f(f^{-1}(-a)) \neq f(-f^{-1}(a))$, but $f(f^{-1}(-a)) = -a$ since f and f^{-1} are inverses and

$$f(-f^{-1}(a)) = -f(f^{-1}(a)) = -a$$

since f is odd. Therefore,

$$f(f^{-1}(-a)) = f(-f^{-1}(a)),$$

which is a contradiction. Therefore, f^{-1} must be an odd function.

SECTION 3.8

Variation and Mathematical Models

You should know the following terms and formulas for variation.

- Direct Variation

  (a) $y = kx$
  (b) $y = kx^n$ (as nth power)

- Inverse Variation

  (a) $y = k/x$
  (b) $y = k/(x^n)$ (as nth power)

- Joint Variation

  (a) $z = kxy$
  (b) $z = kx^n y^m$ (as nth power of x and mth power of y)

- k is called the constant of proportionality.

Solutions to Selected Exercises

5. Find a mathematical model for the statement "z is proportional to the cube root of u".

Solution:

$$z = k\sqrt[3]{u}$$

9. Find a mathematical model for the statement "F varies directly as g and inversely as the square of r".

Solution:

$$F = \frac{kg}{r^2}$$

13. Find a mathematical model for **Newton's Law of Universal Gravitation:** The gravitational attraction F between two objects of masses m_1 and m_2 is proportional to the product of the masses and inversely proportional to the square of the distance r between the objects.

Solution:

$$F = \frac{km_1m_2}{r^2}$$

17. Write a sentence using variation terminology to describe the formula for the volume of a sphere.

$$V = \frac{4}{3}\pi r^3$$

Solution:

$$V = \frac{4}{3}\pi r^3$$

The volume, V, of a sphere is directly proportional to the cube of the radius r. The constant of proportionality is $\frac{4}{3}\pi$.

23. Find a mathematical model for the statement "A varies directly as the square of r". Determine the constant of proportionality given $A = 9\pi$ when $r = 3$.

Solution:

$$\begin{aligned} A &= kr^2 \\ 9\pi &= k(3)^2 \\ \pi &= k \\ A &= \pi r^2 \end{aligned}$$

27. Find a mathematical model for the statement "h is inversely proportional to the third power of t". Determine the constant of proportionality given $h = 3/16$ when $t = 4$.

Solution:

$$\begin{aligned} h &= \frac{k}{t^3} \\ \frac{3}{16} &= \frac{k}{(4)^3} \\ \frac{3}{16} &= \frac{k}{64} \\ k &= 12 \\ h &= \frac{12}{t^3} \end{aligned}$$

31. Find a mathematical model for the statement "F is jointly proportional to r and the third power of s". Determine the constant of proportionality given $F = 4158$ when $r = 11$ and $s = 3$.

Solution:

$$F = krs^3$$
$$4158 = k(11)(3)^3$$
$$k = 14$$
$$F = 14rs^3$$

35. Find a mathematical model for the statement "S varies directly as L and inversely as $L - S$". Determine the constant of proportionality given $S = 4$ when $L = 6$.

Solution:

$$S = \frac{kL}{L-S}$$
$$4 = \frac{k(6)}{6-4}$$
$$4 = 3k$$
$$k = \frac{4}{3}$$
$$S = \frac{4/3L}{L-S} = \frac{4L}{3(L-S)}$$

39. The coiled spring of a toy supports the weight of a child. The spring compresses a distance of 1.9 inches under the weight of a 25-pound child. The toy will not work properly if its spring is compressed more than 3 inches. What is the weight of the heaviest child who should be allowed to use the toy?

Solution:

From Example 1, we have

$$d = kF$$
$$1.9 = k(25) \implies k = 0.076$$
$$d = 0.076F$$

When the distance compressed is 3 inches, we have

$$3 = 0.076F$$
$$F \approx 39.4737$$

No child over 39 pounds should use the toy.

41. A stream with a velocity of 1/4 mile per hour can move coarse sand particles of about 0.02 inch diameter. What must the velocity be to carry particles with a diameter of 0.12 inch? Use the fact that the diameter of a particle moved by a stream varies approximately as the square of the velocity of the stream.

Solution:

$$d = kv^2$$

$$0.02 = k(1/4)^2$$

$$k = 0.32$$

$$d = 0.32v^2$$

$$0.12 = 0.32v^2$$

$$v^2 = 0.12/0.32 = 3/8$$

$$v = \sqrt{3}/(2\sqrt{2}) = \sqrt{6}/4 \approx 0.612 \text{ mi/hr}$$

43. The resistance of a wire carrying electrical current is directly proportional to its length and inversely proportional to its cross-sectional area. If a #28 copper wire (which has a diameter of 0.0126 inch) has a resistance of 66.17 ohms per thousand feet, what length of #28 copper wire will produce a resistance of 33.5 ohms?

Solution:

$$r = \frac{kl}{A}, \quad A = \pi r^2 = \frac{\pi d^2}{4}$$

$$r = \frac{4kl}{\pi d^2}$$

$$66.17 = \frac{4(1000)k}{\pi(0.0126/12)^2}$$

$$k \approx 5.73 \times 10^{-8}$$

$$r = \frac{4(5.73 \times 10^{-8})l}{\pi(0.0126/12)^2}$$

$$33.5 = \frac{4(5.73 \times 10^{-8})l}{\pi(0.0126/12)^2}$$

$$\frac{33.5\pi(0.0126/12)^2}{4(5.73 \times 10^{-8})} = l$$

$$l \approx 506 \text{ feet}$$

49. The illumination from a light source varies inversely as the square of the distance from the light source. When the distance from a light source is doubled, how does the illumination change?

Solution:

$$I = \frac{k}{d^2}$$

When the distance is doubled:

$$I = \frac{k}{(2d)^2}$$

$$I = \frac{1}{4}\left(\frac{k}{d^2}\right)$$

The amount of illumination is 1/4 as bright.

53. Use $k = 1$ to complete the table (shown in the textbook) for the direct variation model $y = kx^2$. Plot the points on the rectangular coordinate system.

Solution:

$k = 1 \quad \Rightarrow \quad y = x^2$

x	2	4	6	8	10
$y = kx^2$	4	16	36	64	100

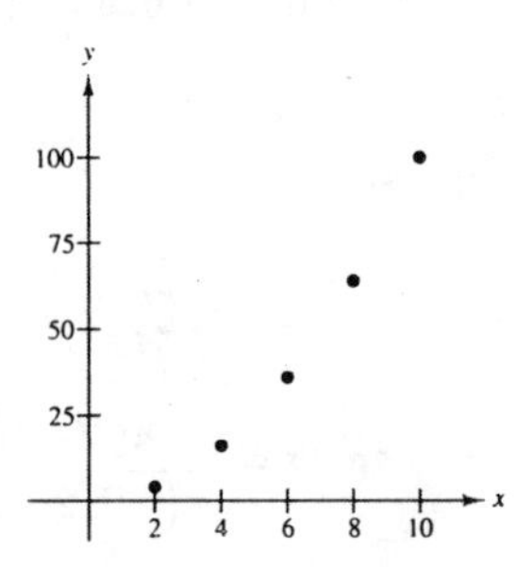

59. Use $k = 10$ to complete the table for the inverse variation model $y = \dfrac{k}{x^2}$. Plot the points on the rectangular coordinate system.

Solution:

$k = 10 \quad \Rightarrow \quad y = \dfrac{10}{x^2}$

x	2	4	6	8	10
$y = \frac{k}{x^2}$	$\frac{5}{2}$	$\frac{5}{8}$	$\frac{5}{18}$	$\frac{5}{32}$	$\frac{1}{10}$

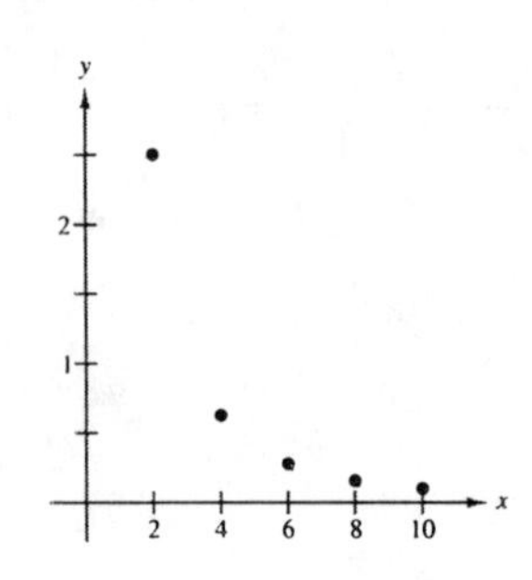

REVIEW EXERCISES FOR CHAPTER 3

Solutions to Selected Exercises

3. For the points (2, 1) and (14, 6), find (a) the distance between the two points, (b) the coordinates of the midpoint of the line segment between the two points, (c) an equation of the line through the two points, and (d) an equation of the circle whose diameter is the line segment between the two points.

Solution:

(a)
$$\begin{aligned} d &= \sqrt{(14-2)^2 + (6-1)^2} \\ &= \sqrt{144+25} \\ &= \sqrt{169} = 13 \end{aligned}$$

(b)
$$\begin{aligned} m &= \left(\frac{2+14}{2}, \frac{1+6}{2}\right) \\ &= \left(8, \frac{7}{2}\right) \end{aligned}$$

(c)
$$\begin{aligned} y - 1 &= \frac{6-1}{14-2}(x-2) \\ &= \frac{5}{12}(x-2) \\ 12y - 12 &= 5x - 10 \\ 5x - 12y + 2 &= 0 \end{aligned}$$

(d) The length of the diameter is 13, so the length of the radius is $\frac{13}{2}$. The midpoint of the line segment is the center of the circle. Center: $(8, \frac{7}{2})$ Radius: $\frac{13}{2}$

$$\begin{aligned} (x-8)^2 + \left(y - \frac{7}{2}\right)^2 &= \left(\frac{13}{2}\right)^2 \\ x^2 - 16x + 64 + y^2 - 7y + \frac{49}{4} &= \frac{169}{4} \\ x^2 + y^2 - 16x - 7y + 34 &= 0 \end{aligned}$$

7. Use the Midpoint Formula to estimate the sales of a company for 1991. Assume the sales followed a linear growth pattern.

Year	1989	1993
Sales	640,000	810,000

Solution:

$$\left(\frac{1989+1993}{2}, \frac{640{,}000+810{,}000}{2}\right) = (1991, \ 725{,}000)$$

The sales for 1991 are estimated to be \$725,000.

9. Find t so that the points $(-2, 5)$, $(0, t)$ and $(1, 1)$ are collinear.

Solution:

The line through $(-2, 5)$ and $(1, 1)$ is

$$y - 5 = \frac{1-5}{1+2}(x+2)$$

$$y - 5 = -\frac{4}{3}(x+2)$$

$$3y - 15 = -4x - 8$$

$$4x + 3y = 7$$

For $(0, t)$ to be on this line also, it must satisfy the equation $4x + 3y = 7$.

$$4(0) + 3(t) = 7$$

Thus, $t = \frac{7}{3}$.

13. Show that the points $(1, 1)$, $(8, 2)$, $(9, 5)$, and $(2, 4)$ form the vertices of a parallelogram.

Solution:

$d_1 = \sqrt{(2-1)^2 + (4-1)^2} = \sqrt{10}$

$d_2 = \sqrt{(9-2)^2 + (5-4)^2} = \sqrt{50} = 5\sqrt{2}$

$d_3 = \sqrt{(8-9)^2 + (2-5)^2} = \sqrt{10}$

$d_4 = \sqrt{(1-8)^2 + (1-2)^2} = \sqrt{50} = 5\sqrt{2}$

Since $d_1 = d_3$ and $d_2 = d_4$, these points are the vertices of a parallelogram.

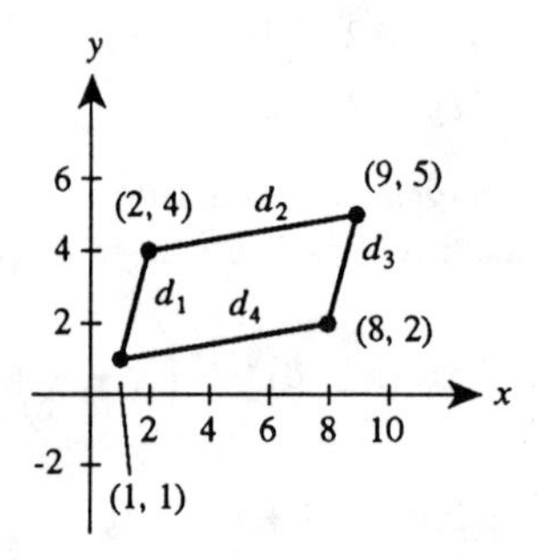

17. Find the intercepts of the graph of $2y^2 = x^3$ and check for symmetry with respect to each of the coordinate axes and the origin.

Solution:

The only intercept is the origin, $(0, 0)$. The graph is symmetric with respect to the x-axis since $2(-y)^2 = x^3$ results in the original equation. Replacing x with $-x$ or replacing both x and y with $-x$ and $-y$ does not yield equivalent equations. Thus, the graph is not symmetric with respect to either the y-axis or the origin.

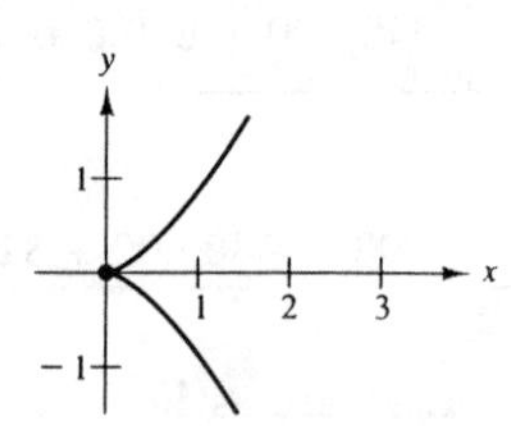

21. Find the intercepts of the graph of $y = x\sqrt{4 - x^2}$ and check for symmetry with respect to each of the coordinate axes and the origin.

Solution:

Let $y = 0$, then $0 = x\sqrt{4 - x^2}$ and $x = 0, \pm 2$.
x-intercepts: $(0, 0)$, $(2, 0)$, $(-2, 0)$
Let $x = 0$, then $y = 0\sqrt{4 - 0^2}$ and $y = 0$.
y-intercept: $(0, 0)$
The graph is symmetric with respect to the origin since

$$-y = -x\sqrt{4 - (-x)^2}$$
$$-y = -x\sqrt{4 - x^2}$$
$$y = x\sqrt{4 - x^2}.$$

The graph is not symmetric with respect to either axis.

25. Find the intercepts of the graph of $y^2 = \dfrac{x^3}{4 - x}$ and check for symmetry with respect to each of the coordinate axes and the origin.

Solution:

Let $y = 0$, then $0^2 = \dfrac{x^3}{4 - x}$ and $x = 0$

Let $x = 0$, then $y^2 = \dfrac{0^3}{4 - 0}$ and $y = 0$

The only intercept is the origin: $(0, 0)$
The graph is symmetric with respect to the x-axis since

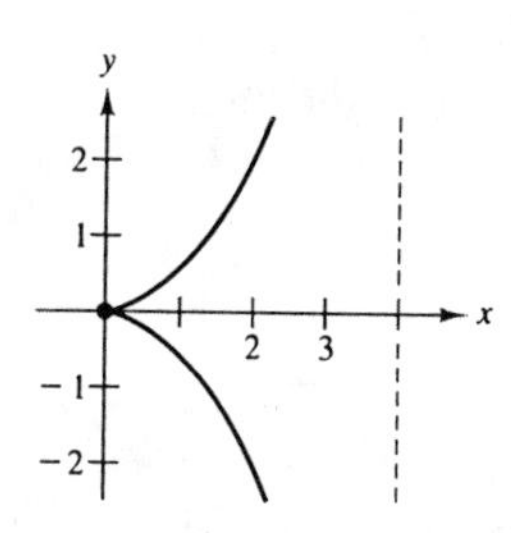

$$(-y)^2 = \frac{x^3}{4 - x}$$
$$y^2 = \frac{x^3}{4 - x}.$$

The graph is not symmetric with respect to the y-axis or the origin.

29. Determine the center and radius of the circle. Then, sketch the graph of $4x^2 + 4y^2 - 4x - 40y + 92 = 0$.

Solution:

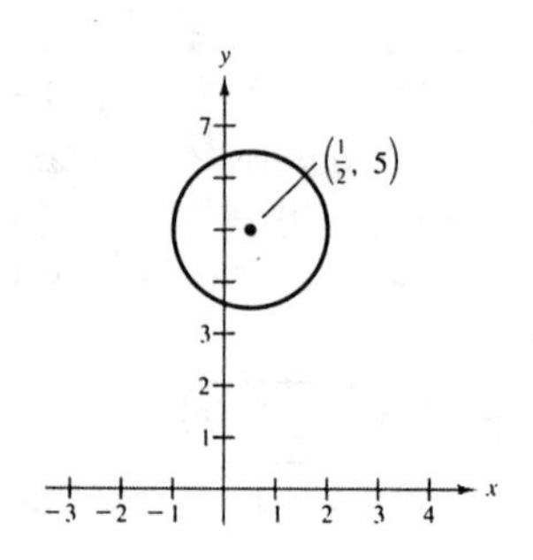

$$4x^2 + 4y^2 - 4x - 40y + 92 = 0$$
$$x^2 + y^2 - x - 10y + 23 = 0$$
$$\left(x^2 - x + \frac{1}{4}\right) + (y^2 - 10y + 25) = -23 + \frac{1}{4} + 25$$
$$\left(x - \frac{1}{2}\right)^2 + (y - 5)^2 = \frac{9}{4}$$

Center: $\left(\frac{1}{2}, 5\right)$ Radius: $\frac{3}{2}$

31. Sketch a graph of the equation $y - 2x - 3 = 0$.

Solution:

$y - 2x - 3 = 0$
$y = 2x + 3$
x-intercept: $\left(-\frac{3}{2}, 0\right)$
y-intercept: $(0, 3)$
The graph is a straight line.

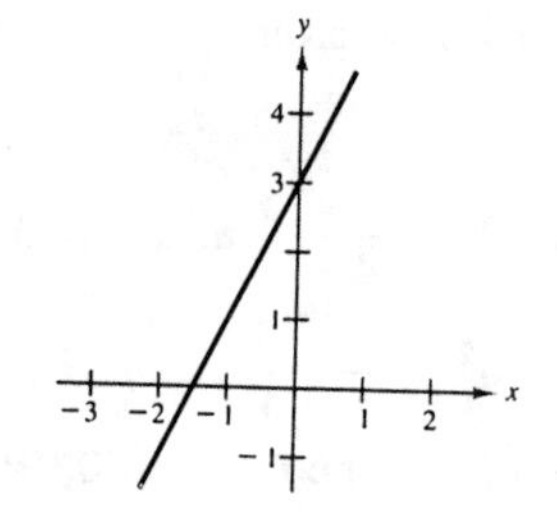

35. Sketch a graph of the equation $y = \sqrt{5 - x}$.

Solution:

$y = \sqrt{5 - x}$
x-intercept: $(5, 0)$
Domain: $(-\infty, 5]$
Range: $[0, \infty)$

x	-4	1	4	5
y	3	2	1	0

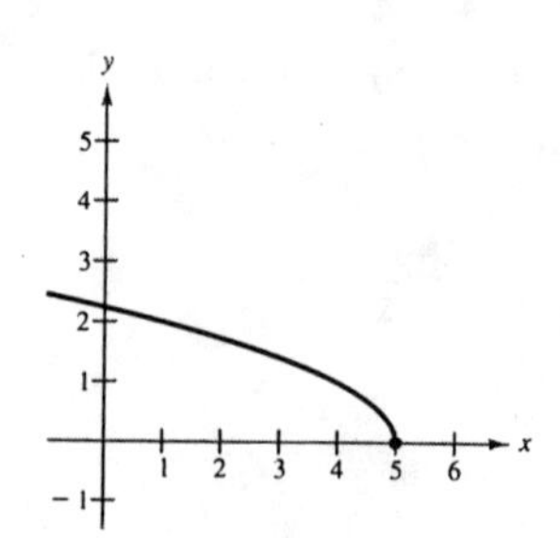

39. Sketch a graph of the equation $y = \sqrt{25 - x^2}$.

Solution:

$y = \sqrt{25 - x^2}$

x-intercepts: $(5, 0)$, $(-5, 0)$

y-intercept: $(0, 5)$

y-axis symmetry

Domain: $[-5, 5]$

Range: $[0, 5]$

x	0	± 3	± 4	± 5
y	5	4	3	0

43. Sketch a graph of the equation $y = \frac{1}{4}(x + 1)^3$.

Solution:

$y = \dfrac{1}{4}(x + 1)^3$

x-intercept: $(-1, 0)$

y-intercept: $\left(0, \dfrac{1}{4}\right)$

Domain: $(-\infty, \infty)$

Range: $(-\infty, \infty)$

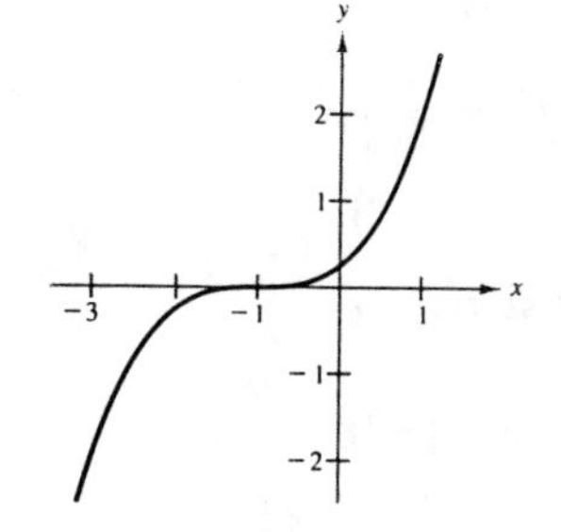

x	-3	-2	-1	0	1
y	-2	$-\frac{1}{4}$	0	$\frac{1}{4}$	2

47. Find an equation of the line that passes through $(3, 0)$ and has a slope of $m = -\frac{2}{3}$. Sketch the graph of the line.

Solution:

$$y - 0 = -\frac{2}{3}(x - 3)$$

$$y = -\frac{2}{3}x + 2$$

OR

$$2x + 3y - 6 = 0$$

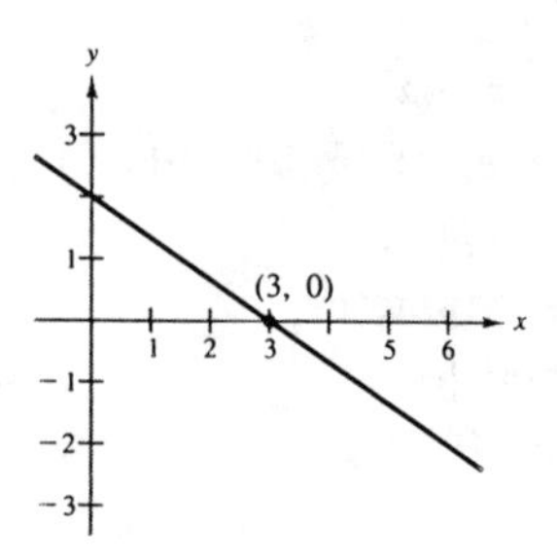

49. Write an equation of the line through the point $(3, -2)$ (a) parallel to $5x - 4y = 8$, and (b) perpendicular to $5x - 4y = 8$.

Solution:

$$5x - 4y = 8$$

$$-4y = -5x + 8$$

$$y = \frac{5}{4}x - 2 \Rightarrow m_1 = \frac{5}{4}$$

(a) Point $(3, -2)$,

Parallel slope $m_2 = \frac{5}{4}$

$$y - (-2) = \frac{5}{4}(x - 3)$$

$$y + 2 = \frac{5}{4}x - \frac{15}{4}$$

$$4y + 8 = 5x - 15$$

$$5x - 4y - 23 = 0$$

(b) Point $(3, -2)$,

Perpendicular slope $m_2 = -\frac{4}{5}$

$$y - (-2) = -\frac{4}{5}(x - 3)$$

$$y + 2 = -\frac{4}{5}x + \frac{12}{5}$$

$$5y + 10 = -4x + 12$$

$$4x + 5y - 2 = 0$$

51. During the second and third quarters of the year, a business had sales of \$160,000 and \$185,000 respectively. If the growth of sales follows a linear pattern, estimate sales during the fourth quarter.

Solution:

$$(2,\ 160{,}000),\quad (3,\ 185{,}000)$$

$$m = \frac{185{,}000 - 160{,}000}{3 - 2} = 25{,}000$$

$$S - 160{,}000 = 25{,}000(t - 2)$$

$$S = 25{,}000t + 110{,}000$$

For the fourth quarter let $t = 4$. Then we have

$$S = 25{,}000(4) + 110{,}000 = \$210{,}000$$

55. Evaluate the function $h(x) = 6 - 5x^2$ at the specified values of the independent variable and simplify your answers.

(a) $h(2)$

(b) $h(x+3)$

(c) $\dfrac{h(4) - h(2)}{4 - 2}$

(d) $\dfrac{h(x + \Delta x) - h(x)}{\Delta x}$

Solution:

(a) $h(2) = 6 - 5(2)^2 = -14$

(b) $$\begin{aligned} h(x+3) &= 6 - 5(x+3)^2 \\ &= 6 - 5(x^2 + 6x + 9) \\ &= -5x^2 - 30x - 39 \end{aligned}$$

(c)
$$h(4) = 6 - 5(4)^2 = -74$$
$$h(2) = -14 \quad \text{from part (a)}$$
$$\frac{h(4) - h(2)}{4 - 2} = \frac{-74 - (-14)}{4 - 2} = \frac{-60}{2} = -30$$

(d)
$$\begin{aligned} h(x + \Delta x) &= 6 - 5(x + \Delta x)^2 \\ &= 6 - 5(x^2 + 2x\Delta x + (\Delta x)^2) \\ &= 6 - 5x^2 - 10x\Delta x - 5(\Delta x)^2 \end{aligned}$$

$$h(x + \Delta x) - h(x)(6 - 5x^2 - 10x\Delta x - 5(\Delta x)^2) - (6 - 5x^2)$$
$$= -10x\Delta x - 5(\Delta x)^2$$

$$\frac{h(x + \Delta x) - h(x)}{\Delta x} = \frac{-10x\Delta x - 5(\Delta x)^2}{\Delta x}$$
$$= -10x - 5\Delta x$$

59. Determine the domain of the function

$$g(s) = \frac{5}{3s - 9}.$$

Solution:

The domain of $g(s) = 5/(3s - 9)$ includes all real numbers except $s = 3$, since this value would yield a zero in the denominator. In interval notation we have $(-\infty,\ 3) \cup (3,\ \infty)$.

65. For $f(x) = \sqrt{x+1}$ (a) find f^{-1}, (b) sketch the graphs of f and f^{-1} on the same coordinate plane, and (c) verify that $f^{-1}(f(x)) = x = f(f^{-1}(x))$.

Solution:

(a)

$$y = \sqrt{x+1}, \quad y \geq 0$$
$$x = \sqrt{y+1}$$
$$x^2 = y + 1$$
$$x^2 - 1 = y$$
$$f^{-1}(x) = x^2 - 1, \quad x \geq 0$$

(b)

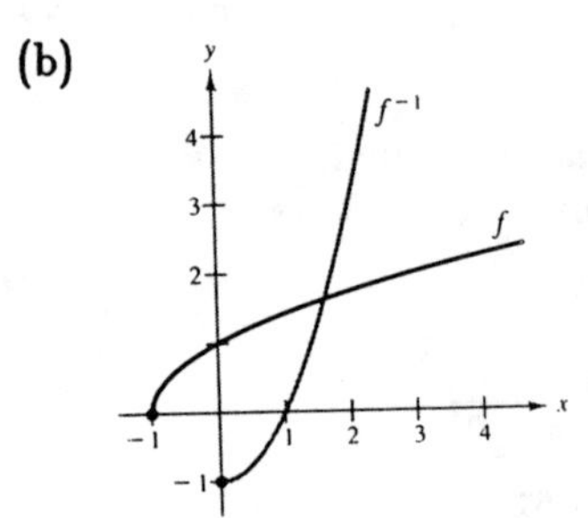

(c)

$$f^{-1}[f(x)] = f^{-1}(\sqrt{x+1})$$
$$= (\sqrt{x+1})^2 - 1 = (x+1) - 1 = x$$
$$f[f^{-1}(x)] = f(x^2 - 1)$$
$$= \sqrt{(x^2 - 1) + 1}$$
$$= \sqrt{x^2} = x, \quad x \geq 0$$

69. Restrict the domain of the function $f(x) = 2(x-4)^2$ to an interval where the function is increasing and determine f^{-1} over that interval.

Solution:

$f(x) = 2(x-4)^2$ is increasing on the interval $[4,\ \infty)$. It is decreasing on the interval $(-\infty,\ 4)$.

$$f(x) = 2(x-4)^2, \ x \geq 4$$
$$y = 2(x-4)^2, \ x \geq 4$$
$$x = 2(y-4)^2$$
$$\sqrt{x} = \sqrt{2}(y-4)$$
$$\frac{\sqrt{x}}{\sqrt{2}} + 4 = y$$
$$f^{-1}(x) = \sqrt{\frac{x}{2}} + 4, \quad x \geq 0$$

73. Let $f(x) = 3 - 2x$, $g(x) = \sqrt{x}$, and $h(x) = 3x^2 + 2$. Find $(f - g)(4)$.

Solution:

$$\begin{aligned}(f-g)(4) &= f(4) - g(4)\\ &= [3 - 2(4)] - \sqrt{4}\\ &= -7\end{aligned}$$

77. Let $f(x) = 3 - 2x$, $g(x) = \sqrt{x}$, and $h(x) = 3x^2 + 2$. Find $(h \circ g)(7)$.

Solution:

$$\begin{aligned}(h \circ g)(7) &= h(g(7))\\ &= h(\sqrt{7})\\ &= 3(\sqrt{7})^2 + 2\\ &= 23\end{aligned}$$

81. The velocity of a ball thrown vertically upward from ground level is

$$v(t) = -32t + 48$$

where t is the time in seconds and v is the velocity in feet per second.

(a) Find the velocity when $t = 1$.

(b) Find the time when the ball reaches its maximum height.
[*Hint:* Find the time when $v(t) = 0$.]

(c) Find the velocity when $t = 2$.

Solution:

(a) $v(1) = -32(1) + 48 = 16$ feet per second

(b)
$$\begin{aligned}v(t) &= 0\\ -32t + 48 &= 0\\ -32t &= -48\\ t &= 1.5 \text{ seconds}\end{aligned}$$

(c) $v(2) = -32(2) + 48 = -16$ feet per second

83. A wire 24 inches long is to be cut into four pieces to form a rectangle whose shortest side has a length of x. Express the area A of the rectangle as a function of x. Determine the domain of the function and sketch its graph over that domain.

Solution:

Let y be the longer side of the rectangle. Then we have $A = xy$. Since the perimeter is 24 inches, we have $2x + 2y = 24$ or $y = (24 - 2x)/2 = 12 - x$. The area equation now becomes: $A = xy = x(12 - x)$. To find the domain of A, we realize that area is a nonnegative quantity. Thus, $x(12 - x) \geq 0$. This gives us the interval $[0,\ 12]$. We also have the further restriction that x is the shortest side. This occurs on the interval $(0,\ 6]$.

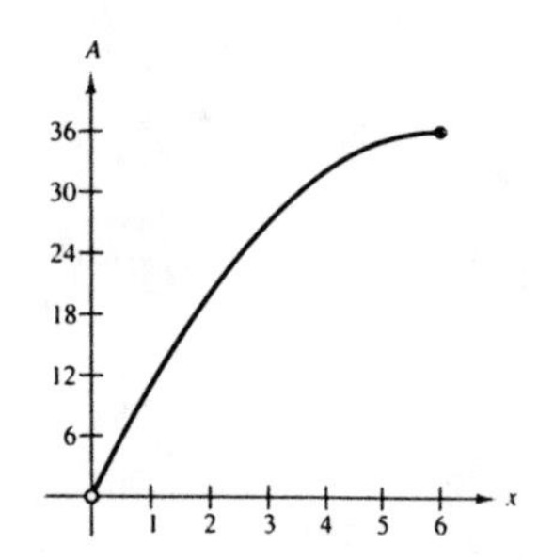

87. Find a mathematical model representing the statement "z varies directly as the square of x and inversely as y". Determine the constant of proportionality if $z = 16$ when $x = 5$ and $y = 2$.

Solution:

$$z = \frac{kx^2}{y}$$

$$16 = \frac{k(5)^2}{2}$$

$$32 = 25k$$

$$k = \frac{32}{25}, \quad \text{therefore, } z = \frac{32x^2}{25y}.$$

Practice Test for Chapter 3

1. Find the distance between $(4, -1)$ and $(0, 3)$.

2. Find the midpoint of the line segment joining $(4, -1)$ and $(0, 3)$.

3. Find x so that the distance from the origin to $(x, -2)$ is 6.

4. Given $y = \dfrac{x-2}{x+3}$, find the intercepts.

5. Given $xy^2 = 6$, list all symmetries.

6. Graph $y = x^3 - 4x$.

7. Find the center and radius of the circle $x^2 + y^2 - 6x + 2y + 6 = 0$.

8. Given $f(x) = x^2 - 2x + 1$, find $f(x-3)$.

9. Given $f(x) = 4x - 11$, find $\dfrac{f(x) - f(3)}{x-3}$.

10. Find the domain and range of $f(x) = \sqrt{36 - x^2}$.

11. Which equations determine y as a function of x?

(a) $6x - 5y + 4 = 0$

(b) $x^2 + y^2 = 9$

(c) $y^3 = x^2 + 6$

12. Sketch the graph of $f(x) = x^2 - 5$.

13. Sketch the graph of $f(x) = |x + 3|$.

14. Sketch the graph of $f(x) = \begin{cases} 2x+1 & \text{if } x \geq 0, \\ x^2 - x & \text{if } x < 0. \end{cases}$

15. Find the equation of the line through $(2, 4)$ and $(3, -1)$.

16. Find the equation of the line with slope $m = 4/3$ and y-intercept $b = -3$.

17. Find the equation of the line through (4, 1) perpendicular to the line $2x + 3y = 0$.

18. If it costs a company $32 to produce 5 units of a product and $44 to produce 9 units, how much does it cost to produce 20 units? (Assume that the cost function is linear.)

19. Given $f(x) = x^2 - 2x + 16$ and $g(x) = 2x + 3$, find $f(g(x))$.

20. Given $f(x) = x^3 + 7$, find $f^{-1}(x)$.

21. Which of the following functions are one-to-one?

(a) $f(x) = |x - 6|$

(b) $f(x) = ax + b,\ a \neq 0$

(c) $f(x) = x^3 - 19$

22. Given $f(x) = \sqrt{\dfrac{3 - x}{x}}$, $0 < x \leq 3$, find $f^{-1}(x)$.

23. Find the equation: y varies directly as x and $y = 30$ when $x = 5$.

24. Find the equation: y varies inversely as x and $y = 0.5$ when $x = 14$.

25. z varies directly as the square of x and inversely as y, and $z = 3$ when $x = 3$ and $y = -6$. Find the equation relating z to x and y.

CHAPTER 4
Polynomial Functions: Graphs and Zeros

SECTION 4.1

Quadratic Functions

You should know the following facts about parabolas.

- $f(x) = ax^2 + bx + c$, $a \neq 0$, is a quadratic function, and its graph is a parabola.
- If $a > 0$, the parabola opens upward and the vertex is the minimum point. If $a < 0$, the parabola opens downward and the vertex is the maximum point.
- The vertex is $(-b/2a,\ f(-b/2a))$.
- To find the x-intercepts (if any), solve

  $$ax^2 + bx + c = 0.$$

- The standard form of the equation of a parabola is

  $$f(x) = a(x - h)^2 + k$$

  where $a \neq 0$.
  (a) The vertex is $(h,\ k)$.
  (b) The axis is the vertical line $x = h$.

Solutions to Selected Exercises

3. Match the quadratic function $f(x) = x^2 - 4$ with the correct graph.

Solution:

$f(x) = x^2 - 4$

Vertex: $(0,\ -4)$

$a = 1 > 0$ opens upward

x-intercepts: $(\pm 2,\ 0)$

Matches graph (c)

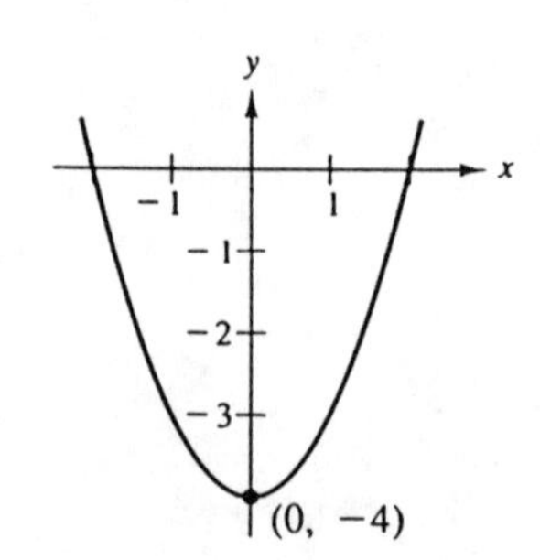

7. Find an equation for the given parabola.

Solution:

The vertex is (2, 0) and the parabola passes through the point (0, 4).

$f(x) = a(x-2)^2 + 0$

$f(x) = a(x-2)^2$

$4 = a(0-2)^2 \implies a = 1$

$f(x) = (x-2)^2$

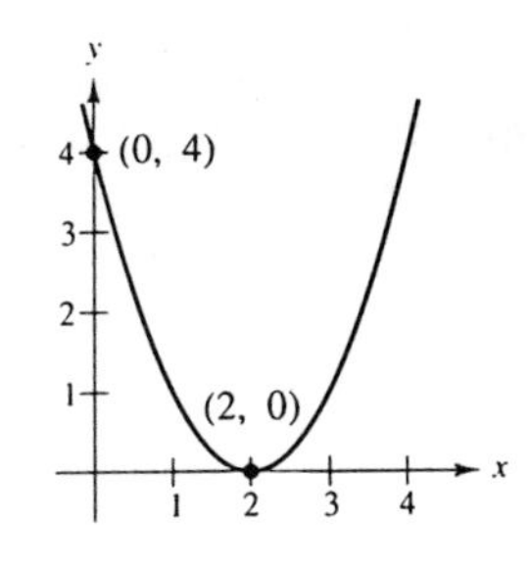

11. Find an equation for the given parabola.

Solution:

The vertex is (−3, 3) and the parabola passes through (−2, 1).

$f(x) = a[x-(-3)]^2 + 3$

$f(x) = a(x+3)^2 + 3$

$1 = a(-2+3)^2 + 3 \implies a = -2$

$f(x) = -2(x+3)^2 + 3$

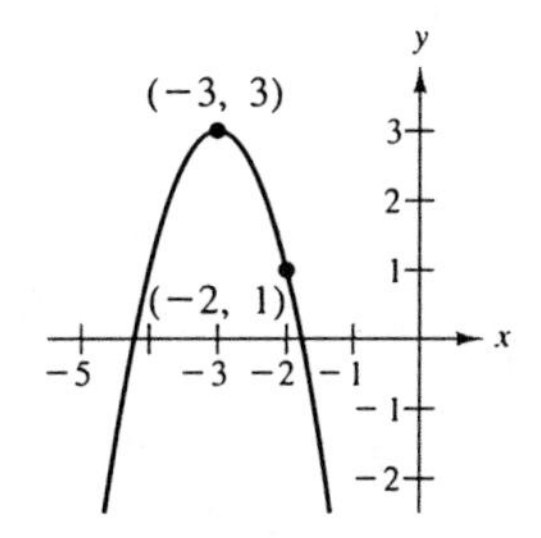

17. Sketch the graph of $f(x) = (x+5)^2 - 6$. Identify the vertex and intercepts.

Solution:

Vertex: (−5, −6)

x-intercepts:

$0 = (x+5)^2 - 6$

$(x+5)^2 = 6$

$x + 5 = \pm\sqrt{6}$

$x = -5 \pm \sqrt{6}$

$(-5-\sqrt{6}, 0)$, $(-5+\sqrt{6}, 0)$

y-intercept: (0, 19) Let $x = 0$ and solve for y.

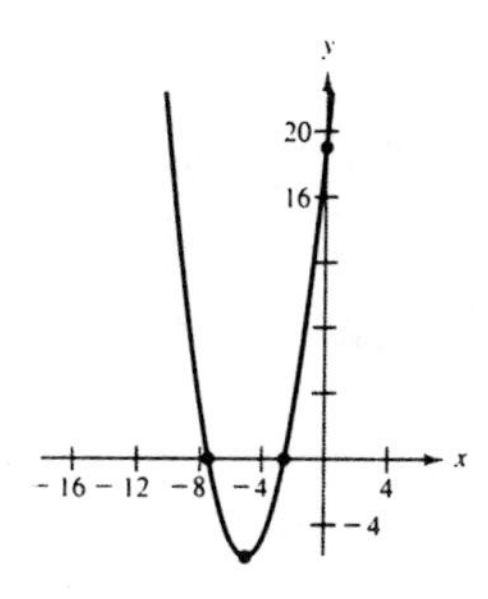

19. Sketch the graph of $h(x) = x^2 - 8x + 16$. Identify the vertex and intercepts.

Solution:

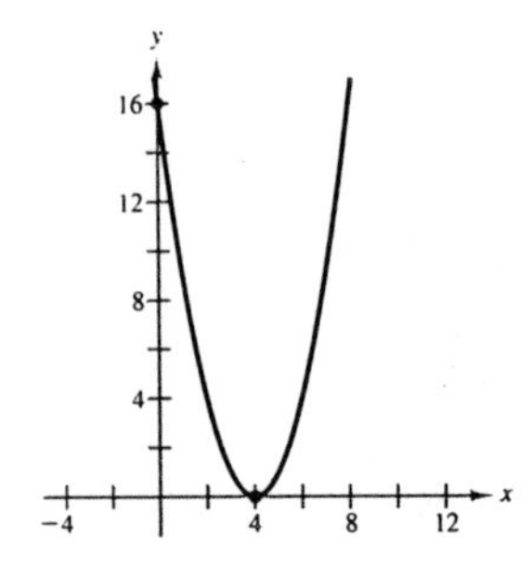

$h(x) = x^2 - 8x + 16$
$h(x) = (x - 4)^2$

Vertex: $(4, 0)$
x-intercept: $(4, 0)$ Let $y = 0$ and solve for x.
y-intercept: $(0, 16)$ Let $x = 0$ and solve for y.

23. Sketch the graph of $f(x) = x^2 - x + \frac{5}{4}$. Identify the vertex and intercepts.

Solution:

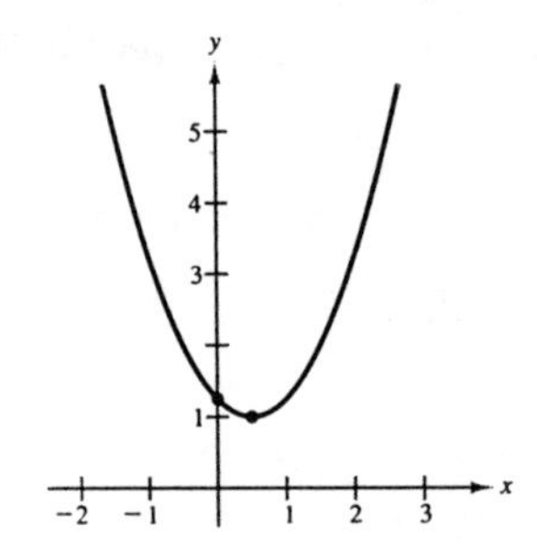

$$\begin{aligned} f(x) &= x^2 - x + \tfrac{5}{4} \\ &= x^2 - x + \tfrac{1}{4} - \tfrac{1}{4} + \tfrac{5}{4} \\ &= (x - \tfrac{1}{2})^2 + 1 \end{aligned}$$

Vertex: $(\frac{1}{2}, 1)$

x-intercept: None since

$$0 = (x - \tfrac{1}{2})^2 + 1$$
$$-1 = (x - \tfrac{1}{2})^2$$

$\pm\sqrt{-1} = x - \frac{1}{2}$ has no real solutions.

y-intercept: $(0, \frac{5}{4})$ Let $x = 0$ and solve for y.

27. Sketch the graph of $h(x) = 4x^2 - 4x + 21$. Identify the vertex and intercepts.

Solution:

$$\begin{aligned} h(x) &= 4x^2 - 4x + 21 \\ &= 4(x^2 - x + \tfrac{1}{4} - \tfrac{1}{4}) + 21 \\ &= 4(x^2 - x + \tfrac{1}{4}) - 1 + 21 \\ &= 4(x - \tfrac{1}{2})^2 + 20 \end{aligned}$$

–CONTINUED–

27. –CONTINUED–

Vertex: $(\frac{1}{2}, 20)$
x-intercept: None since

$$0 = 4(x - \tfrac{1}{2})^2 + 20$$
$$-20 = 4(x - \tfrac{1}{2})^2$$
$$-5 = (x - \tfrac{1}{2})^2$$

$\pm\sqrt{-5} = x - \frac{1}{2}$ has no real solutions.
y-intercept: $(0, 21)$ Let $x = 0$ and solve for y.

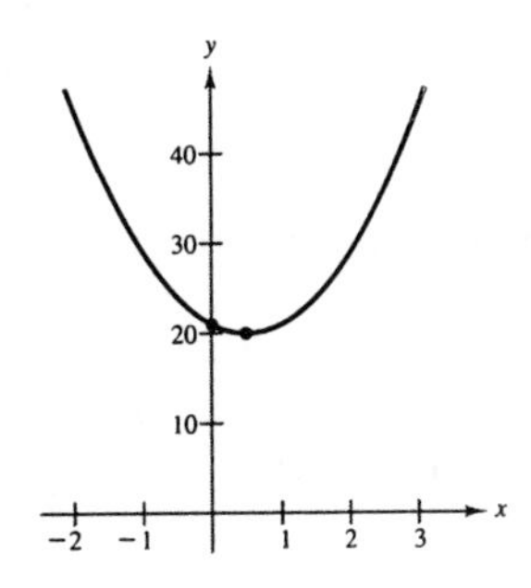

33. Find the quadratic function with a vertex of $(5, 12)$ and whose graph passes through the point $(7, 15)$.

Solution:

$(5, 12)$ is the vertex.

$f(x) = a(x - 5)^2 + 12$

Since the graph passes through the point $(7, 15)$, we have

$$15 = a(7 - 5)^2 + 12$$
$$3 = 4a \implies a = \tfrac{3}{4}$$
$$f(x) = \tfrac{3}{4}(x - 5)^2 + 12.$$

37. Find two quadratic functions whose graphs have the x-intercepts $(0, 0)$ and $(10, 0)$. (One function has a graph that opens upward and the other has a graph that opens downward.)

Solution:

$f(x) = (x - 0)(x - 10)$ opens upward
$= x^2 - 10x$
$g(x) = -(x - 0)(x - 10)$ opens downward
$= -x^2 + 10x$

Note: $f(x) = a(x - 0)(x - 10) = ax(x - 10)$ has x-intercepts $(0, 0)$ and $(10, 0)$ for all real numbers a.

43. Find two positive real numbers satisfying the requirements "the sum of the first and twice the second is 24 and the product is a maximum."

Solution:

Let $x =$ the first number and $y =$ the second number. Then the sum is

$$x + 2y = 24 \quad \Longrightarrow \quad y = \frac{24 - x}{2}.$$

The product is $P(x) = xy = x\left(\dfrac{24 - x}{2}\right)$.

$$\begin{aligned} P(x) &= \frac{1}{2}(-x^2 + 24x) \\ &= -\frac{1}{2}(x^2 - 24x + 144 - 144) \\ &= -\frac{1}{2}[(x - 12)^2 - 144] = -\frac{1}{2}(x - 12)^2 + 72 \end{aligned}$$

The maximum value of the product occurs at the vertex of $P(x)$ and is 72. This happens when $x = 12$ and $y = (24 - 12)/2 = 6$. Thus, the numbers are 12 and 6.

47. A rancher has 200 feet of fencing to enclose two adjacent rectangular corrals, as shown in the figure. What dimensions will produce a maximum enclosed area?

Solution:

Since the rancher has 200 feet of fencing, we have the equation $4x + 3y = 200$ or $y = (200 - 4x)/3$. The area is

$$\begin{aligned} A = 2xy &= 2x\left(\frac{200 - 4x}{3}\right) \\ &= \frac{2}{3}(-4x^2 + 200x) \\ &= -\frac{8}{3}(x^2 - 50x) \\ &= -\frac{8}{3}(x^2 - 50x + 625 - 625) \\ &= -\frac{8}{3}[(x - 25)^2 - 625] \\ &= -\frac{8}{3}(x - 25)^2 + \frac{5000}{3} \end{aligned}$$

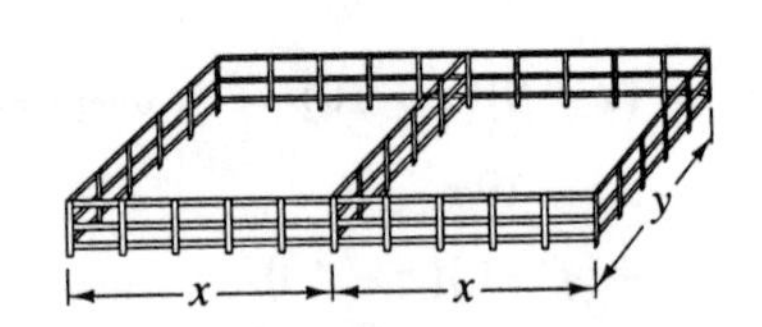

The maximum area occurs at the vertex and is 5000/3 square feet. This happens when $x = 25$ feet and $y = (200 - 4(25))/3 = 100/3$ feet. The dimensions are $2x = 50$ feet by $33\frac{1}{3}$ feet.

55. The number of board feet in a 16-foot log is approximated by the model

$$V = 0.77x^2 - 1.32x - 9.31, \quad 5 \le x \le 40$$

where V is the number of board feet and x is the diameter of the log at the small end in inches. (One board foot is a measure of volume equivalent to a board that is 12 inches wide, 12 inches long, and 1 inch thick.)

(a) Sketch a graph of the function.

(b) Estimate the number of board feet in a 16–foot log with a diameter of 16 inches.

(c) Estimate the diameter of a 16–foot log that scaled 500 board feet when the lumber was sold.

Solution:

(a) $V = 0.77x^2 - 1.32x - 9.31,$

$5 \le x \le 40$

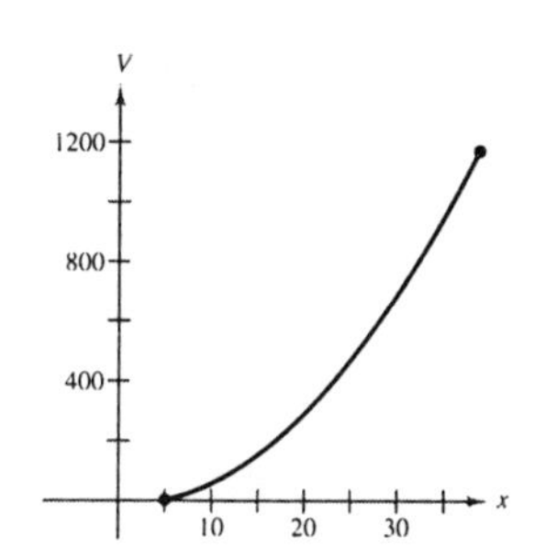

(b) $V(16) = 166.69$ board feet

(c) $500 = 0.77x^2 - 1.32x - 9.31$

$$0 = 0.77x^2 - 1.32x - 509.31$$

$$x = \frac{1.32 \pm \sqrt{1570.4172}}{1.54}$$

Since x is positive, we have

$x \approx 26.59$ inches.

59. Assume that the function $f(x) = ax^2 + bx + c \ \ (a \ne 0)$ has two real zeros. Show that the x-coordinate of the vertex of the graph is the average of the zeros of f. [*Hint:* Use the Quadratic Formula.]

Solution:

If $f(x) = ax^2 + bx + c$ has two real zeros, then by the Quadratic Formula they are

$$x = \frac{-b \pm \sqrt{b^2 - 4ac}}{2a}.$$

The average of the zeros of f is

$$\frac{\dfrac{-b - \sqrt{b^2 - 4ac}}{2a} + \dfrac{-b + \sqrt{b^2 - 4ac}}{2a}}{2} = \frac{\dfrac{-2b}{2a}}{2} = -\frac{b}{2a}.$$

This is the x-coordinate of the vertex of the graph.

SECTION 4.2

Polynomial Functions of Higher Degree

You should know the following basic principles about polynomials.

- $f(x) = a_n x^n + a_{n-1}x^{n-1} + \cdots + a_2x^2 + a_1x + a_0$ is a polynomial function of degree n.
- If f is of odd degree and
  (a) $a_n > 0$, then
    1. $f(x) \to \infty$ as $x \to \infty$
    2. $f(x) \to -\infty$ as $x \to -\infty$

  (b) $a_n < 0$, then
    1. $f(x) \to -\infty$ as $x \to \infty$
    2. $f(x) \to \infty$ as $x \to -\infty$
- If f is of even degree and
  (a) $a_n > 0$, then
    1. $f(x) \to \infty$ as $x \to \infty$
    2. $f(x) \to \infty$ as $x \to -\infty$

  (b) $a_n < 0$, then
    1. $f(x) \to -\infty$ as $x \to \infty$
    2. $f(x) \to -\infty$ as $x \to -\infty$
- The following are equivalent for a polynomial function.
  (a) $x = a$ is a zero of a function.
  (b) $x = a$ is a solution of the polynomial equation $f(x) = 0$.
  (c) $(x - a)$ is a factor of the polynomial.
  (d) $(a, 0)$ is an x-intercept of the graph of f.
- A polynomial of degree n has at most n distinct zeros.
- If f is a polynomial function such that $a < b$ and $f(a) \neq f(b)$, then f takes on every value between $f(a)$ and $f(b)$ in the interval $[a, b]$.
- If you can find a value where a polynomial is positive and another value where it is negative, then there is at least one real zero between the values.

Solutions to Selected Exercises

5. Match the polynomial function $f(x) = -2x^2 - 8x - 9$.

Solution:

$$f(x) = -2x^2 - 8x - 9$$
$$f(x) \to -\infty \text{ as } x \to \infty$$
$$f(x) \to -\infty \text{ as } x \to -\infty$$

Parabola opens downward

Matches (b)

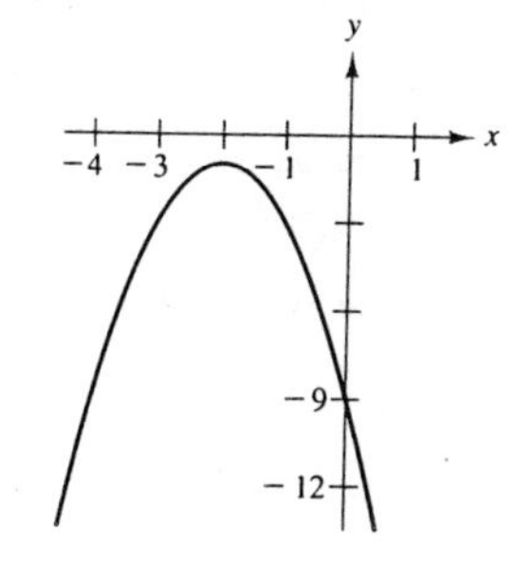

9. Match the polynomial function $f(x) = 3x^4 + 4x^3$.

Solution:

$$f(x) = 3x^4 + 4x^3 = x^3(3x + 4)$$

Zeros: $0, \ -\frac{4}{3}$

$f(x) \to \infty$ as $x \to \infty$

$f(x) \to \infty$ as $x \to -\infty$

Matches (d)

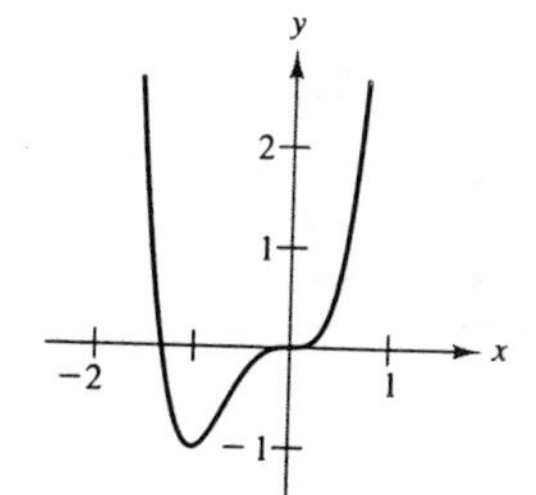

13. Determine the right-hand and left-hand behavior of the graph of $g(x) = 5 - \frac{7}{2}x - 3x^2$.

Solution:

$$g(x) = 5 - \tfrac{7}{2}x - 3x^2 = -3x^2 - \tfrac{7}{2}x + 5$$

Even degree with leading coefficient of -3

Left: $g(x) \to -\infty$ as $x \to -\infty$, so the graph moves down to the left.

Right: $g(x) \to -\infty$ as $x \to +\infty$, so the graph moves down to the right.

17. Determine the right-hand and left-hand behavior of the graph of $f(x) = 6 - 2x + 4x^2 - 5x^3$.

Solution:

$$f(x) = 6 - 2x + 4x^2 - 5x^3 = -5x^3 + 4x^2 - 2x + 6$$

Odd degree with a negative leading coefficient of -5

Left: $f(x) \to \infty$ as $x \to -\infty$, so the graph moves up to the left.

Right: $f(x) \to -\infty$ as $x \to \infty$, so the graph moves down to the right.

23. Find all the real zeros of $h(t) = t^2 - 6t + 9$.

Solution:

$$\begin{aligned} h(t) &= t^2 - 6t + 9 \\ 0 &= t^2 - 6t + 9 \\ 0 &= (t-3)^2 \\ t &= 3 \end{aligned}$$

27. Find all the real zeros of $f(x) = 3x^2 - 12x + 3$.

Solution:

$$\begin{aligned} f(x) &= 3x^2 - 12x + 3 \\ 0 &= 3(x^2 - 4x + 1) \\ x &= \frac{4 \pm \sqrt{12}}{2} \quad \text{by the Quadratic Formula} \\ &= \frac{4 \pm 2\sqrt{3}}{2} \\ &= 2 \pm \sqrt{3} \end{aligned}$$

31. Find all the real zeros of $g(t) = \frac{1}{2}t^4 - \frac{1}{2}$.

Solution:

$$\begin{aligned} g(t) &= \tfrac{1}{2}t^4 - \tfrac{1}{2} \\ 0 &= \tfrac{1}{2}(t^4 - 1) \\ 0 &= \tfrac{1}{2}(t^2 + 1)(t^2 - 1) \\ 0 &= \tfrac{1}{2}(t^2 + 1)(t + 1)(t - 1) \\ t &= \pm 1 \end{aligned}$$

35. Find all the real zeros of $f(x) = 5x^4 + 15x^2 + 10$.

Solution:

$$f(x) = 5x^4 + 15x^2 + 10$$
$$0 = 5(x^4 + 3x^2 + 2)$$
$$0 = 5(x^2 + 1)(x^2 + 2)$$

No real zeros

37. Find a polynomial function that has the zeros 0 and 10.

Solution:

$$f(x) = (x - 0)(x - 10)$$
$$f(x) = x^2 - 10x$$

Note: $f(x) = a(x - 0)(x - 10) = ax(x - 10)$ has the zeros 0 and 10 for all real numbers a.

43. Find a polynomial function that has the zeros 4, −3, 3 and 0.

Solution:

$$f(x) = (x - 4)(x + 3)(x - 3)(x - 0)$$
$$= (x - 4)(x^2 - 9)x$$
$$= x^4 - 4x^3 - 9x^2 + 36x$$

Note: $f(x) = a(x^4 - 4x^3 - 9x^2 + 36x)$ has these zeros for all real numbers a.

47. Sketch the graph of $f(x) = -\frac{3}{2}$.

Solution:

$f(x) = -\frac{3}{2}$ is a horizontal line.
y-intercept: $(0, -\frac{3}{2})$

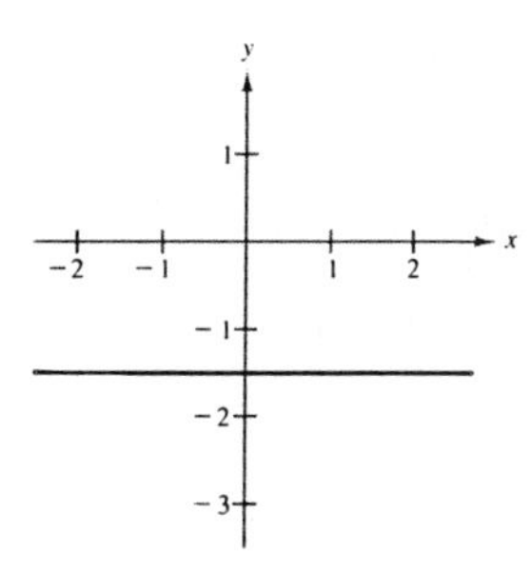

51. Sketch the graph of $f(x) = x^3 - 3x^2$.

Solution:

$$f(x) = x^3 - 3x^2 = x^2(x - 3)$$

Zeros: 0 and 3

Right: Moves up

Left: Moves down

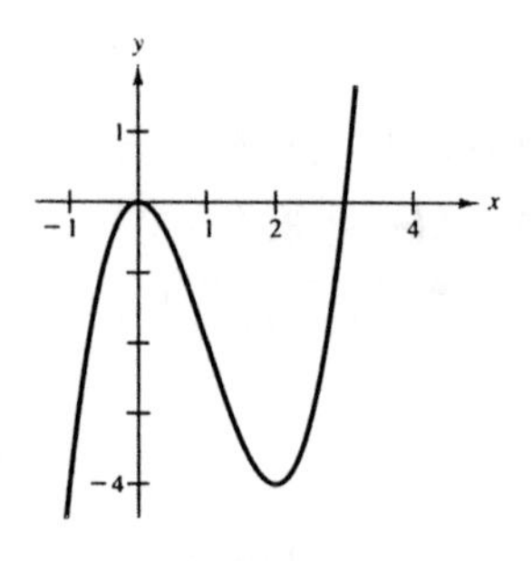

x	0	1	2	3	-1
$f(x)$	0	-2	-4	0	-4

55. Sketch the graph of $g(t) = -\frac{1}{4}(t-2)^2(t+2)^2$.

Solution:

$$g(t) = -\tfrac{1}{4}(t-2)^2(t+2)^2$$

Zeros: 2 and -2

Right: Moves down

Left: Moves down

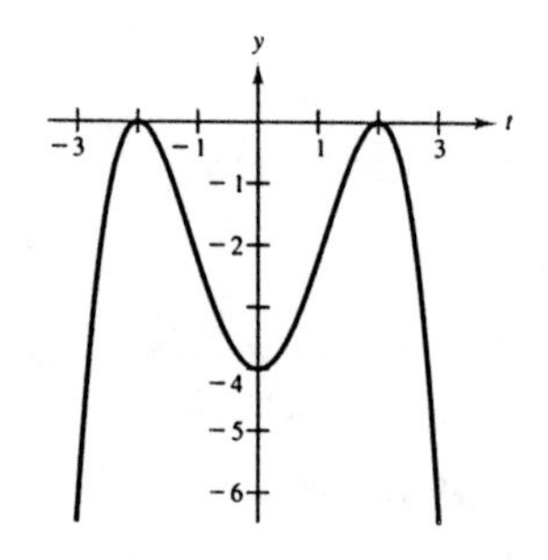

t	-3	-2	-1	0	1	2	3
$g(t)$	$-\frac{25}{4}$	0	$-\frac{9}{4}$	-4	$-\frac{9}{4}$	0	$-\frac{25}{4}$

59. Sketch the graph of $h(x) = \frac{1}{3}x^3(x-4)^2$.

Solution:

$$h(x) = \tfrac{1}{3}x^3(x-4)^2$$

Zeros: 0 and 4

Right: Moves up

Left: Moves down

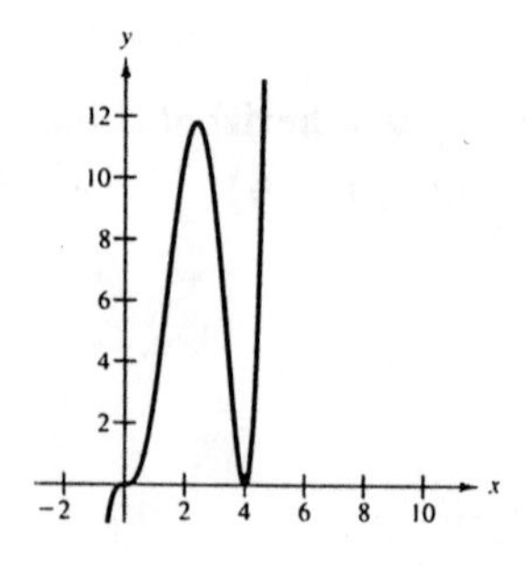

x	-1	0	1	2	3	4	5
$h(x)$	$-\frac{25}{3}$	0	3	$\frac{32}{3}$	9	0	$\frac{125}{3}$

65. Follow the procedure given in Section 4.2 in the textbook to estimate the zero of $f(x) = x^4 - 10x^2 - 11$ in the interval $[3, 4]$. (Give your approximation to the nearest tenth.)

Solution:

$$f(x) = x^4 - 10x^2 - 11, \quad [3, 4]$$

x	3	3.1	3.2	3.3	3.4	3.5
$f(x)$	−20	−14.748	−8.542	−1.308	7.034	16.563

x	3.6	3.7	3.8	3.9	4.0
$f(x)$	27.362	39.516	53.114	68.244	85

The zero lies between 3.3 and 3.4. It is closer to 3.3.

69. The total revenue for a soft drink company is related to its advertising expense by the function

$$R = \frac{1}{50{,}000}(-x^3 + 600x^2), \quad 0 \le x \le 400$$

where R is the total revenue in millions of dollars and x is the amount spent on advertising (in 10,000's of dollars). Use the graph of this function shown on the figure to estimate the point on the graph at which the function is increasing most rapidly. This is the *point of diminishing returns* because any expense above this amount will yield less return per dollar invested in advertising.

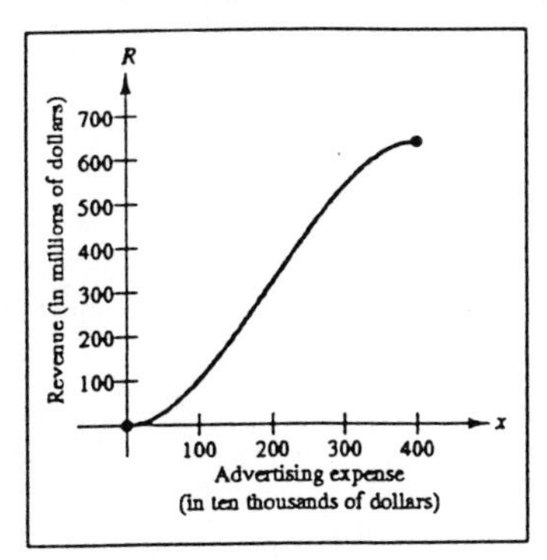

Solution:

The point of diminishing returns (where the graph changes from curving upward to curving downward) occurs when $x = 200$. The point is $(200, \ 320)$ which corresponds to spending \$2,000,000 on advertising to obtain a revenue of \$320 million.

SECTION 4.3

Polynomial Division and Synthetic Division

You should know the following basic techniques and principles of polynomial division.

- The Division Algorithm (Long Division of Polynomials)
- Synthetic Division
- $f(k)$ is equal to the remainder of $f(x)$ divided by $(x - k)$.
- $f(k) = 0$ if and only if $(x - k)$ is a factor of $f(x)$.

Solutions to Selected Exercises

5. Divide $x^4 + 5x^3 + 6x^2 - x - 2$ by $x + 2$ using long division.

Solution:

$$\begin{array}{r}
x^3 + 3x^2 \qquad\quad - 1 \\
x + 2 \overline{\big) x^4 + 5x^3 + 6x^2 - x - 2} \\
\underline{x^4 + 2x^3 \qquad\qquad\qquad\quad} \\
3x^3 + 6x^2 \qquad\quad \\
\underline{3x^3 + 6x^2 \qquad\quad} \\
- x - 2 \\
\underline{- x - 2} \\
0
\end{array}$$

Thus, $\dfrac{x^4 + 5x^3 + 6x^2 - x - 2}{x + 2} = x^3 + 3x^2 - 1$.

7. Divide $7x+3$ by $x+2$ using long division.

Solution:

$$
\begin{array}{r}
7 \\
x+2\;\overline{)\;7x+\ \ 3} \\
7x+14 \\
\hline
-\ 11
\end{array}
$$

Thus, $\dfrac{7x+3}{x+2} = 7 - \dfrac{11}{x+2}$.

11. Divide $x^4 + 3x^2 + 1$ by $x^2 - 2x + 3$ using long division.

Solution:

$$
\begin{array}{r}
x^2 + \ 2x + \ \ \ 4 \\
x^2-2x+3\;\overline{)\;x^4 + 0x^3 + 3x^2 + 0x + \ 1} \\
x^4 - 2x^3 + 3x^2 \\
\hline
2x^3 + 0x^2 + 0x \\
2x^3 - 4x^2 + 6x \\
\hline
4x^2 - 6x + \ 1 \\
4x^2 - 8x + 12 \\
\hline
2x - 11
\end{array}
$$

Thus, $\dfrac{x^4 + 3x^2 + 1}{x^2 - 2x + 3} = x^2 + 2x + 4 + \dfrac{2x - 11}{x^2 - 2x + 3}$.

15. Divide $3x^3 - 17x^2 + 15x - 25$ by $x - 5$ using synthetic division.

Solution:

$$
\begin{array}{r|rrrr}
5 & 3 & -17 & 15 & -25 \\
 & & 15 & -10 & 25 \\
\hline
 & 3 & -2 & 5 & 0
\end{array}
$$

Thus, $\dfrac{3x^3 - 17x^2 + 15x - 25}{x - 5} = 3x^2 - 2x + 5$.

19. Divide $-x^3 + 75x - 250$ by $x + 10$ using synthetic division.

Solution:

$$\begin{array}{r|rrrr} -10 & -1 & 0 & 75 & -250 \\ & & 10 & -100 & 250 \\ \hline & -1 & 10 & -25 & 0 \end{array}$$

Thus, $\dfrac{-x^3 + 75x - 250}{x + 10} = -x^2 + 10x - 25.$

23. Divide $10x^4 - 50x^3 - 800$ by $x - 6$ using synthetic division.

Solution:

$$\begin{array}{r|rrrrr} 6 & 10 & -50 & 0 & 0 & -800 \\ & & 60 & 60 & 360 & 2160 \\ \hline & 10 & 10 & 60 & 360 & 1360 \end{array}$$

Thus, $\dfrac{10x^4 - 50x^3 - 800}{x - 6} = 10x^3 + 10x^2 + 60x + 360 + \dfrac{1360}{x - 6}.$

27. Divide $-3x^4$ by $x - 2$ using synthetic division.

Solution:

$$\begin{array}{r|rrrrr} 2 & -3 & 0 & 0 & 0 & 0 \\ & & -6 & -12 & -24 & -48 \\ \hline & -3 & -6 & -12 & -24 & -48 \end{array}$$

Thus, $\dfrac{-3x^4}{x - 2} = -3x^3 - 6x^2 - 12x - 24 - \dfrac{48}{x - 2}.$

31. Divide $4x^3 + 16x^2 - 23x - 15$ by $x + \frac{1}{2}$ using synthetic division.

Solution:

$$\begin{array}{r|rrrr} -\frac{1}{2} & 4 & 16 & -23 & -15 \\ & & -2 & -7 & 15 \\ \hline & 4 & 14 & -30 & 0 \end{array}$$

Thus, $\dfrac{4x^3 + 16x^2 - 23x - 15}{x + \frac{1}{2}} = 4x^2 + 14x - 30.$

35. Use synthetic division to show that $x = \frac{1}{2}$ is a solution of $2x^3 - 15x^2 + 27x - 10 = 0$, and use the result to factor the polynomial completely.

Solution:

$$\begin{array}{r|rrrr} \frac{1}{2} & 2 & -15 & 27 & -10 \\ & & 1 & -7 & 10 \\ \hline & 2 & -14 & 20 & 0 \end{array}$$

$$\begin{aligned} 2x^3 - 15x^2 + 27x - 10 &= (x - \tfrac{1}{2})(2x^2 - 14x + 20) \\ &= (x - \tfrac{1}{2})2(x^2 - 7x + 10) \\ &= (2x - 1)(x - 2)(x - 5) \end{aligned}$$

39. Use synthetic division to show that $x = 1 + \sqrt{3}$ is a solution of $x^3 - 3x^2 + 2 = 0$, and use the result to factor the polynomial completely.

Solution:

$$\begin{array}{r|rrrr} 1+\sqrt{3} & 1 & -3 & 0 & 2 \\ & & 1+\sqrt{3} & 1-\sqrt{3} & -2 \\ \hline & 1 & -2+\sqrt{3} & 1-\sqrt{3} & 0 \end{array}$$

$$\begin{aligned} x^3 - 3x^2 + 2 &= \left[x - (1 + \sqrt{3})\right]\left[x^2 + (-2 + \sqrt{3})x + 1 - \sqrt{3}\right] \\ &= (x - 1 - \sqrt{3})(x^2 - 2x + \sqrt{3}x + 1 - \sqrt{3}) \\ &= (x - 1 - \sqrt{3})\left[(x^2 - 2x + 1) + \sqrt{3}(x - 1)\right] \\ &= (x - 1 - \sqrt{3})\left[(x - 1)^2 + \sqrt{3}(x - 1)\right] \\ &= (x - 1 - \sqrt{3})(x - 1)\left[(x - 1) + \sqrt{3}\right] \\ &= \left(x - 1 - \sqrt{3}\right)\left(x - 1 + \sqrt{3}\right)(x - 1) \end{aligned}$$

41. Express the function $f(x) = x^3 - x^2 - 14x + 11$ in the form $f(x) = (x - k)q(x) + r$ for $k = 4$, and demonstrate that $f(k) = r$.

Solution:

$$\begin{array}{r|rrrr} 4 & 1 & -1 & -14 & 11 \\ & & 4 & 12 & -8 \\ \hline & 1 & 3 & -2 & 3 \end{array}$$

$$f(x) = (x - 4)(x^2 + 3x - 2) + 3$$
$$r = 3$$

$$\begin{aligned} f(4) &= 4^3 - 4^2 - 14(4) + 11 \\ &= 64 - 16 - 56 + 11 \\ &= 3 \end{aligned}$$

45. Use synthetic division to find the required function values of $f(x) = 4x^3 - 13x + 10$.

(a) $f(1)$ (b) $f(-2)$ (c) $f(1/2)$ (d) $f(8)$

Solution:

(a)
$$\begin{array}{r|rrrr} 1 & 4 & 0 & -13 & 10 \\ & & 4 & 4 & -9 \\ \hline & 4 & 4 & -9 & 1 \end{array}$$

Thus, $f(1) = 1$.

(b)
$$\begin{array}{r|rrrr} -2 & 4 & 0 & -13 & 10 \\ & & -8 & 16 & -6 \\ \hline & 4 & -8 & 3 & 4 \end{array}$$

Thus, $f(-2) = 4$.

(c)
$$\begin{array}{r|rrrr} \frac{1}{2} & 4 & 0 & -13 & 10 \\ & & 2 & 1 & -6 \\ \hline & 4 & 2 & -12 & 4 \end{array}$$

Thus, $f(1/2) = 4$.

(d)
$$\begin{array}{r|rrrr} 8 & 4 & 0 & -13 & 10 \\ & & 32 & 256 & 1944 \\ \hline & 4 & 32 & 243 & 1954 \end{array}$$

Thus, $f(8) = 1954$.

49. Use synthetic division to find the required function values of $f(x) = x^3 - 2x^2 - 11x + 52$.

(a) $f(5)$ (b) $f(-4)$ (c) $f(1.2)$ (d) $f(2)$

Solution:

(a)
$$\begin{array}{r|rrrr} 5 & 1 & -2 & -11 & 52 \\ & & 5 & 15 & 20 \\ \hline & 1 & 3 & 4 & 72 \end{array}$$

Thus, $f(5) = 72$.

(b)
$$\begin{array}{r|rrrr} -4 & 1 & -2 & -11 & 52 \\ & & -4 & 24 & -52 \\ \hline & 1 & -6 & 13 & 0 \end{array}$$

Thus, $f(-4) = 0$.

(c)
$$\begin{array}{r|rrrr} 1.2 & 1 & -2 & -11 & 52 \\ & & 1.2 & -0.96 & -14.352 \\ \hline & 1 & -0.8 & -11.96 & 37.648 \end{array}$$

Thus, $f(1.2) = 37.648$.

(d)
$$\begin{array}{r|rrrr} 2 & 1 & -2 & -11 & 52 \\ & & 2 & 0 & -22 \\ \hline & 1 & 0 & -11 & 30 \end{array}$$

Thus, $f(2) = 30$.

53. Simplify the rational expression $\dfrac{x^3 + 3x^2 - x - 3}{x + 1}$.

Solution:

$$\begin{array}{r|rrrr} -1 & 1 & 3 & -1 & -3 \\ & & -1 & -2 & 3 \\ \hline & 1 & 2 & -3 & 0 \end{array}$$

Thus, $\dfrac{x^3 + 3x^2 - x - 3}{x + 1} = x^2 + 2x - 3 = (x + 3)(x - 1)$.

57. The horsepower y developed by a compact car engine is approximated by the model

$$y = -1.42x^3 + 5.04x^2 + 32.45x - 0.75, \quad 1 \le x \le 5$$

where x is the engine speed in thousands of revolutions per minute. Note on the graph that there are two engine speeds that develop 110 horsepower, one of which is 5000 rpm. Approximate the other engine speed.

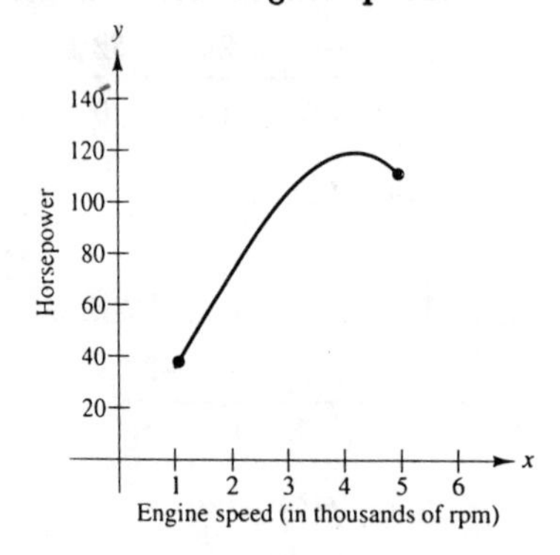

Solution:

Since $y = 110$ when $x = 5$ we have $x = 5$ as a zero to the equation $y - 110$.

$$\begin{aligned} f(x) &= y - 110, \quad 1 \le x \le 5 \\ &= -1.42x^3 + 5.04x^2 + 32.45x - 110.75 \end{aligned}$$

$$\begin{array}{r|rrrr} 5 & -1.42 & 5.04 & 32.45 & -110.75 \\ & & -7.10 & -10.30 & 110.75 \\ \hline & -1.42 & -2.06 & 22.15 & 0 \end{array}$$

Thus, $f(x) = (x-5)(-1.42x^2 - 2.06x + 22.15)$. Using the Quadratic Formula to find the other two zeros yields

$$\begin{aligned} x &= \frac{-(-2.06) \pm \sqrt{(-2.06)^2 - 4(-1.42)(22.15)}}{2(-1.42)} \\ &= \frac{2.06 \pm \sqrt{130.0556}}{-2.84} \end{aligned}$$

$x \approx -4.74$ or $x \approx 3.29$.

Since $1 \le x \le 5$, we choose $x = 3.29$ which corresponds to 3290 rpm.

SECTION 4.4

Real Zeros of Polynomial Functions

- You should know Descartes's Rule of Signs.
  (a) The number of positive real zeros of f is either equal to the number of variations of sign of f or is less than that number by an even integer.
  (b) The number of negative real zeros of f is either equal to the number of variations in sign of $f(-x)$ or is less than that number by an even integer.
  (c) When there is only one variation in sign, there is exactly one positive (or negative) real zero.

- You should know the Rational Zero Test.

- You should know shortcuts for the Rational Zero Test.
  (a) Use a programmable calculator.
  (b) Sketch a graph.
  (c) After finding a root, use synthetic division to reduce the degree of the polynomial.

- You should be able to observe the last row obtained from synthetic division in order to determine upper or lower bounds.
  (a) If the test value is positive and all of the entries in the last row are positive or zero, then the test value is an upper bound.
  (b) If the test value is negative and the entries in the last row alternate from positive to negative, then the test value is a lower bound. (Zero entries count as positive or negative.)

Solutions to Selected Exercises

5. Use Descartes's Rule of Signs to determine the possible number of positive and negative zeros of $g(x) = 2x^3 - 3x^2 - 3$.

Solution:

$$g(x) = 2x^3 - 3x^2 - 3$$
$$g(-x) = -2x^3 - 3x^2 - 3$$

Since $g(x)$ has one variation in sign, g has exactly one positive real zero. Since $g(-x)$ has no variations in sign, there are no negative real zeros.

9. Use Descartes's Rule of Signs to determine the possible number of positive and negative real zeros of $h(x) = 4x^2 - 8x + 3$.

Solution:

$$h(x) = 4x^2 - 8x + 3$$
$$h(-x) = 4x^2 + 8x + 3$$

Since $h(x)$ has two variations in sign, h has either two or zero positive real zeros. Since $h(-x)$ has no variations in sign, there are no negative real zeros.

13. Use the Rational Zero Test to list all the possible rational zeros of $f(x) = -4x^3 + 15x^2 - 8x - 3$ and verify that the zeros of f shown on the graph are contained in the list.

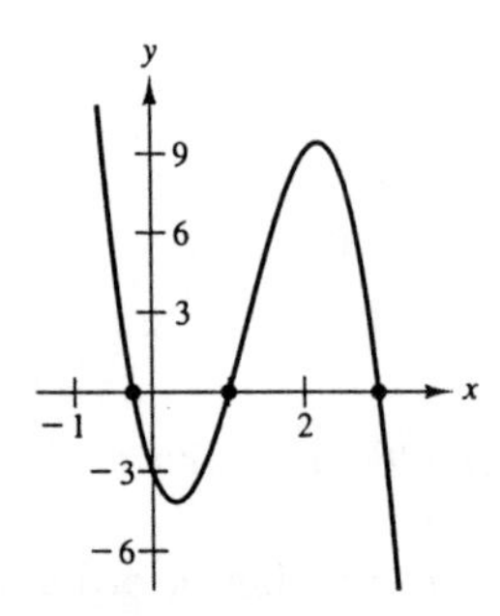

Solution:

Since the leading coefficient is -4 and the constant term is -3, the possible rational zeros of f are

$$\frac{\text{factors of } -3}{\text{factors of } -4} = \frac{\pm 1, \ \pm 3}{\pm 1, \ \pm 2, \ \pm 4} = \pm 1, \ \pm 3, \ \pm\frac{1}{2}, \ \pm\frac{3}{2}, \ \pm\frac{1}{4}, \ \pm\frac{3}{4}$$

The zeros shown on the graph are $-\frac{1}{4}$, 1 and 3 and are contained in the list.

17. Use synthetic division to determine if the given x-value is an upper bound or lower bound of the zeros of $f(x) = x^4 - 4x^3 + 15$.

(a) $x = 4$ (b) $x = -1$ (c) $x = 3$

Solution:

$$\text{(a) } 4 \begin{array}{|rrrrr} 1 & -4 & 0 & 0 & 15 \\ & 4 & 0 & 0 & 0 \\ \hline 1 & 0 & 0 & 0 & 15 \end{array}$$

Since the test value is positive and all the entries in the last row are positive, $x = 4$ is an upper bound.

$$\text{(b) } -1 \begin{array}{|rrrrr} 1 & -4 & 0 & 0 & 15 \\ & -1 & 5 & -5 & 5 \\ \hline 1 & -5 & 5 & -5 & 20 \end{array}$$

Since the test value is negative and the entries in the last row alternate in sign, $x = -1$ is a lower bound.

$$\text{(c) } 3 \begin{array}{|rrrrr} 1 & -4 & 0 & 0 & 15 \\ & 3 & -3 & -9 & -27 \\ \hline 1 & -1 & -3 & -9 & -12 \end{array}$$

$x = 3$ is neither an upper nor a lower bound.

21. Find the real zeros of $f(x) = x^3 - 6x^2 + 11x - 6$.

Solution:

Possible rational zeros: ± 1, ± 2, ± 3, ± 6

$$1 \begin{array}{|rrrr} 1 & -6 & 11 & -6 \\ & 1 & -5 & 6 \\ \hline 1 & -5 & 6 & 0 \end{array}$$

$$x^3 - 6x^2 + 11x - 6 = (x-1)(x^2 - 5x + 6) = (x-1)(x-2)(x-3)$$

The zeros are 1, 2, and 3.

25. Find the real zeros of $h(t) = t^3 + 12t^2 + 21t + 10$.

Solution:

Possible rational zeros: $\pm 1,\ \pm 2,\ \pm 5,\ \pm 10$

$$\begin{array}{r|rrrr} -1 & 1 & 12 & 21 & 10 \\ & & -1 & -11 & -10 \\ \hline & 1 & 11 & 10 & 0 \end{array}$$

$$t^3 + 12t^2 + 21t + 10 = (t+1)(t^2 + 11t + 10) = (t+1)(t+1)(t+10)$$

Thus, the zeros are -1 and -10.

27. Find the real zeros of $f(x) = x^3 - 4x^2 + 5x - 2$.

Solution:

Possible rational zeros: $\pm 1,\ \pm 2$

$$\begin{array}{r|rrrr} 1 & 1 & -4 & 5 & -2 \\ & & 1 & -3 & 2 \\ \hline & 1 & -3 & 2 & 0 \end{array}$$

$$x^3 - 4x^2 + 5x - 2 = (x-1)(x^2 - 3x + 2) = (x-1)(x-1)(x-2)$$

Thus, the zeros are 1 and 2.

31. Find the real zeros of $f(x) = 4x^3 - 3x - 1$.

Solution:

Possible rational zeros: $\pm 1,\ \pm\frac{1}{2},\ \pm\frac{1}{4}$

$$\begin{array}{r|rrrr} 1 & 4 & 0 & -3 & -1 \\ & & 4 & 4 & 1 \\ \hline & 4 & 4 & 1 & 0 \end{array}$$

$$4x^3 - 3x - 1 = (x-1)(4x^2 + 4x + 1) = (x-1)(2x+1)^2$$

Thus, the zeros are 1 and $-\frac{1}{2}$.

33. Find the real zeros of $f(y) = 4y^3 + 3y^2 + 8y + 6$.

Solution:

Possible rational zeros: $\pm 1, \pm 2, \pm 3, \pm 6, \pm\frac{1}{2}, \pm\frac{3}{2}, \pm\frac{1}{4}, \pm\frac{3}{4}$

$-\frac{3}{4}$	4	3	8	6
		-3	0	-6
	4	0	8	0

$$4y^3 + 3y^2 + 8y + 6 = (y + \tfrac{3}{4})(4y^2 + 8) = (y + \tfrac{3}{4})4(y^2 + 2) = (4y + 3)(y^2 + 2)$$

Thus, the only zero is $-\frac{3}{4}$.

35. Find the real zeros of $f(x) = x^4 - 3x^2 + 2$.

Solution:

$$\begin{aligned} f(x) &= x^4 - 3x^2 + 2 \\ &= (x^2 - 1)(x^2 - 2) \\ &= (x + 1)(x - 1)(x + \sqrt{2})(x - \sqrt{2}) \end{aligned}$$

Thus, the zeros are ± 1 and $\pm\sqrt{2}$.

39. Find all the real solutions of $x^4 - 13x^2 - 12x = 0$.

Solution:

$$f(x) = x^4 - 13x^2 - 12x = x(x^3 - 13x - 12)$$

0 is a zero.

Possible rational zeros: $\pm 1, \pm 2, \pm 3, \pm 4, \pm 6, \pm 12$

-1	1	0	-13	-12
		-1	1	12
	1	-1	-12	0

$$x^4 - 13x^2 - 12x = x(x^3 - 13x - 12) = x(x + 1)(x^2 - x - 12) = x(x + 1)(x + 3)(x - 4)$$

Thus, the zeros are $0, -1, -3,$ and 4.

43. Find all the real solutions of $x^5 - 7x^4 + 10x^3 + 14x^2 - 24x = 0$.

Solution:

$$f(x) = x^5 - 7x^4 + 10x^3 + 14x^2 - 24x = x(x^4 - 7x^3 + 10x^2 + 14x - 24)$$

0 is a zero.

Possible rational zeros: ±1, ±2, ±3, ±4, ±6, ±8, ±12, ±24

$$\begin{array}{c|ccccc} 4 & 1 & -7 & 10 & 14 & -24 \\ & & 4 & -12 & -8 & 24 \\ \hline \end{array}$$

$$\begin{array}{c|ccccc} 3 & 1 & -3 & -2 & 6 & 0 \\ & & 3 & 0 & -6 & \\ \hline & 1 & 0 & -2 & 0 & \end{array}$$

$$\begin{aligned} x^5 - 7x^4 + 10x^3 + 14x^2 - 24x &= x(x^4 - 7x^3 + 10x^2 + 14x - 24) \\ &= x(x-4)(x-3)(x^2-2) \\ &= x(x-4)(x-3)(x+\sqrt{2})(x-\sqrt{2}) \end{aligned}$$

Thus, the zeros are 0, 4, 3, and $\pm\sqrt{2}$.

47. For $f(x) = 4x^3 + 7x^2 - 11x - 18$

(a) list all the possible rational zeros of f,

(b) sketch the graph of f so that some of the possible zeros in part (a) can be disregarded, and

(c) then determine all the real zeros of f.

Solution:

(a) Possible rational roots: ±1, ±2, ±3, ±6, ±9, ±18, $\pm\frac{1}{2}$, $\pm\frac{3}{2}$, $\pm\frac{9}{2}$, $\pm\frac{1}{4}$, $\pm\frac{3}{4}$, $\pm\frac{9}{4}$

(b)

x	0	1	-1	$\frac{1}{2}$
$f(x)$	-18	-18	-4	-21.25

x	-2	-3	$-\frac{3}{2}$
$f(x)$	0	-30	0.75

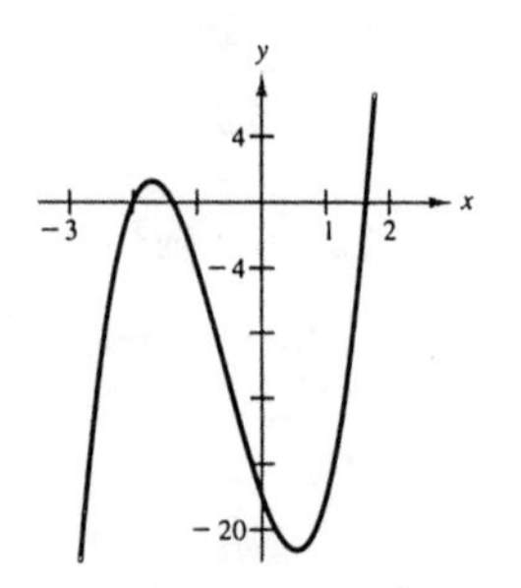

By testing values using synthetic division, we find that 2 is an upper bound and -3 is a lower bound. This eliminates 3, ±6, ±9, ±18, $\pm\frac{9}{2}$, and $\frac{9}{4}$ as possible zeros.

47. --CONTINUED--

(c) -2 is a zero.

$4x^3 + 7x^2 - 11x - 18 = (x+2)(4x^2 - x - 9) = 0$

$4x^2 - x - 9$ does not factor, so by the Quadratic Formula

$x = \dfrac{1 \pm \sqrt{145}}{8}$ are also zeros.

51. Find all the rational zeros of $f(x) = x^3 - \frac{1}{4}x^2 - x + \frac{1}{4}$.

Solution:

$$f(x) = x^3 - \tfrac{1}{4}x^2 - x + \tfrac{1}{4} = \tfrac{1}{4}(4x^3 - x^2 - 4x + 1)$$

Possible rational zeros: ± 1, $\pm \frac{1}{4}$, $\pm \frac{1}{2}$

By testing these values, we see that $x = \pm 1$ and $x = \frac{1}{4}$ work.

55. Match the cubic equation $f(x) = x^3 - x$ with the number of rational and irrational zeros (a), (b), (c) or (d).

(a) Rational zeros: 0
Irrational zeros: 1

(b) Rational zeros: 3
Irrational zeros: 0

(c) Rational zeros: 1
Irrational zeros: 2

(d) Rational zeros: 1
Irrational zeros: 0

Solution:

$$f(x) = x^3 - x = x(x^2 - 1) = x(x+1)(x-1)$$

Zeros: 0, ± 1

Three rational zeros/no irrational zeros

Matches (b)

59. A rectangular package to be sent by a postal service can have a maximum combined length and girth (perimeter of a cross section) of 108 inches. (see figure) Find the dimensions of the package, given that the volume is to be 11,664 cubic inches.

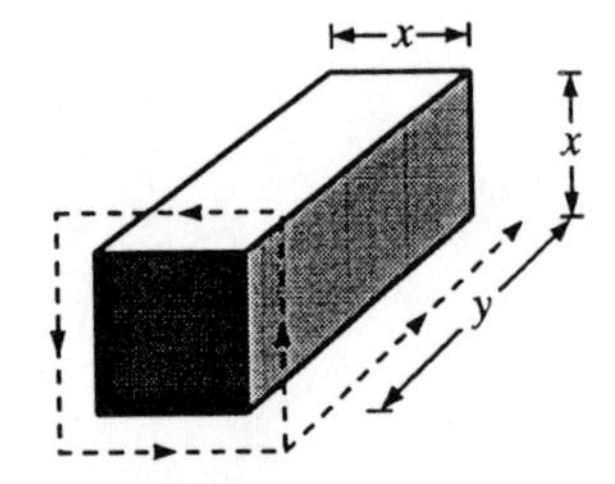

Solution:

The combined length and girth is

$$4x + y = 108$$
$$y = 108 - 4x.$$

The volume is

$$V = x^2y = x^2(108 - 4x).$$

Since the volume is 11,664 cubic inches, we have

$$11{,}664 = x^2(108 - 4x)$$
$$11{,}664 = 108x^2 - 4x^3$$
$$4x^3 - 108x^2 + 11{,}664 = 0$$
$$4(x^3 - 27x^2 + 2916) = 0$$
$$4(x - 18)^2(x + 2) = 0$$

Using the methods of this section yields

Thus, $x = 18$in. and $y = 36$in.

The dimensions of the package are 18 in. × 18 in. × 36 in.

SECTION 4.5

The Fundamental Theorem of Algebra

- You should know that if f is a polynomial of degree $n > 0$, then f has exactly n zeros (roots) in the complex number system.
- You should know that if $a + bi$ is a complex zero of a polynomial f, with real coefficients, then $a - bi$ is also a complex zero of f.
- You should know the difference between a factor that is irreducible over the rationals (such as $x^2 - 7$) and a factor that is irreducible over the reals (such as $x^2 + 9$).

Solutions to Selected Exercises

3. Find all the zeros of $h(x) = x^2 - 4x + 1$ and write the polynomial as a product of linear factors.

Solution:

h has no rational zeros.

By the Quadratic Formula, the zeros are $x = \dfrac{4 \pm \sqrt{16 - 4}}{2} = 2 \pm \sqrt{3}$.

$$h(x) = [x - (2 + \sqrt{3})][x - (2 - \sqrt{3})] = (x - 2 - \sqrt{3})(x - 2 + \sqrt{3})$$

7. Find all the zeros of $f(z) = z^2 - 2z + 2$ and write the polynomial as a product of linear factors.

Solution:

f has no rational zeros.

By the Quadratic Formula, the zeros are $z = \dfrac{2 \pm \sqrt{4 - 8}}{2} = 1 \pm i$.

$$f(z) = [z - (1 + i)][z - (1 - i)] = (z - 1 - i)(z - 1 + i)$$

9. Find all the zeros of $g(x) = x^3 - 6x^2 + 13x - 10$ and write the polynomial as a product of linear factors.

Solution:

Possible rational zeros: $\pm 1,\ \pm 2,\ \pm 5,\ \pm 10$

$$\begin{array}{r|rrrr} 2 & 1 & -6 & 13 & -10 \\ & & 2 & -8 & 10 \\ \hline & 1 & -4 & 5 & 0 \end{array}$$

$g(x) = (x-2)(x^2 - 4x + 5)$

$x = 2$ is a zero, and by the Quadratic Formula $x = \dfrac{4 \pm \sqrt{16-20}}{2} = 2 \pm i$ are also zeros.

$g(x) = (x-2)[x-(2+i)][x-(2-i)] = (x-2)(x-2-i)(x-2+i)$

15. Find all the zeros of $f(x) = 16x^3 - 20x^2 - 4x + 15$ and write the polynomial as a product of linear factors.

Solution:

Possible rational zeros: $\pm 1,\ \pm 3,\ \pm 5,\ \pm 15,\ \pm\frac{1}{2},\ \pm\frac{3}{2},\ \pm\frac{5}{2},\ \pm\frac{15}{2},\ \pm\frac{1}{4},\ \pm\frac{3}{4},\ \pm\frac{5}{4},\ \pm\frac{15}{4},$
$\pm\frac{1}{8},\ \pm\frac{3}{8},\ \pm\frac{5}{8},\ \pm\frac{15}{8},\ \pm\frac{1}{16},\ \pm\frac{3}{16},\ \pm\frac{5}{16},\ \pm\frac{15}{16}$

$$\begin{array}{r|rrrr} -\frac{3}{4} & 16 & -20 & -4 & 15 \\ & & -12 & 24 & -15 \\ \hline & 16 & -32 & 20 & 0 \end{array}$$

$$f(x) = \left(x + \frac{3}{4}\right)(16x^2 - 32x + 20) = 4\left(x + \frac{3}{4}\right)(4x^2 - 8x + 5)$$
$$= (4x+3)(4x^2 - 8x + 5)$$

$x = -\dfrac{3}{4}$ is a zero, and by the Quadratic Formula $x = \dfrac{8 \pm \sqrt{64-80}}{8} = 1 \pm \dfrac{1}{2}i$ are also zeros.

$$f(x) = 16\left(x + \frac{3}{4}\right)\left[x - \left(1 + \frac{1}{2}i\right)\right]\left[x - \left(1 - \frac{1}{2}i\right)\right] = (4x+3)(2x-2-i)(2x-2+i)$$

21. Find all the zeros of $g(x) = x^4 - 4x^3 + 8x^2 - 16x + 16$ and write the polynomial as a product of linear factors.

Solution:

Possible rational zeros: $\pm 1, \pm 2, \pm 4, \pm 8, \pm 16$

$$\begin{array}{r|rrrrr} 2 & 1 & -4 & 8 & -16 & 16 \\ & & 2 & -4 & 8 & -16 \\ \hline \end{array}$$

$$\begin{array}{r|rrrrr} 2 & 1 & -2 & 4 & -8 & 0 \\ & & 2 & 0 & 8 & \\ \hline & 1 & 0 & 4 & 0 & \end{array}$$

$$g(x) = (x-2)(x-2)(x^2+4) = (x-2)^2(x+2i)(x-2i)$$

The zeros of g are 2 and $\pm 2i$.

27. Find a polynomial with integer coefficients that has the zeros 1, $5i$, and $-5i$.

Solution:

$$\begin{aligned} f(x) &= (x-1)(x-5i)(x+5i) \\ &= (x-1)(x^2+25) \\ &= x^3 - x^2 + 25x - 25 \end{aligned}$$

Note: $f(x) = a(x^3 - x^2 + 25x - 25)$, where a is any real number, has the zeros 1 and $\pm 5i$.

31. Find a polynomial with integer coefficients that has the zeros i, $-i$, $6i$, and $-6i$.

Solution:

$$\begin{aligned} f(x) &= (x-i)(x+i)(x-6i)(x+6i) \\ &= (x^2+1)(x^2+36) \\ &= x^4 + 37x^2 + 36 \end{aligned}$$

Note: $f(x) = a(x^4 + 37x^2 + 36)$, where a is any real number, has the zeros $\pm i$, and $\pm 6i$.

35. Find a polynomial with integer coefficients that has the zeros $\frac{3}{4}$, -2, and $-\frac{1}{2}+i$.

Solution:

Since $-\frac{1}{2}+i$ is a zero, so is $-\frac{1}{2}-i$.

$$\begin{aligned} f(x) &= 16\left(x-\tfrac{3}{4}\right)(x+2)\left[x-\left(-\tfrac{1}{2}+i\right)\right]\left[x-\left(-\tfrac{1}{2}-i\right)\right] \quad \text{Multiply by 16 to clear all the fractions.} \\ &= 4(4x-3)(x+2)\left[x^2+x+\left(\tfrac{1}{4}+1\right)\right] \\ &= (4x^2+5x-6)(4x^2+4x+5) \\ &= 16x^4+36x^3+16x^2+x-30 \end{aligned}$$

Note: $f(x) = a(16x^4+36x^3+16x^2+x-30)$, where a is any real number, has the zeros $\frac{3}{4}$, -2, and $-\frac{1}{2} \pm i$.

39. Write $f(x) = x^4 - 4x^3 + 5x^2 - 2x - 6$

(a) as the product of factors that are irreducible over the rationals,

(b) as the product of linear and quadratic factors that are irreducible over the reals, and

(c) in completely factored form. [*Hint:* One factor is $x^2 - 2x - 2$.]

Solution:

$$\begin{array}{r|l} & x^2 - 2x + 3 \\ \hline x^2-2x-2 & x^4 - 4x^3 + 5x^2 - 2x - 6 \\ & x^4 - 2x^3 - 2x^2 \\ \hline & -2x^3 + 7x^2 - 2x \\ & -2x^3 + 4x^2 + 4x \\ \hline & 3x^2 - 6x - 6 \\ & 3x^2 - 6x - 6 \\ \hline & 0 \end{array}$$

$$f(x) = (x^2-2x+3)(x^2-2x-2)$$

(a) $f(x) = (x^2-2x+3)(x^2-2x-2)$

(b) $f(x) = (x^2-2x+3)(x-1+\sqrt{3})(x-1-\sqrt{3})$

(c) $f(x) = (x-1+\sqrt{2}i)(x-1-\sqrt{2}i)(x-1+\sqrt{3})(x-1-\sqrt{3})$

Note: Use the Quadratic Formula for (b) and (c).

43. Use the zero, $r = 2i$, to find all the zeros of $f(x) = 2x^4 - x^3 + 7x^2 - 4x - 4$.

Solution:

Since $2i$ is a zero of f, so is $-2i$.

$$\begin{array}{r|rrrrr} 2i & 2 & -1 & 7 & -4 & -4 \\ & & 0+4i & -8-2i & 4-2i & 4 \\ \hline \end{array}$$

$$\begin{array}{r|rrrrr} -2i & 2 & -1+4i & -1-2i & -2i & 0 \\ & & 0-4i & 0+2i & 2i & \\ \hline & 2 & -1 & -1 & 0 & \end{array}$$

$$f(x) = (x - 2i)(x + 2i)(2x^2 - x - 1) = (x - 2i)(x + 2i)(2x + 1)(x - 1)$$

The zeros of f are $\pm 2i$, $-\frac{1}{2}$, and 1.

49. Use the zero $r = (1 - \sqrt{5}\,i)/2$ to find all the zeros of $h(x) = 8x^3 - 14x^2 + 18x - 9$.

Solution:

Since $(1 - \sqrt{5}\,i)/2$ is a zero, so is $(1 + \sqrt{5}\,i)/2$.

$$\begin{array}{r|rrrr} \frac{1-\sqrt{5}\,i}{2} & 8 & -14 & 18 & -9 \\ & & 4-4\sqrt{5}\,i & -15+3\sqrt{5}\,i & 9 \\ \hline \end{array}$$

$$\begin{array}{r|rrrr} \frac{1+\sqrt{5}\,i}{2} & 8 & -10-4\sqrt{5}\,i & 3+3\sqrt{5}\,i & 0 \\ & & 4+4\sqrt{5}\,i & -3-3\sqrt{5}\,i & \\ \hline & 8 & -6 & 0 & \end{array}$$

$$f(x) = \left[x - \left(\frac{1}{2} - \frac{\sqrt{5}}{2}i\right)\right]\left[x - \left(\frac{1}{2} + \frac{\sqrt{5}}{2}i\right)\right](8x - 6)$$

The zeros of f are: $\frac{3}{4}, \frac{1}{2} \pm \frac{\sqrt{5}}{2}i$.

53. Find a quadratic function f (with integer coefficients) that has $\pm\sqrt{b}\,i$ as zeros. Assume that b is a positive integer.

Solution:

$$f(x) = (x - \sqrt{b}\,i)(x + \sqrt{b}\,i) = x^2 + b$$

Note: $f(x) = a(x^2 + b)$, where a is any real number, has the zeros $\pm\sqrt{b}\,i$.

SECTION 4.6

Approximating Zeros of a Polynomial Function

- You should be able to use the Bisection Method to approximate the zeros of a polynomial f.
  (a) Sketch f and roughly approximate x_0, a point where the graph crosses the x-axis.
  (b) Find an x-value a such that $f(a) < 0$.
  (c) Find an x-value b such that $f(b) > 0$.
  (d) Find the midpoint $c = (a+b)/2 = x_1$.
  (e) Approximate the zero.
    1. If $f(c) = 0$, c is a zero of f.
    2. If $f(c) > 0$, the zero is between a and c.
    3. If $f(c) < 0$, the zero is between b and c.
  (f) The maximum error is $|b - a|/2$.
  (g) Continue this process until the error falls below the desired bound.

- You should know that the smaller the initial interval about the zero is, the fewer the number of iterations you will have.

Solutions to Selected Exercises

3. The value of the function $f(x) = -x^3 + 4x^2 - 5x + 3$ is negative at one of the x values and positive for the other. Determine which is positive and which is negative.

x-values: $x = 2.4$, $x = 2.5$

Solution:

$$f(x) = -x^3 + 4x^2 - 5x + 3$$
$$f(2.4) = 0.216 \quad \text{Positive}$$
$$f(2.5) = -0.125 \quad \text{Negative}$$

7. Use the Bisection Method to approximate the real zero of $f(x) = 2x^3 - 6x^2 + 6x - 1$ to two decimal places.

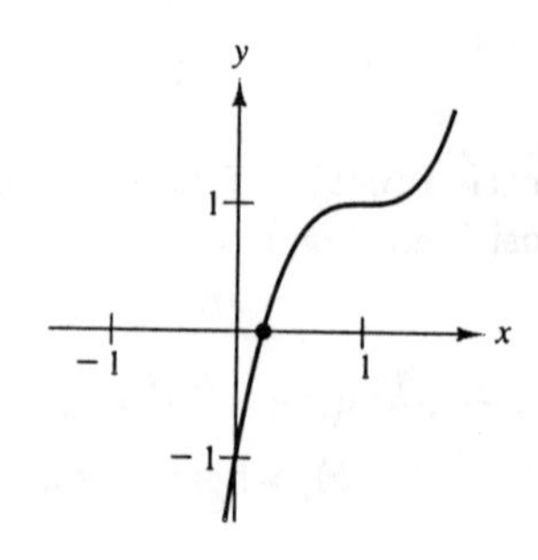

7. —CONTINUED—

Solution:

Iteration	a	c	b	$f(a)$	$f(c)$	$f(b)$	Error
1	0.2000	0.2500	0.3000	−0.0240	0.1563	0.3140	0.0500
2	0.2000	0.2250	0.2500	−0.0240	0.0690	0.1563	0.0250
3	0.2000	0.2125	0.2250	−0.0240	0.0233	0.0690	0.0125
4	0.2000	0.2063	0.2125	−0.0240	−0.0002	0.0233	0.0063
5	0.2063	0.2094	0.2125	−0.0002	0.0116	0.0233	0.0031

Accurate to two decimal places, the zero of f is 0.21.

11. Approximate the real zeros of $f(x) = x^4 - x - 3$ to two decimal places.

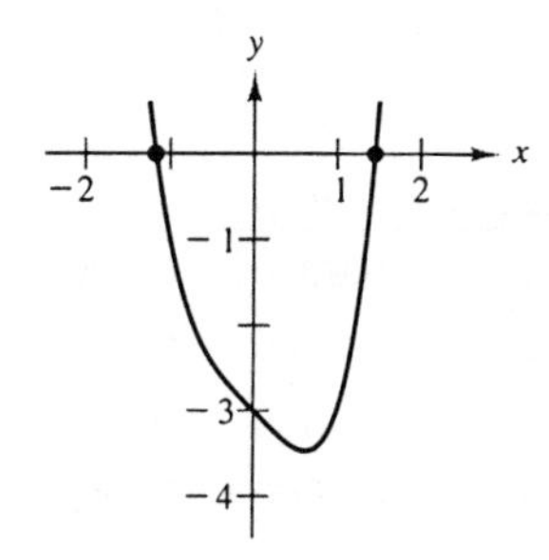

Solution:

Iteration	a	c	b	$f(a)$	$f(c)$	$f(b)$	Error
1	−1.5000	−1.2500	−1.0000	3.5625	0.6914	−1.0000	0.2500
2	−1.2500	−1.1250	−1.0000	0.6914	−0.2732	−1.0000	0.1250
3	−1.2500	−1.1875	−1.1250	0.6914	0.1760	−0.2732	0.0625
4	−1.1875	−1.1563	−1.1250	0.1760	−0.0564	−0.2732	0.0313
5	−1.1875	−1.1719	−1.1563	0.1760	0.0578	−0.0564	0.0156
6	−1.1719	−1.1641	−1.1563	0.0578	0.0002	−0.0564	0.0078
7	−1.1641	−1.1602	−1.1563	0.0002	−0.0282	−0.0564	0.0039

Accurate to two decimal places, one zero of f is −1.16.

–CONTINUED ON NEXT PAGE–

11. –CONTINUED–

Iteration	a	c	b	$f(a)$	$f(c)$	$f(b)$	Error
1	1.3000	1.4500	1.6000	−1.4439	−0.0295	1.9536	0.1500
2	1.4500	1.5250	1.6000	−0.0295	0.8835	1.9536	0.0750
3	1.4500	1.4875	1.5250	−0.0295	0.4083	0.8835	0.0375
4	1.4500	1.4688	1.4875	−0.0295	0.1849	0.4083	0.0188
5	1.4500	1.4594	1.4688	−0.0295	0.0766	0.1849	0.0094
6	1.4500	1.4547	1.4594	−0.0295	0.0233	0.0766	0.0047
7	1.4500	1.4523	1.4547	−0.0295	−0.0032	0.0233	0.0023

Accurate to two decimal places, the other zero of f is 1.45.

15. Find the zero of $f(x) = 4x^3 + 14x - 8$ in $[0,\ 1]$ correct to two decimal places.

Solution:

Iteration	a	c	b	$f(a)$	$f(c)$	$f(b)$	Error
1	0.0000	0.5000	1.0000	−8.0000	−0.5000	10.0000	0.5000
2	0.5000	0.7500	1.0000	−0.5000	4.1875	10.0000	0.2500
3	0.5000	0.6250	0.7500	−0.5000	1.7266	4.1875	0.1250
4	0.5000	0.5625	0.6250	−0.5000	0.5869	1.7266	0.0625
5	0.5000	0.5313	0.5625	−0.5000	0.0372	0.5869	0.0313
6	0.5000	0.5156	0.5313	−0.5000	−0.2329	0.0372	0.0156
7	0.5156	0.5234	0.5313	−0.2329	−0.0982	0.0372	0.0078
8	0.5234	0.5273	0.5313	−0.0982	−0.0306	0.0372	0.0039
9	0.5273	0.5293	0.5313	−0.0306	0.0033	0.0372	0.0020

The zero of f is approximately 0.53.

19. Find the zero of $f(x) = 7x^4 - 42x^3 + 43x^2 + 216x - 324$ in $[1,\ 2]$ to two correct decimal places.

Solution:

Iteration	a	c	b	$f(a)$	$f(c)$	$f(b)$	Error
1	1.0000	1.5000	2.0000	−100.0000	−9.5625	56.0000	0.5000
2	1.5000	1.7500	2.0000	−9.5625	26.2461	56.0000	0.2500
3	1.5000	1.6250	1.7500	−9.5625	9.1345	26.2461	0.1250
4	1.5000	1.5625	1.6250	−9.5625	−0.0135	9.1345	0.0625
5	1.5625	1.5938	1.6250	−0.0135	4.6104	9.1345	0.0313
6	1.5625	1.5781	1.5938	−0.0135	2.3109	4.6104	0.0156
7	1.5625	1.5703	1.5781	−0.0135	1.1518	2.3109	0.0078
8	1.5625	1.5664	1.5703	−0.0135	0.5699	1.1518	0.0039
9	1.5625	1.5645	1.5664	−0.0135	0.2784	0.5699	0.0020
10	1.5625	1.5635	1.5645	−0.0135	0.1324	0.2784	0.0010

The zero of f is approximately 1.56.

23. The concentration of a certain chemical in the bloodstream t hours after injection into muscle tissue is given by

$$C = \frac{3t^2 + t}{t^3 + 50}.$$

The concentration is greater when $3t^4 + 2t^3 - 300t - 50 = 0$. Approximate this time to the nearest tenth of an hour.

Solution:

$$f(t) = 3t^4 + 2t^3 - 300t - 50$$

Since $f(4) = -354$ and $f(5) = 575$, a zero is in the interval $[4,\ 5]$.

Iteration	a	c	b	$f(a)$	$f(c)$	$f(b)$	Error
1	4.000	4.500	5.000	−354.000	12.438	575.000	0.500
2	4.000	4.250	4.500	−354.000	−192.707	12.438	0.250
3	4.250	4.375	4.500	−192.707	−95.929	12.438	0.125
4	4.375	4.438	4.500	−95.929	−43.234	12.438	0.063
5	4.438	4.469	4.500	−43.234	−15.776	12.438	0.031
6	4.469	4.484	4.500	−15.776	−1.764	12.438	0.016

The zero of f is approximately 4.5 hours.

REVIEW EXERCISES FOR CHAPTER 4

Solutions to Selected Exercises

3. Sketch the graph of the quadratic function $f(x) = \frac{1}{3}(x^2 + 5x - 4)$. Identify the vertex and intercepts.

Solution:

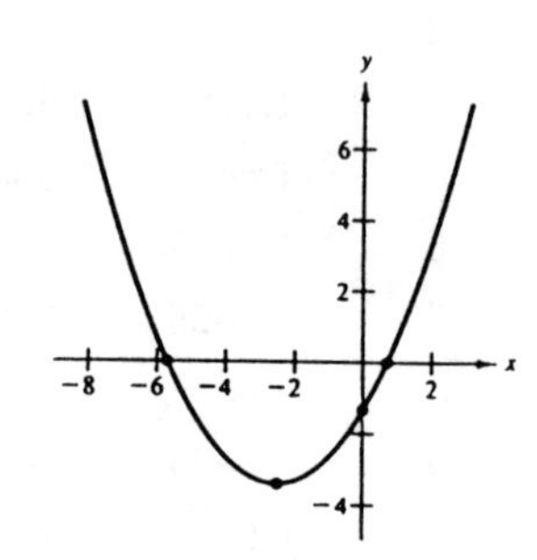

$$
\begin{aligned}
f(x) &= \frac{1}{3}(x^2 + 5x - 4) \\
&= \frac{1}{3}\left(x^2 + 5x + \frac{25}{4} - \frac{25}{4} - 4\right) \\
&= \frac{1}{3}\left[\left(x + \frac{5}{2}\right)^2 - \frac{41}{4}\right] \\
&= \frac{1}{3}\left(x + \frac{5}{2}\right)^2 - \frac{41}{12}
\end{aligned}
$$

Vertex: $\left(-\frac{5}{2}, \frac{41}{12}\right)$

Intercepts: $\left(0, -\frac{4}{3}\right)$, $\left(\frac{-5 \pm \sqrt{41}}{2}, 0\right)$

Use the Quadratic Formula for the x-intercepts.

7. Find the maximum or minimum value of $g(x) = x^2 - 2x$.

Solution:

$$
\begin{aligned}
g(x) &= x^2 - 2x && \text{has a minimum since } a > 0 \\
&= x^2 - 2x + 1 - 1 \\
&= (x - 1)^2 - 1
\end{aligned}
$$

The minimum is at the vertex $(1, -1)$.

11. Find the maximum or minimum value of $f(t) = -2t^2 + 4t + 1$.

Solution:

$$f(t) = -2t^2 + 4t + 1 \qquad \text{has a maximum since } a < 0$$

$$= -2\left(t^2 - 2t - \frac{1}{2}\right)$$

$$= -2\left(t^2 - 2t + 1 - 1 - \frac{1}{2}\right)$$

$$= -2\left[(t-1)^2 - \frac{3}{2}\right]$$

$$= -2(t-1)^2 + 3$$

The maximum is at the vertex $(1,\ 3)$.

15. A rectangle is inscribed in the region bounded by the x-axis, the y-axis, and the graph of $x + 2y - 6 = 0$, as shown in the figure. Find the coordinates $(x,\ y)$ that yield a maximum area for the rectangle.

Solution:

$$x + 2y - 6 = 0 \quad \Longrightarrow \quad y = \frac{6 - x}{2}$$

The area of the rectangle is

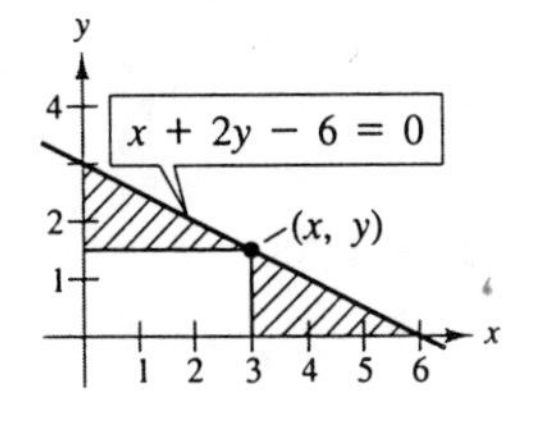

$$A = xy = x\left(\frac{6-x}{2}\right) = -\frac{1}{2}x^2 + 3x$$

$$= -\frac{1}{2}(x^2 - 6x + 9 - 9)$$

$$= -\frac{1}{2}[(x-3)^2 - 9]$$

$$= -\frac{1}{2}(x-3)^2 + \frac{9}{2}$$

The maximum value of the area is 9/2 square units and this occurs when $x = 3$ and $y = (6-3)/2 = 3/2$.

17. A textile manufacturer has daily production costs of

$$C = 10{,}000 - 10x + 0.045x^2$$

where C is the total cost in dollars and x is the number of units produced. How many fixtures should be produced each day to yield a minimum cost?

Solution:

$$\begin{aligned} C &= 10{,}000 - 10x + 0.045x^2 \\ &= 0.045\left(x^2 - \frac{2000}{9}x\right) + 10{,}000 \\ &= 0.045\left(x - \frac{1000}{9}\right)^2 + \frac{85{,}000}{9} \end{aligned}$$

The cost is minimum when $x = \dfrac{1000}{9} \approx 111$ units.

21. Determine the right hand and left hand behavior of the graph of $f(x) = -x^2 + 6x + 9$.

Solution:

$$f(x) = -x^2 + 6x + 9$$

The degree is even and the leading coefficient is negative, therefore

$f(x) \to -\infty$ as $x \to \infty$ Falls to the right

$f(x) \to -\infty$ as $x \to -\infty$ Falls to the left.

27. Sketch the graph of $g(x) = x^4 - x^3 - 2x^2$.

Solution:

$$\begin{aligned} g(x) &= x^4 - x^3 - 2x^2 \\ &= x^2(x^2 - x - 2) \\ &= x^2(x+1)(x-2) \end{aligned}$$

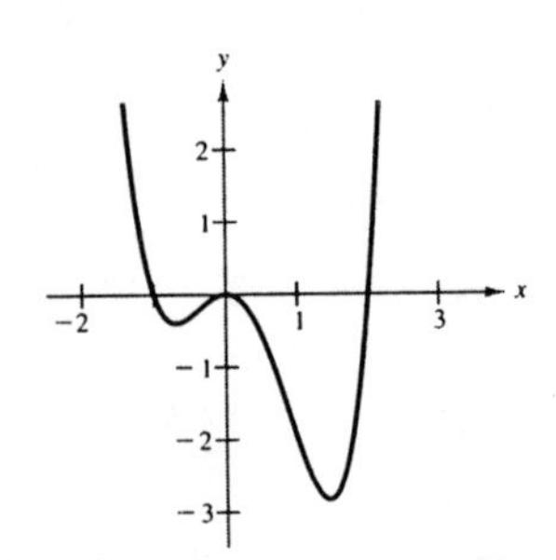

The zeros of g are 0, -1, and 2, so the graph of g crosses the x-axis at these points. Since the degree of g is even and the leading coefficient is positive, the graph moves up to the right and left.

x	0	-1	2	1	$-\frac{1}{2}$	$\frac{1}{2}$
$g(x)$	0	0	0	-2	$-\frac{5}{16}$	$-\frac{9}{16}$

31. Sketch the graph of $f(x) = x(x+3)^2$.

Solution:

$$f(x) = x(x+3)^2$$

The zeros of f are 0 and -3, so the graph of f crosses the x-axis at these points. Since the degree of f is odd and the leading coefficient is positive, the graph moves up to the right and down to the left.

x	-4	-3	-2	-1	0	1	2
$f(x)$	-4	0	-2	-4	0	16	50

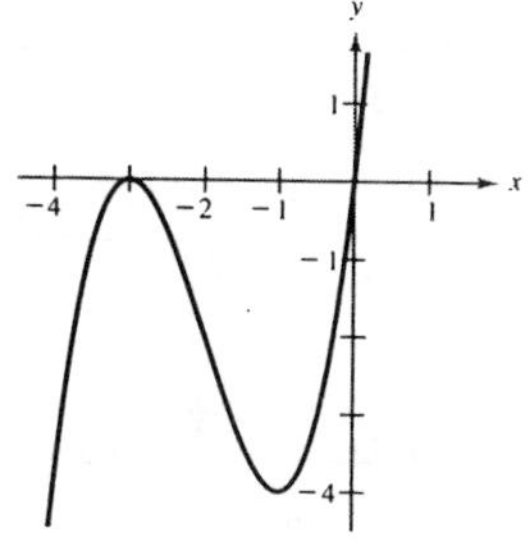

35. Perform the indicated division.

$$\frac{x^4 + x^3 - x^2 + 2x}{x^2 + 2x}$$

Solution:

$$\begin{array}{r}
x^2 - x + 1 \\
x^2 + 2x \overline{\big) x^4 + x^3 - x^2 + 2x} \\
\underline{x^4 + 2x^3 } \\
-x^3 - x^2 \\
\underline{-x^3 - 2x^2 } \\
x^2 + 2x \\
\underline{x^2 + 2x} \\
0
\end{array}$$

Thus, $\dfrac{x^4 + x^3 - x^2 + 2x}{x^2 + 2x} = x^2 - x + 1.$

39. Perform the indicated division.

$$\frac{x^4 - 3x^3 + 4x^2 - 6x + 3}{x^2 + 2}$$

Solution:

$$\begin{array}{r}
x^2 - 3x + 2 \\
x^2 + 2 \overline{\big) x^4 - 3x^3 + 4x^2 - 6x + 3} \\
\underline{x^4 + 2x^2 } \\
-3x^3 + 2x^2 - 6x \\
\underline{-3x^3 - 6x } \\
2x^2 + 3 \\
\underline{2x^2 + 4} \\
-1
\end{array}$$

Thus, $\dfrac{x^4 - 3x^3 + 4x^2 - 6x + 3}{x^2 + 2}$

$$= x^2 - 3x + 2 - \frac{1}{x^2 + 2}.$$

43. Use synthetic division to perform the indicated division.

$$\frac{6x^4 - 4x^3 - 27x^2 + 18x}{x - (2/3)}$$

Solution:

$\frac{2}{3}$	6	−4	−27	18	0
		4	0	−18	0
	6	0	−27	0	0

Thus, $\dfrac{6x^4 - 4x^3 - 27x^2 + 18x}{x - (2/3)} = 6x^3 - 27x.$

49. Use synthetic division to determine whether the given values of x are zeros of

$$f(x) = 2x^3 + 7x^2 - 18x - 30.$$

(a) $x = 1$ (b) $x = \frac{5}{2}$ (c) $x = -3 + \sqrt{3}$ (d) $x = 0$

Solution:

(a) 1	2	7	−18	−30
		2	9	−9
	2	9	−9	−39

$x = 1$ is *not* a zero of f.

(b) $\frac{5}{2}$	2	7	−18	−30
		5	30	30
	2	12	12	0

$x = \frac{5}{2}$ *is* a zero of f.

(c) $-3 + \sqrt{3}$	2	7	−18	−30
		$-6 + 2\sqrt{3}$	$3 - 5\sqrt{3}$	30
	2	$1 + 2\sqrt{3}$	$-15 - 5\sqrt{3}$	0

$x = -3 + \sqrt{3}$ *is* a zero of f.

(d) 0	2	7	−18	−30
		0	0	0
	2	7	−18	−30

$x = 0$ is *not* a zero of f.

53. Use synthetic division to find the specified value of $f(x) = x^4 + 10x^3 - 24x^2 + 20x + 44$.

(a) $f(-3)$ (b) $f(\sqrt{2}\,i)$

Solution:

(a)
$$\begin{array}{r|rrrrr} -3 & 1 & 10 & -24 & 20 & 44 \\ & & -3 & -21 & 135 & -465 \\ \hline & 1 & \text{[illegible]} & -45 & 155 & -421 \end{array}$$

Thus, $f(-3) = -421$.

(b)
$$\begin{array}{r|ccccc} \sqrt{2}\,i & 1 & 10 & -24 & 20 & 44 \\ & & 0 + \sqrt{2}\,i & -2 + 10\sqrt{2}\,i & -20 - 26\sqrt{2}\,i & 52 \\ \hline & 1 & 10 + \sqrt{2}\,i & -26 + 10\sqrt{2}\,i & -26\sqrt{2}\,i & 96 \end{array}$$

Thus, $f(\sqrt{2}\,i) = 96$.

55. Find a fourth degree polynomial with the zeros -1, -1, $\frac{1}{3}$, and $-\frac{1}{2}$.

Solution:

$$f(x) = 6(x+1)^2\left(x - \frac{1}{3}\right)\left(x + \frac{1}{2}\right) \qquad \text{Multiply by 6 to clear the fractions.}$$

$$= (x+1)^2 3\left(x - \frac{1}{3}\right) 2\left(x + \frac{1}{2}\right)$$

$$= (x^2 + 2x + 1)(3x - 1)(2x + 1)$$

$$= (x^2 + 2x + 1)(6x^2 + x - 1)$$

$$= 6x^4 + 13x^3 + 7x^2 - x - 1$$

Note: $f(x) = a(6x^4 + 13x^3 + 7x^2 - x - 1)$, where a is any real number, has zeros -1, -1, $\frac{1}{3}$, and $-\frac{1}{2}$.

59. Use Descartes's Rule of Signs to determine the possible number of positive and negative zeros of the function $f(x) = 5x^3 + 3x^2 - 6x + 9$.

Solution:

$$f(x) = 5x^3 + 3x^2 - 6x + 9$$

$f(x)$ has two variations in sign so f has either two or zero positive real zeros.

$$f(-x) = -5x^3 + 3x^2 + 6x + 9$$

$f(-x)$ has one variation in sign so f has one negative real zero.

65. Find all the zeros of $f(x) = 6x^3 - 5x^2 + 24x - 20$.

Solution:

Possible rational zeros: ±1, ±2, ±4, ±5, ±10, ±20, $\pm\frac{1}{2}$, $\pm\frac{5}{2}$, $\pm\frac{1}{3}$, $\pm\frac{2}{3}$, $\pm\frac{4}{3}$, $\pm\frac{5}{3}$, $\pm\frac{10}{3}$, $\pm\frac{20}{3}$, $\pm\frac{1}{6}$, $\pm\frac{5}{6}$

$$\begin{array}{r|rrrr} \frac{5}{6} & 6 & -5 & 24 & -20 \\ & & 5 & 0 & 20 \\ \hline & 6 & 0 & 24 & 0 \end{array}$$

$$f(x) = \left(x - \tfrac{5}{6}\right)\left(6x^2 + 24\right) = 6\left(x - \tfrac{5}{6}\right)\left(x^2 + 4\right) = 6\left(x - \tfrac{5}{6}\right)(x + 2i)(x - 2i)$$

The zeros of f are $\frac{5}{6}$ and $\pm 2i$.

Note: This problem can also be solved by factoring by grouping.

$$\begin{aligned} f(x) &= 6x^3 - 5x^2 + 24x - 20 = x^2(6x - 5) + 4(6x - 5) \\ &= (6x - 5)(x^2 + 4) = (6x - 5)(x + 2i)(x - 2i) \end{aligned}$$

The zeros of f are $\frac{5}{6}$ and $\pm 2i$.

69. Use the Bisection Method to find the zero of $f(x) = x^4 + 2x - 1$ in the interval $[0, 1]$ to the nearest hundredth.

Solution:

Iteration	a	c	b	$f(a)$	$f(c)$	$f(b)$	Error
1	0.0000	0.5000	1.0000	−1.0000	0.0625	2.0000	0.5000
2	0.0000	0.2500	0.5000	−1.0000	−0.4961	0.0625	0.2500
3	0.2500	0.3750	0.5000	−0.4961	−0.2302	0.0625	0.1250
4	0.3750	0.4375	0.5000	−0.2302	−0.0884	0.0625	0.0625
5	0.4375	0.4688	0.5000	−0.0884	−0.0142	0.0625	0.0313
6	0.4688	0.4844	0.5000	−0.0142	0.0238	0.0625	0.0156
7	0.4688	0.4766	0.4844	−0.0142	0.0047	0.0238	0.0078
8	0.4688	0.4727	0.4766	−0.0142	−0.0048	0.0047	0.0039
9	0.4727	0.4746	0.4766	−0.0048	−0.0000	0.0047	0.0020
10	0.4746	0.4756	0.4766	−0.0000	0.0023	0.0047	0.0010
11	0.4746	0.4751	0.4756	−0.0000	0.0011	0.0023	0.0005
12	0.4746	0.4749	0.4751	−0.0000	0.0006	0.0011	0.0002
13	0.4746	0.4747	0.4749	−0.0000	0.0003	0.0006	0.0001

The zero of f is approximately 0.47.

73. A spherical tank of radius 50 feet (see figure) will be two-thirds full when the depth of the fluid is $x + 50$ feet, where

$$3x^3 - 22{,}550x + 250{,}000 = 0.$$

Use the Bisection Method to approximate x to within 0.01 unit.

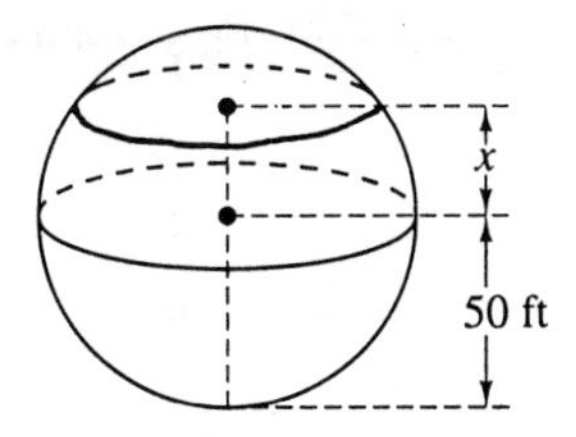

Solution:

Let $f(x) = 3x^3 - 22{,}500x + 250{,}000$

Since $f(11) = 6493$ and $f(12) = -14{,}816$, the zero is in the interval $[11, \ 12]$.

Iteration	a	c	b	$f(a)$	$f(c)$	$f(b)$	Error
1	11.0000	11.5000	12.0000	6493.0000	−4187.3750	−14816.0000	0.5000
2	11.0000	11.2500	11.5000	6493.0000	1146.4840	−4187.3750	0.2500
3	11.2500	11.3750	11.5000	1146.4840	−1522.0470	−4187.3750	0.1250
4	11.2500	11.3125	11.3750	1146.4840	−188.1719	−1522.0470	0.0625
5	11.2500	11.2813	11.3125	1146.4840	479.0625	−188.1719	0.0313
6	11.2813	11.2969	11.3125	479.0625	145.4063	−188.1719	0.0156
7	11.2969	11.3047	11.3125	145.4063	−21.3906	−188.1719	0.0078

The zero is approximately $x \approx 11.30$.

The depth of the tank is $x + 50 = 61.30$ feet when it is two-thirds full.

Practice Test for Chapter 4

1. Sketch the graph of $f(x) = x^2 - 6x + 5$ and identify the vertex and the intercepts.

2. Find the number of units x that produce a minimum cost C if $C = 0.01x^2 - 90x + 15{,}000$.

3. Find the quadratic function that has a maximum at (1, 7) and passes through the point (2, 5).

4. Find two quadratic functions that have x-intercepts (2, 0) and $(\frac{4}{3}, 0)$.

5. Use the Leading Coefficient Test to determine the right-hand and left hand behavior of the graph of the polynomial function $f(x) = -3x^5 + 2x^3 - 17$.

6. Find all the real zeros of $f(x) = x^5 - 5x^3 + 4x$.

7. Find a polynomial function with 0, 3, and -2 as zeros.

8. Sketch $f(x) = x^3 - 12x$.

9. Divide $3x^4 - 7x^2 + 2x - 10$ by $x - 3$ using long division.

10. Divide $x^3 - 11$ by $x^2 + 2x - 1$.

11. Use synthetic division to divide $3x^5 + 13x^4 + 12x - 1$ by $x + 5$.

12. Use synthetic division to find $f(-6)$ when $f(x) = 7x^3 + 40x^2 - 12x + 15$.

13. Find the real zeros of $f(x) = x^3 - 19x - 30$.

14. Find the real zeros of $f(x) = x^4 + x^3 - 8x^2 - 9x - 9$.

15. List all the possible rational zeros of the function $f(x) = 6x^3 - 5x^2 + 4x - 15$.

16. Find the rational zeros of the polynomial $f(x) = x^3 - \frac{20}{3}x^2 + 9x - \frac{10}{3}$.

17. Write $f(x) = x^4 + x^3 + 3x^2 + 5x - 10$ as a product of linear factors.

18. Find a polynomial with real coefficients that has 2, $3 + i$, and $3 - i$ as zeros.

19. Use synthetic division to show that $3i$ is a zero of $f(x) = x^3 + 4x^2 + 9x + 36$.

20. Find the zero of the function $f(x) = x^3 + 2x - 1$ in the interval [0, 1], accurate to within 0.001 unit.

CHAPTER 5

Rational Functions and Conic Sections

SECTION 5.1

Rational Functions

- You should know the following basic facts about rational functions.
  (a) A function of the form $f(x) = P(x)/Q(x)$, $Q(x) \neq 0$, where $P(x)$ and $Q(x)$ are polynomials, is called a rational function.
  (b) The domain of a rational function is the set of all real numbers except those which make the denominator zero.
  (c) If $f(x) = P(x)/Q(x)$ is in reduced form, and a is a value such that $Q(a) = 0$, then the line $x = a$ is a vertical asymptote of the graph of f.
  (d) The line $y = b$ is a horizontal asymptote of the graph of f if $f(x) \to b$ as $x \to \infty$ or $x \to -\infty$.
  (e) If $f(x) = P(x)/Q(x) = mx + b + R(x)/Q(x)$, then the line $y = mx + b$ is a slant asymptote of the graph of f.
- Be able to graph rational functions.

Solutions to Selected Exercises

3. Match $f(x) = (x+1)/x$ with its graph.

Solution:

$$f(x) = \frac{x+1}{x}$$

Vertical asymptote: $x = 0$

Horizontal asymptote: $y = 1$

x-intercept: $(-1, 0)$

Matches graph (a)

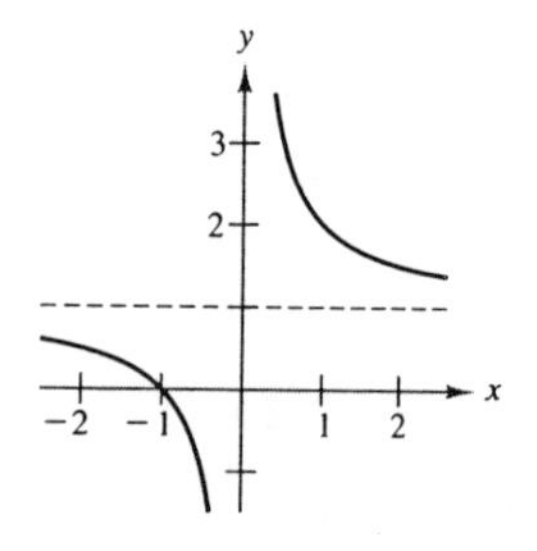

7. Match $f(x) = (x^2+1)/x$ with its graph.

Solution:

$$f(x) = \frac{x^2+1}{x} = x + \frac{1}{x}$$

Vertical asymptote: $x = 0$

Slant asymptote: $y = x$

No intercepts

Matches graph (h)

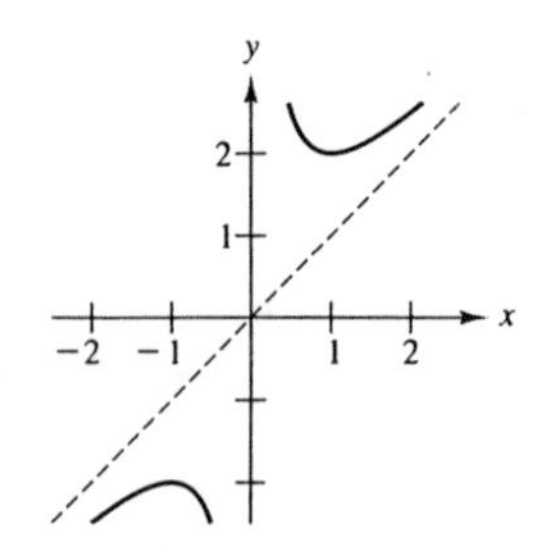

11. Find the domain of the following function and identify any horizontal, vertical, or slant asymptotes.

$$f(x) = \frac{2+x}{2-x}$$

Solution:

$$f(x) = \frac{2+x}{2-x} = \frac{x+2}{-x+2}$$

Domain: all real numbers except 2

Vertical asymptote: $x = 2$

Horizontal asymptote: $y = -1$

[degree of $p(x)$ = degree of $q(x)$]

15. Find the domain of the following function and identify any horizontal, vertical, or slant asymptotes.

$$f(x) = \frac{3x^2+1}{x^2+9}$$

Solution:

$$f(x) = \frac{3x^2+1}{x^2+9}$$

Domain: all real numbers

Horizontal asymptotes: $y = 3$

[degree of $p(x)$ = degree of $q(x)$]

19. Use the graph of $y = \frac{1}{x}$ to sketch the graph of $y = f(x) + 1 = \frac{1}{x} + 1$.

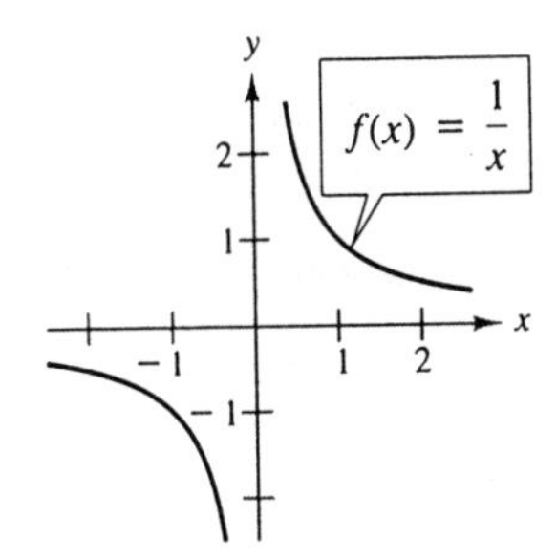

Solution:

$$y = f(x) + 1 = \frac{1}{x} + 1$$

Vertical shift one unit upward

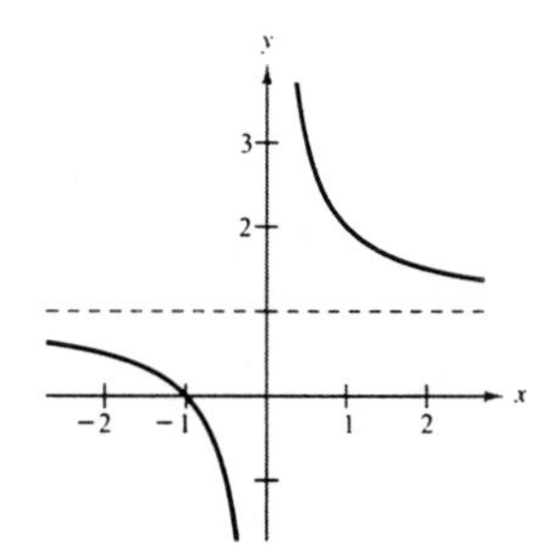

23. Use the graph of $f(x) = \frac{4}{x^2}$ to sketch the graph of $y = f(x) - 2 = \frac{4}{x^2} - 2$.

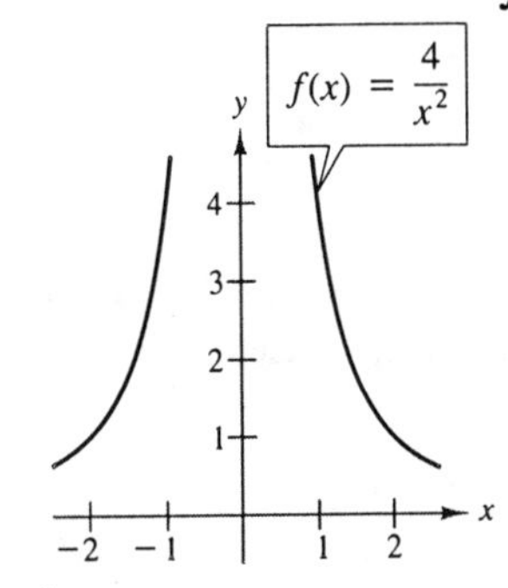

Solution:

$$y = f(x) - 2 = \frac{4}{x^2} - 2$$

Vertical shift two units downward

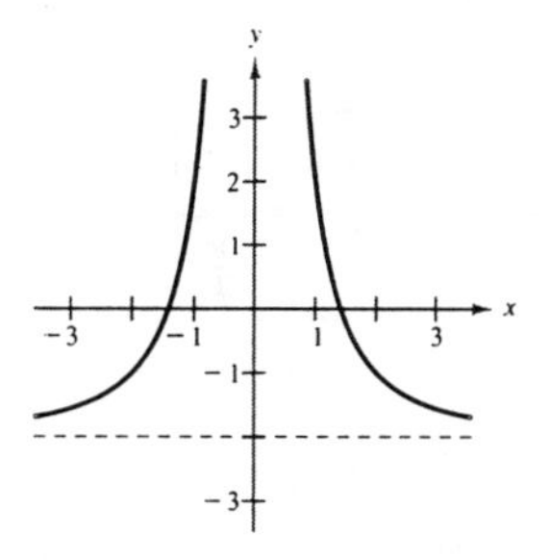

27. Use the graph of $g(x) = \dfrac{8}{x^3}$ to sketch the graph of $y = g(x+2) = \dfrac{8}{(x+2)^3}$.

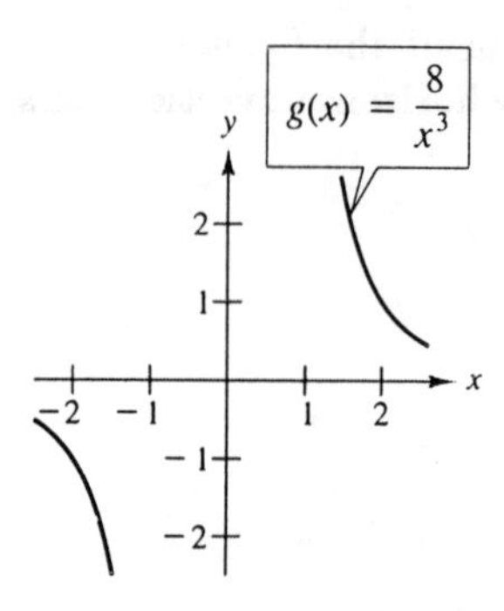

Solution:

$$y = g(x+2) = \frac{8}{(x+2)^3}$$

Horizontal shift two units to the left

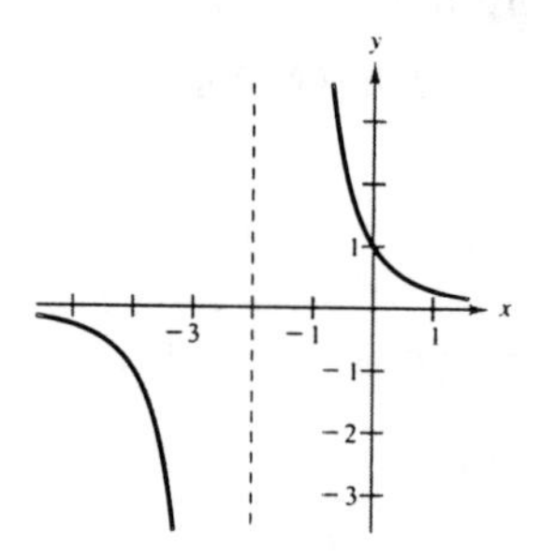

33. Sketch the graph of the following rational function. As sketching aids, check for intercepts, symmetry, vertical asymptotes, and horizontal asymptotes.

$$h(x) = \frac{-1}{x+2}$$

Solution:

Vertical asymptote: $x = -2$

Horizontal asymptote: $y = 0$

[degree $p(x)$ < degree $q(x)$]

y-intercept: $(0, -\frac{1}{2})$

x	-4	-3	-1	0
y	$\frac{1}{2}$	1	-1	$-\frac{1}{2}$

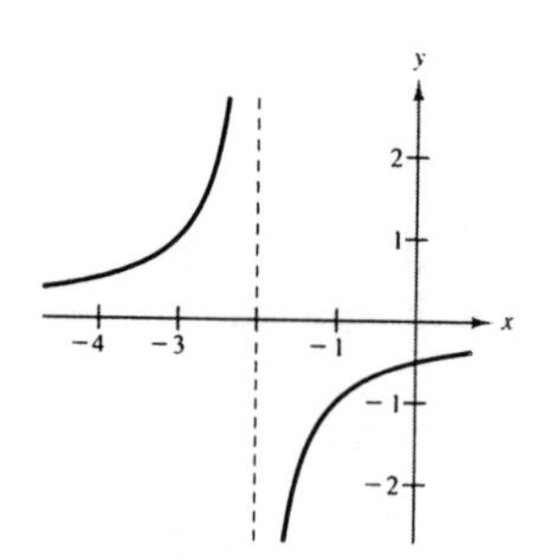

35. Sketch the graph of the following rational function. As sketching aids, check for intercepts, symmetry, vertical asymptotes, and horizontal asymptotes.

$$f(x) = \frac{x+1}{x+2}$$

Solution:

$$f(x) = \frac{x+1}{x+2}$$

Vertical asymptote: $x = -2$

Horizontal asymptote: $y = 1$

$[\text{degree } p(x) = \text{degree } q(x)]$

x-intercept: $(-1, 0)$

y-intercept: $(0, \frac{1}{2})$

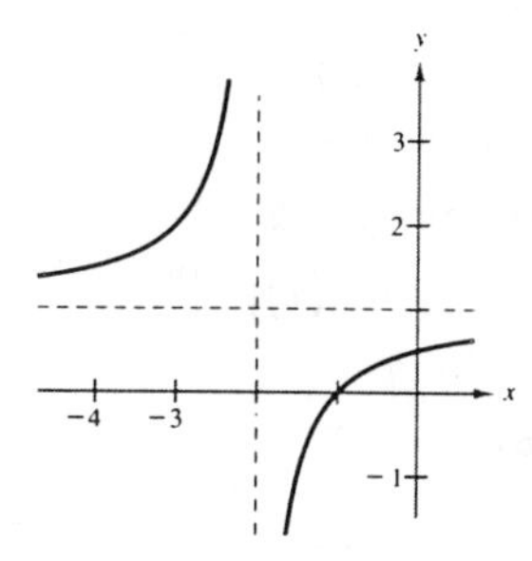

x	-4	-3	-1	0	1
y	$\frac{3}{2}$	2	0	$\frac{1}{2}$	$\frac{2}{3}$

39. Sketch the graph of the following rational function. As sketching aids, check for intercepts, symmetry, vertical asymptotes, and horizontal asymptotes.

$$f(t) = \frac{3t+1}{t}$$

Solution:

$$f(t) = \frac{3t+1}{t}$$

Vertical asymptote: $t = 0$

Horizontal asymptote: $y = 3$

$[\text{degree } p(x) = \text{degree } q(x)]$

t-intercept: $(-\frac{1}{3}, 0)$

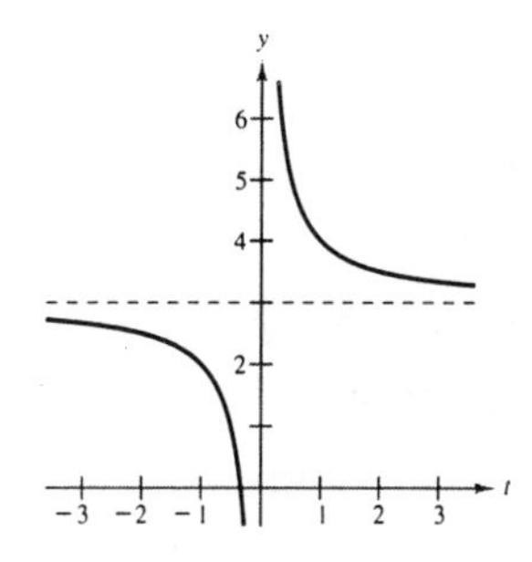

x	-2	-1	1	2
y	$\frac{5}{2}$	2	4	$\frac{7}{2}$

43. Sketch the graph of the following rational function. As sketching aids, check for intercepts, symmetry, vertical asymptotes, and horizontal asymptotes.

$$C(x) = \frac{5+2x}{1+x}$$

Solution:

$$C(x) = \frac{5+2x}{1+x} = \frac{2x+5}{x+1}$$

Vertical asymptote: $x = -1$

Horizontal asymptote: $y = 2$

[degree $p(x)$ = degree $q(x)$]

x-intercept: $(-\frac{5}{2}, 0)$

y-intercept: $(0, 5)$

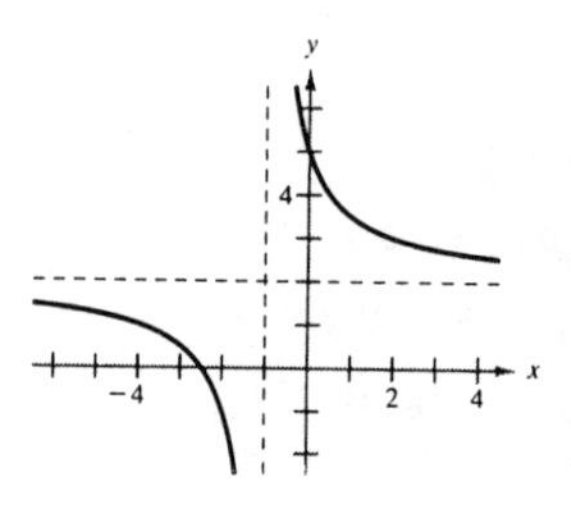

x	-3	-2	0	1	2
y	$\frac{1}{2}$	-1	5	$\frac{7}{2}$	3

47. Sketch the graph of the following rational function. As sketching aids, check for intercepts, symmetry, vertical asymptotes, and horizontal asymptotes.

$$h(x) = \frac{x^2}{x^2-9}$$

Solution:

$$h(x) = \frac{x^2}{x^2-9}$$

Vertical asymptotes: $x = \pm 3$

Horizontal asymptote: $y = 1$

[degree $p(x)$ = degree $q(x)$]

Intercept: $(0, 0)$

y-axis symmetry

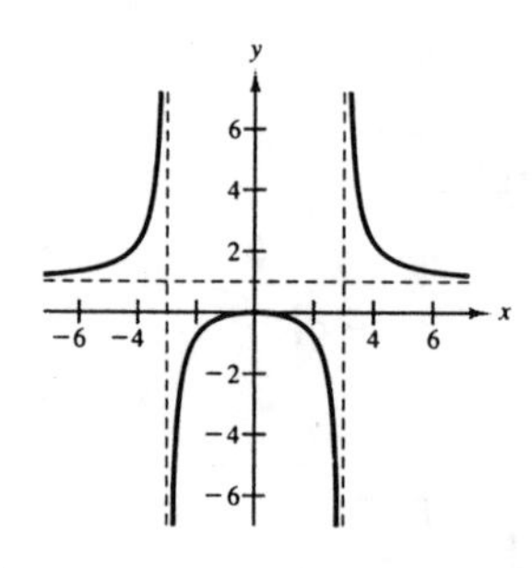

x	± 5	± 4	± 2	± 1	0
y	$\frac{25}{16}$	$\frac{16}{7}$	$-\frac{4}{5}$	$-\frac{1}{8}$	0

51. Sketch the graph of the following rational function. As sketching aids, check for intercepts, symmetry, vertical asymptotes, and horizontal asymptotes.

$$f(x) = -\frac{1}{(x-2)^2} + 3$$

Solution:

$$f(x) = \frac{-1 + 3(x-2)^2}{(x-2)^2} = \frac{3x^2 - 12x + 11}{x^2 - 4x + 4}$$

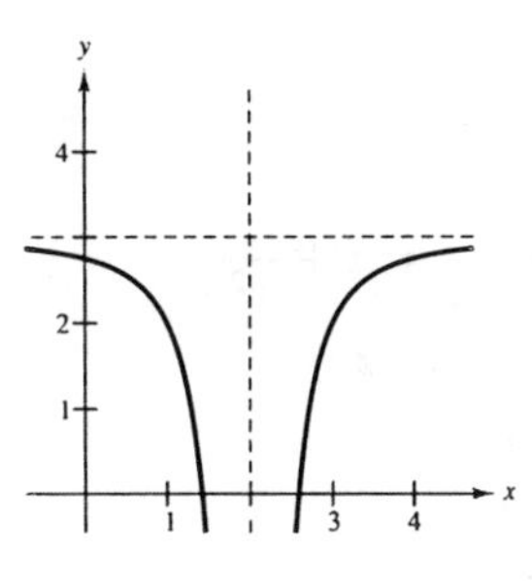

Vertical asymptote: $x = 2$

Horizontal asymptote: $y = 3$

[degree $p(x)$ = degree $q(x)$]

y-intercept: $(0, \frac{11}{4})$

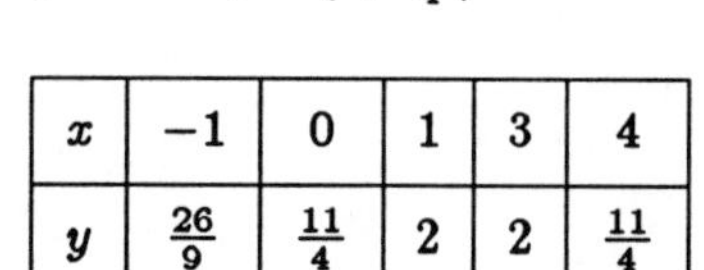

x	-1	0	1	3	4
y	$\frac{26}{9}$	$\frac{11}{4}$	2	2	$\frac{11}{4}$

55. Sketch the graph of the following rational function. As sketching aids, check for intercepts, symmetry, vertical asymptotes, and slant asymptotes.

$$f(x) = \frac{2x^2 + 1}{x}$$

Solution:

$$f(x) = \frac{2x^2 + 1}{x} = 2x + \frac{1}{x}$$

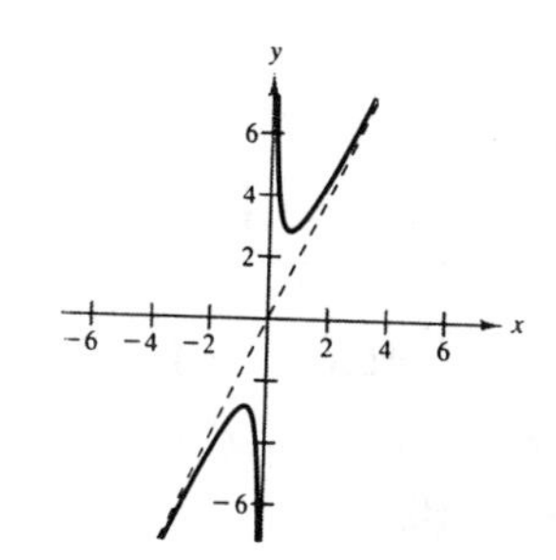

Vertical asymptote: $x = 0$

Slant asymptote: $y = 2x$

Origin symmetry

x	-3	-2	-1	$-\frac{1}{2}$	$\frac{1}{2}$	1	2	3
y	$-\frac{19}{3}$	$-\frac{9}{2}$	-3	-3	3	3	$\frac{9}{2}$	$\frac{19}{3}$

59. Sketch the graph of the following rational function. As sketching aids, check for intercepts, symmetry, vertical asymptotes, and slant asymptotes.

$$f(x) = \frac{x^3}{x^2 - 1} \quad \text{[Long Division of Polynomials]}$$

Solution:

$$f(x) = \frac{x^3}{x^2 - 1} = x + \frac{x}{x^2 - 1}$$

Vertical asymptotes: $x = \pm 1$

Slant asymptote: $y = x$

Intercept: (0, 0)

Origin symmetry

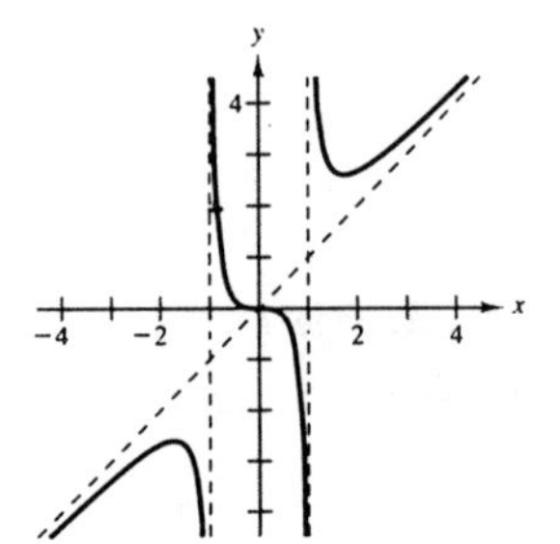

x	-3	-2	$-\frac{1}{2}$	0	$\frac{1}{2}$	2	3
y	$-\frac{27}{8}$	$-\frac{8}{3}$	$\frac{1}{6}$	0	$-\frac{1}{6}$	$\frac{8}{3}$	$\frac{27}{8}$

61. Sketch the graph of the following rational function. As sketching aids, check for intercepts, symmetry, vertical asymptotes, and slant asymptotes.

$$f(x) = \frac{x^2 - x + 1}{x - 1} \quad \text{[Long Division of Polynomials]}$$

Solution:

$$f(x) = \frac{x^2 - x + 1}{x - 1} = x + \frac{1}{x - 1}$$

Vertical asymptote: $x = 1$

Slant asymptote: $y = x$

y-intercept: (0, −1)

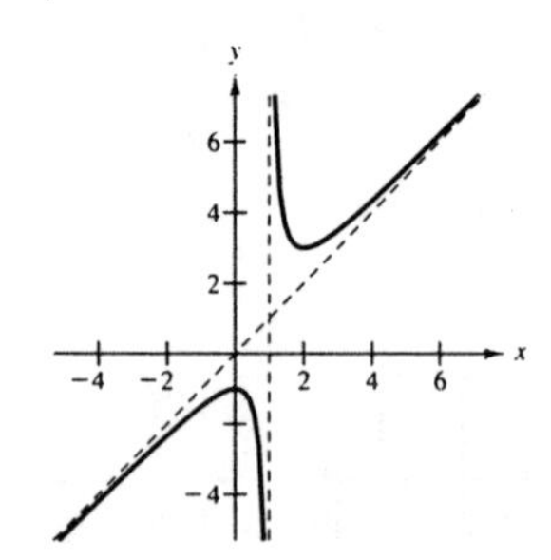

x	-2	-1	0	2	3	4
y	$-\frac{7}{3}$	$-\frac{3}{2}$	-1	3	$\frac{7}{2}$	$\frac{13}{3}$

67. The game commission introduces 50 deer into newly acquired state game lands. The population of the herd is

$$N = \frac{10(5 + 3t)}{1 + 0.04t}, \quad 0 \le t$$

where t is time in years (see figure).

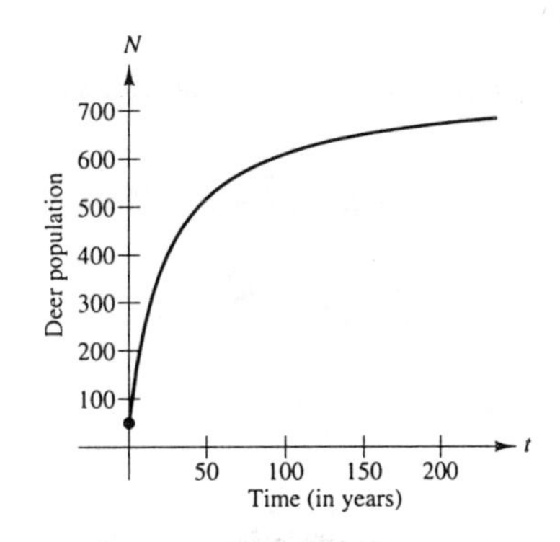

(a) Find the population when t is 5, 10, and 25.

(b) What is the limiting size of the herd as time increases?

Solution:

$$N = \frac{10(5 + 3t)}{1 + 0.04t} = \frac{30t + 50}{0.04t + 1}$$

(a) $N(5) \approx 167$ deer

$N(10) = 250$ deer

$N(25) = 400$ deer

(b) The herd is limited by a horizontal asymptote:

$$N = \frac{30}{0.04} = 750 \text{ deer.}$$

71. A right triangle is formed in the first quadrant by the x-axis, the y-axis, and a line segment through the point $(2, 3)$, as shown in the figure.

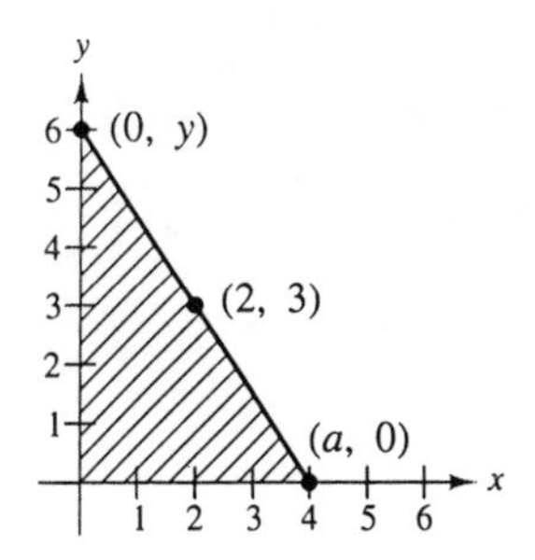

(a) Show that an equation of the line segment is

$$y = \frac{3(x - a)}{2 - a}, \quad 0 \le x \le a.$$

(b) Show that the area of the triangle is

$$A = \frac{-3a^2}{2(2 - a)}.$$

(c) Sketch the graph of the area function of part (b), and from the graph estimate the value of a that yields a minimum area.

71. --CONTINUED--

Solution:

(a) The line passes through the points $(a, 0)$ and $(2, 3)$ and has a slope of

$$m = \frac{3-0}{2-a} = \frac{3}{2-a}$$

$$y - 0 = \frac{3}{2-a}(x-a) \qquad \text{By the point slope equation}$$

$$y = \frac{3(x-a)}{2-a}$$

(b) The area of a triangle is $A = \frac{1}{2}bh$.

$b = a$

$h = y$ when $x = 0$, so $h = \frac{3(0-a)}{2-a} = \frac{-3a}{2-a}$

$$A = \left(\frac{1}{2}a\right)\left(\frac{-3a}{2-a}\right) = \frac{-3a^2}{2(2-a)}$$

(c) $A = \frac{-3a^2}{2(2-a)} = \frac{-3a^2}{-2a+4} = \frac{3}{2}a + 3 + \frac{12}{2a-4} = \frac{3}{2}a + 3 + \frac{6}{a-2}, \quad a > 2$

Vertical asymptote: $a = 2$

Slant asymptote: $A = \frac{3}{2}a + 3$

A is minimum when $a \approx 4$.

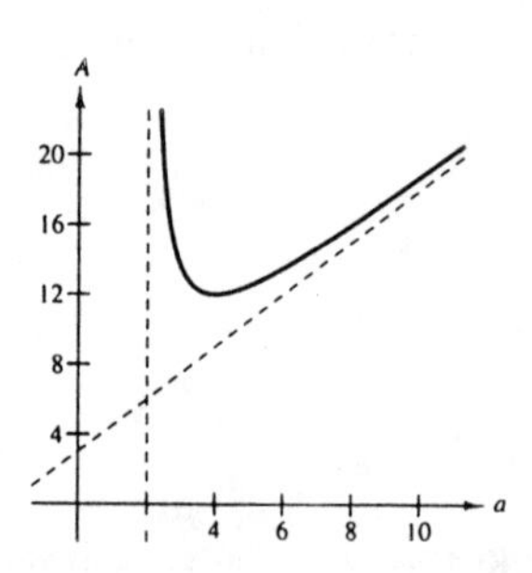

SECTION 5.2

Partial Fractions

- You should know how to decompose a rational function $\dfrac{N(x)}{D(x)}$ into partial fractions.

  (a) If the fraction is improper, divide to obtain

  $$\frac{N(x)}{D(x)} = p(x) + \frac{N_1(x)}{D(x)}$$

  where $p(x)$ is a polynomial.

  (b) Factor the denominator completely into linear and irreducible (over the reals) quadratic factors.

  (c) For each factor of the form $(px+q)^m$, the partial fraction decomposition includes the terms

  $$\frac{A_1}{(px+q)} + \frac{A_2}{(px+q)^2} + \cdots + \frac{A_m}{(px+q)^m}.$$

  (d) For each factor of the form $(ax^2+bx+c)^n$, the partial fraction decomposition includes the terms

  $$\frac{B_1x+C_1}{ax^2+bx+c} + \frac{B_2x+C_2}{(ax^2+bx+c)^2} + \cdots + \frac{B_nx+C_n}{(ax^2+bx+c)^n}.$$

- You should know how to determine the values of the constants in the numerators.

  (a) Set $\dfrac{N_1(x)}{D(x)} =$ partial fraction decomposition.

  (b) Multiply both sides by $D(x)$. This is called the basic equation.

  (c) For distinct linear factors, substitute the roots of the distinct linear factors into the basic equation.

  (d) For repeated linear factors, use the coefficients found in part (c) to rewrite the basic equation. Then use other values of x to solve for the remaining coefficients.

  (e) For quadratic factors, expand the basic equation, collect like terms, and then equate the coefficients of like powers.

Solutions to Selected Exercises

3. Write the partial fraction decomposition for the rational expression

$$\frac{1}{x^2+x}.$$

Solution:

Since $x^2+x=x(x+1)$,

$$\frac{1}{x^2+x} = \frac{A}{x}+\frac{B}{x+1}$$

$$1 = A(x+1)+Bx. \qquad \text{Basic Equation}$$

Let $x=0$: $\quad 1=A$

Let $x=-1$: $\quad 1=-B \Rightarrow B=-1$

Thus, $\dfrac{1}{x^2+x}=\dfrac{1}{x}-\dfrac{1}{x+1}.$

7. Write the partial fraction decomposition for the rational expression

$$\frac{3}{x^2+x-2}.$$

Solution:

Since $x^2+x-2=(x-1)(x+2)$,

$$\frac{3}{x^2+x-2} = \frac{A}{x-1}+\frac{B}{x+2}$$

$$3 = A(x+2)+B(x-1). \qquad \text{Basic Equation}$$

Let $x=1$: $\quad 3=3A$

$\quad 1=A$

Let $x=-2$: $\quad 3=-3B$

$\quad -1=B$

Thus, $\dfrac{3}{x^2+x-2}=\dfrac{1}{x-1}-\dfrac{1}{x+2}.$

11. Write the partial fraction decomposition for the rational expression $\dfrac{x^2+12x+12}{x^3-4x}$.

Solution:

Since $x^3-4x = x(x+2)(x-2)$,

$$\frac{x^2+12x+12}{x^3-4x} = \frac{A}{x} + \frac{B}{x+2} + \frac{C}{x-2}$$

$$x^2+12x+12 = A(x+2)(x-2) + Bx(x-2) + Cx(x+2). \quad \text{Basic Equation}$$

Let $x=0$: $\quad 12 = -4A$

$-3 = A$

Let $x=-2$: $\quad -8 = 8B$

$-1 = B$

Let $x=2$: $\quad 40 = 8C$

$5 = C$

Thus, $\dfrac{x^2+12x+12}{x^3-4x} = -\dfrac{3}{x} - \dfrac{1}{x+2} + \dfrac{5}{x-2}$.

13. Write the partial fraction decomposition for the rational expression $\dfrac{4x^2+2x-1}{x^2(x+1)}$.

Solution:

$$\frac{4x^2+2x-1}{x^2(x+1)} = \frac{A}{x} + \frac{B}{x^2} + \frac{C}{x+1}$$

$$4x^2+2x-1 = Ax(x+1) + B(x+1) + Cx^2 \quad \text{Basic Equation}$$

Let $x=0$: $\quad -1 = B$

Let $x=-1$: $\quad 1 = C$

Let $x=1$: $\quad 5 = 2A+2B+C$

$5 = 2A-2+1$

$6 = 2A$

$3 = A$

Thus, $\dfrac{4x^2+2x-1}{x^2(x+1)} = \dfrac{3}{x} - \dfrac{1}{x^2} + \dfrac{1}{x+1}$.

Note: $x^2 = (x-0)^2$ and is a linear factor squared. It is not an irreducible quadratic factor. Do not write $\dfrac{Bx+C}{x^2}$ in the partial fraction decomposition.

19. Write the partial fraction decomposition for the rational expression

$$\frac{x^2-1}{x(x^2+1)}.$$

Solution:

$$\frac{x^2-1}{x(x^2+1)} = \frac{A}{x} + \frac{Bx+C}{x^2+1}$$

$$x^2-1 = A(x^2+1) + (Bx+C)x \qquad \text{Basic Equation}$$

Let $x = 0$: $\quad -1 = A$

$$x^2-1 = Ax^2 + A + Bx^2 + Cx = -x^2 - 1 + Bx^2 + Cx = x^2(B-1) + Cx - 1$$

Equating coefficients of like powers,

$$1 = B - 1$$

$$2 = B \quad \text{and} \quad 0 = C$$

Thus, $\dfrac{x^2-1}{x(x^2+1)} = -\dfrac{1}{x} + \dfrac{2x}{x^2+1}.$

23. Write the partial fraction decomposition for the rational expression

$$\frac{x}{16x^4-1}.$$

Solution:

Since $16x^4 - 1 = (4x^2+1)(2x+1)(2x-1)$,

$$\frac{x}{16x^4-1} = \frac{A}{2x+1} + \frac{B}{2x-1} + \frac{Cx+D}{4x^2+1}$$

$$x = A(2x-1)(4x^2+1) + B(2x+1)(4x^2+1)$$
$$+ (Cx+D)(2x+1)(2x-1). \qquad \text{Basic Equation}$$

Let $x = -\frac{1}{2}$: $\quad -\frac{1}{2} = -4A$

$$\frac{1}{8} = A$$

Let $x = \frac{1}{2}$: $\quad \frac{1}{2} = 4B$

$$\frac{1}{8} = B$$

Let $x = 0$: $\quad 0 = -A + B - D$

$$0 = -\frac{1}{8} + \frac{1}{8} - D$$

$$0 = D$$

–CONTINUED ON NEXT PAGE–

23. –CONTINUED–

Let $x = 1$:
$$\begin{aligned} 1 &= 5A + 15B + 3C + 3D \\ 1 &= \frac{5}{8} + \frac{15}{8} + 3C + 0 \\ 1 &= \frac{20}{8} + 3C \\ -\frac{3}{2} &= 3C \\ -\frac{1}{2} &= C \end{aligned}$$

Thus, $\dfrac{x}{16x^4 - 1} = \dfrac{1/8}{2x+1} + \dfrac{1/8}{2x-1} - \dfrac{x/2}{4x^2+1} = \dfrac{1}{8}\left[\dfrac{1}{2x+1} + \dfrac{1}{2x-1} - \dfrac{4x}{4x^2+1}\right].$

27. Write the partial fraction decomposition for the rational expression

$$\frac{x^2+5}{(x+1)(x^2-2x+3)}.$$

Solution:

$$\frac{x^2+5}{(x+1)(x^2-2x+3)} = \frac{A}{x+1} + \frac{Bx+C}{x^2-2x+3}$$

$$x^2 + 5 = A(x^2 - 2x + 3) + (Bx + C)(x + 1) \qquad \text{Basic Equation}$$

Let $x = -1$: $6 = 6A$

$1 = A$

$$\begin{aligned} x^2 + 5 &= x^2 - 2x + 3 + Bx^2 + Bx + Cx + C \\ &= x^2(1 + B) + x(-2 + B + C) + (3 + C) \end{aligned}$$

Equating coefficients of like powers,

$1 = 1 + B$, $\quad 0 = -2 + B + C$, $\quad$ and $\quad 5 = 3 + C$

$0 = B \qquad 0 = -2 + 0 + C \qquad 2 = C$

$2 = C$

Thus, $\dfrac{x^2+5}{(x+1)(x^2-2x+3)} = \dfrac{1}{x+1} + \dfrac{2}{x^2-2x+3}.$

31. Write the partial fraction decomposition for the rational expression

$$\frac{x^4}{(x-1)^3}.$$

Solution:

$$\frac{x^4}{(x-1)^3} = \frac{x^4}{x^3-3x^2+3x-1}$$

$$= x+3+\frac{6x^2-8x+3}{(x-1)^3} \qquad \text{By long division of polynomials}$$

$$\frac{6x^2-8x+3}{(x-1)^3} = \frac{A}{x-1}+\frac{B}{(x-1)^2}+\frac{C}{(x-1)^3}$$

$$6x^2-8x+3 = A(x-1)^2+B(x-1)+C \qquad \text{Basic Equation}$$

Let $x=1$: $\quad 1 = C$

$$6x^2-8x+3 = Ax^2-2Ax+A+Bx-B+1$$

$$6x^2-8x+3 = Ax^2+(-2A+B)x+(A-B+1)$$

Equating coefficients of like powers

$$6 = A \quad -8 = -2A+B \text{ and } 3 = A-B+1$$

$$-8 = -12+B \qquad 3 = 6-B+1$$

$$4 = B \qquad 4 = B$$

Thus, $\dfrac{x^4}{(x-1)^3} = x+3+\dfrac{6}{x-1}+\dfrac{4}{(x-1)^2}+\dfrac{1}{(x-1)^3}.$

35. **Write the partial fraction decomposition for the rational expression**

$$\frac{1}{y(L-y)}, \qquad L \text{ is a constant.}$$

Solution:

$$\frac{1}{y(L-y)} = \frac{A}{y} + \frac{B}{L-y}$$

$$1 = A(L-y) + By \qquad \text{Basic Equation}$$

Let $y = 0$: $1 = LA$

$1/L = A$

Let $y = L$: $1 = LB$

$1/L = B$

Thus, $\dfrac{1}{y(L-y)} = \dfrac{1/L}{y} + \dfrac{1/L}{L-y} = \dfrac{1}{L}\left(\dfrac{1}{y} + \dfrac{1}{L-y}\right).$

SECTION 5.3

Conic Sections

You should know the following basic definitions of conic sections.

- A circle is the set of all points $(x,\ y)$ that are equidistant from a fixed point $(h,\ k)$.

  (a) Standard Equation: $(x-h)^2+(y-k)^2=r^2$

  (b) Center: $(h,\ k)$

  (c) Radius: r

- A parabola is the set of all points $(x,\ y)$ that are equidistant from a fixed line (directrix) and a fixed point (focus) not on the line.

  (a) Standard Equation with Vertex $(0, 0)$ and Directrix $y=-p$ (vertical axis): $x^2=4py$

  (b) Standard Equation with Vertex $(0, 0)$ and Directrix $x=-p$ (horizontal axis): $y^2=4px$

  (c) The focus lies on the axis p units (directed distance) from the vertex.

- An ellipse is the set of all points $(x,\ y)$ the sum of whose distances from two distinct fixed points (foci) is constant.

  (a) Standard Equation of an Ellipse with Center $(0, 0)$, Major Axis Length $2a$, and Minor Axis Length $2b$:

    1. Horizontal Major Axis: $\dfrac{x^2}{a^2}+\dfrac{y^2}{b^2}=1$

    2. Vertical Major Axis: $\dfrac{x^2}{b^2}+\dfrac{y^2}{a^2}=1$

  (b) The foci lie on the major axis, c units from the center, where a, b, and c are related by the equation $c^2=a^2-b^2$.

  (c) The vertices and endpoints of the minor axis are

    1. Horizontal Axis: $(\pm a,\ 0)$ and $(0,\ \pm b)$

    2. Vertical Axis: $(0,\ \pm a)$ and $(\pm b,\ 0)$

- A hyperbola is the set of all points (x, y) the difference of whose distances from two distinct fixed points (foci) is constant.

  (a) Standard Equation of a Hyperbola with Center (0, 0)

  1. Horizontal Transverse Axis: $\dfrac{x^2}{a^2} - \dfrac{y^2}{b^2} = 1$
  2. Vertical Transverse Axis: $\dfrac{y^2}{a^2} - \dfrac{x^2}{b^2} = 1$

  (b) The vertices and foci are a and c units from the center and $b^2 = c^2 - a^2$.

  (c) The asymptotes of the hyperbola are

  1. Horizontal Transverse Axis: $y = \pm\dfrac{b}{a}x$
  2. Vertical Transverse Axis: $y = \pm\dfrac{a}{b}x$

Solutions to Selected Exercises

7. Match the following equation with its graph.

$$\frac{x^2}{1} - \frac{y^2}{4} = 1$$

Solution:

This is the standard equation of a hyperbola vertices $(\pm 1, 0)$ and thus matches the graph shown in (e).

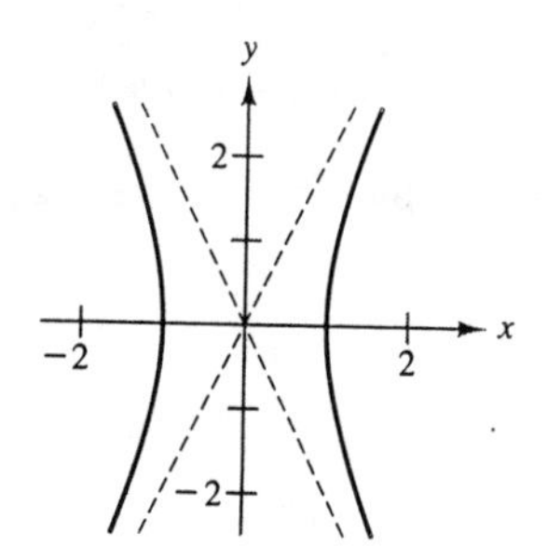

11. Find the vertex and focus of $y^2 = -6x$ and sketch its graph.

Solution:

$y^2 = -6x$ Horizontal axis

$y^2 = 4(-\frac{3}{2})x;\ \ p = -\frac{3}{2}$

Vertex: (0, 0)

Focus: $(-\frac{3}{2}, 0)$

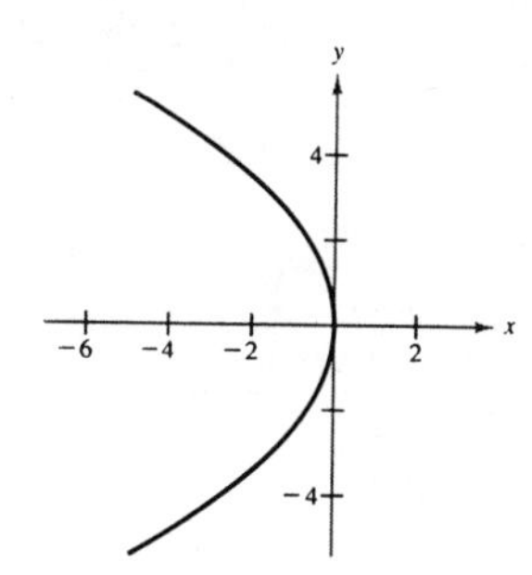

15. Find the vertex and focus of $y^2 - 8x = 0$ and sketch its graph.

Solution:

$$y^2 - 8x = 0$$

$$y^2 = 8x \qquad \text{Horizontal axis}$$

$$y^2 = 4(2)x; \quad p = 2$$

Vertex: $(0, 0)$

Focus: $(2, 0)$

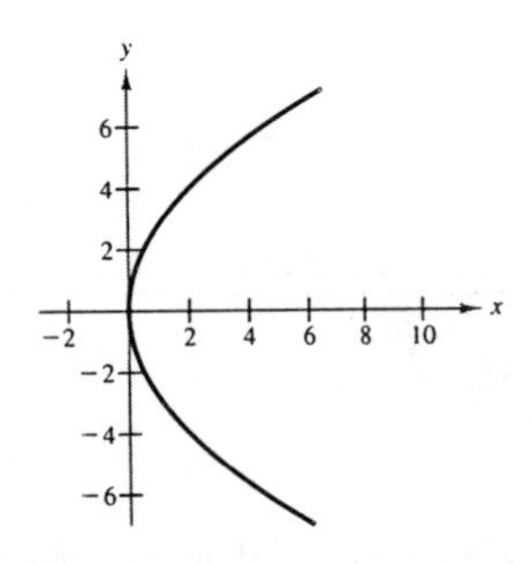

17. Find an equation of the parabola with vertex at the origin and focus $(0, -\frac{3}{2})$.

Solution:

Vertex: $(0, 0)$

Focus: $(0, -\frac{3}{2})$

$$p = -\frac{3}{2} \qquad \text{Vertical axis}$$

$$x^2 = 4py$$

$$x^2 = 4(-\tfrac{3}{2})y$$

$$x^2 = -6y$$

21. Find an equation of the parabola with vertex at the origin and directrix $y = -1$.

Solution:

Vertex: $(0, 0)$

Directrix: $y = -1$

$$p = 1, \qquad \text{Vertical axis}$$

$$x^2 = 4(1)y$$

$$x^2 = 4y$$

25. Find an equation of the parabola with vertex at the origin, horizontal axis, and passes through the point $(4, 6)$.

Solution:

Vertex: $(0, 0)$

Horizontal axis, passes through the point $(4, 6)$

$$y^2 = 4px \qquad \text{Since the axis is horizontal}$$

$$6^2 = 4p(4) \qquad \text{Since } (4, 6) \text{ is on the graph.}$$

$$36 = 16p$$

$$p = \frac{36}{16}$$

$$p = \frac{9}{4}$$

Thus, $y^2 = 4(\frac{9}{4})x$

$y^2 = 9x$.

29. Find the center and vertices of the following ellipse and sketch its graph.

$$\frac{x^2}{16} + \frac{y^2}{25} = 1$$

Solution:

Vertical major axis

$a = 5, \; b = 4$

Center: $(0, 0)$

Vertices: $(0, \pm 5)$

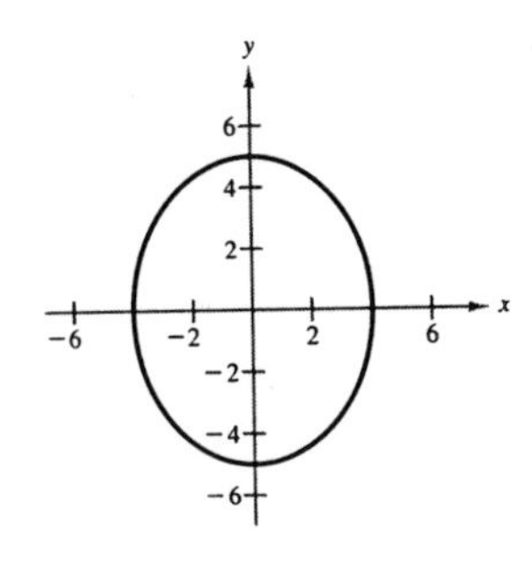

33. Find the center and vertices of $5x^2 + 3y^2 = 15$ and sketch its graph.

Solution:

$$5x^2 + 3y^2 = 15$$

$$\frac{x^2}{3} + \frac{y^2}{5} = 1$$

Vertical major axis

$a = \sqrt{5}, \; b = \sqrt{3}$

Center: $(0, 0)$

Vertices $(0, \pm\sqrt{5})$

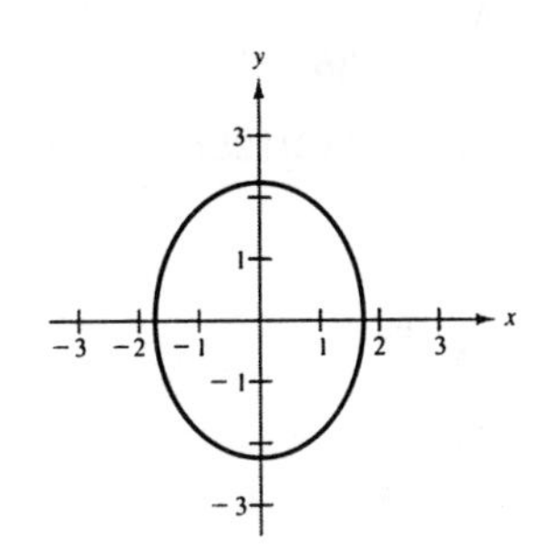

37. Find an equation of the ellipse with center at the origin, vertices at $(\pm 5, 0)$, and foci at $(\pm 2, 0)$.

Solution:

Vertices: $(5, 0), (-5, 0)$

Foci: $(2, 0), (-2, 0)$

$a = 5, \; c = 2, \; b = \sqrt{25 - 4} = \sqrt{21}$

Horizontal major axis

$$\frac{x^2}{a^2} + \frac{y^2}{b^2} = 1$$

$$\frac{x^2}{25} + \frac{y^2}{21} = 1$$

41. Find an equation of the ellipse with center at the origin, vertices at $(0, \pm 5)$, and passes through the point $(4, 2)$.

Solution:

Vertices: $(0, \pm 5)$; passes through $(4, 2)$

$a = 5$ Vertical major axis

$$\frac{x^2}{b^2} + \frac{y^2}{a^2} = 1$$

$$\frac{x^2}{b^2} + \frac{y^2}{25} = 1$$

$$\frac{16}{b^2} + \frac{4}{25} = 1 \qquad \text{Since } (4, 2) \text{ is on the graph.}$$

$$\frac{16}{b^2} = \frac{21}{25}$$

$$b^2 = \frac{(16)(25)}{21} = \frac{400}{21}$$

$$\frac{x^2}{400/21} + \frac{y^2}{25} = 1$$

$$\frac{21x^2}{400} + \frac{y^2}{25} = 1$$

45. Find the center and vertices of the following hyperbola and sketch its graph, using asymptotes as an aid.

$$\frac{y^2}{1} - \frac{x^2}{4} = 1$$

Solution:

$\dfrac{y^2}{1} - \dfrac{x^2}{4} = 1$; Vertical transverse axis

$a = 1,\ b = 2$

Center: $(0, 0)$

Vertices: $(0, \pm 1)$

Asymptotes: $y = \pm \dfrac{1}{2}x$

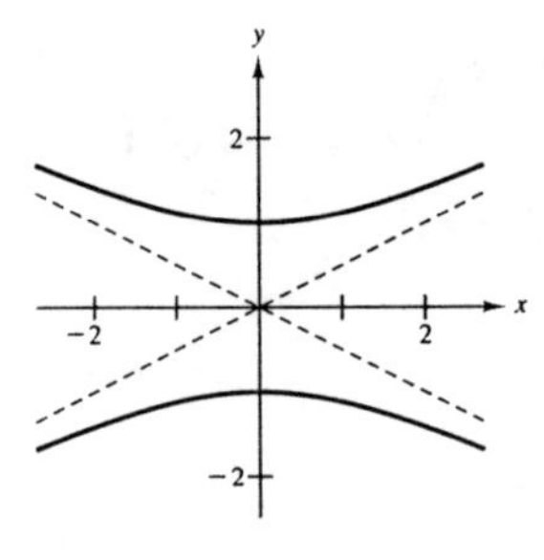

49. Find the center and vertices of $2x^2 - 3y^2 = 6$ and sketch its graph, using asymptotes as an aid.

Solution:

$$2x^2 - 3y^2 = 6$$

$$\frac{x^2}{3} - \frac{y^2}{2} = 1; \text{ Horizontal transverse axis}$$

$$a = \sqrt{3},\ b = \sqrt{2}$$

Center: $(0, 0)$

Vertices: $(\pm\sqrt{3}, 0)$

Asymptotes: $y = \pm\sqrt{2/3}x = \pm(\sqrt{6}/3)x$

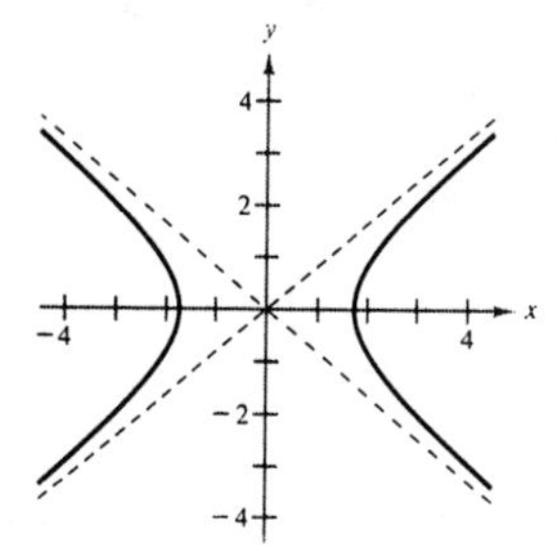

53. Find an equation of the hyperbola with center at the origin, vertices at $(\pm 1, 0)$ and asymptotes $y = \pm 3x$.

Solution:

Vertices: $(\pm 1, 0)$

Asymptotes: $y = \pm 3x$

Horizontal transverse axis

$a = 1$

$$3 = \frac{b}{a} = \frac{b}{1},\quad b = 3$$

$$\frac{x^2}{a^2} - \frac{y^2}{b^2} = 1$$

$$\frac{x^2}{1} - \frac{y^2}{9} = 1$$

57. Find an equation of the hyperbola with center at the origin, vertices at $(0, \pm 3)$ and passes through the point $(-2, 5)$.

Solution:

Vertices: $(0, \pm 3)$; passes through $(-2, 5)$

Vertical transverse axis; $a = 3$

$$\frac{y^2}{a^2} - \frac{x^2}{b^2} = 1$$

$$\frac{y^2}{9} - \frac{x^2}{b^2} = 1$$

$$\frac{25}{9} - \frac{4}{b^2} = 1 \qquad \text{Since } (-2, 5) \text{ is on the graph.}$$

$$-\frac{4}{b^2} = -\frac{16}{9}$$

$$b^2 = \frac{(4)(9)}{16} = \frac{9}{4}$$

$$\frac{y^2}{9} - \frac{x^2}{9/4} = 1$$

$$\frac{y^2}{9} - \frac{4x^2}{9} = 1$$

61. A fireplace arch is to be constructed in the shape of a semi-ellipse. The opening is to have a height of 2 feet at the center and a width of 5 feet along the base, as shown in the figure. The contractor draws the outline of the ellipse by the method shown in Example 2 (see textbook). Where should the tacks be placed and what should be the length of the piece of string?

Solution:

$2a = 5 \rightarrow a = 2.5, \quad b = 2,$

$c^2 = (2.5)^2 - (2)^2, \quad c = \sqrt{2.25} = 1.5$

The tacks should be placed at $(\pm 1.5, 0)$.

The length of the string should be $2a = 5$ feet.

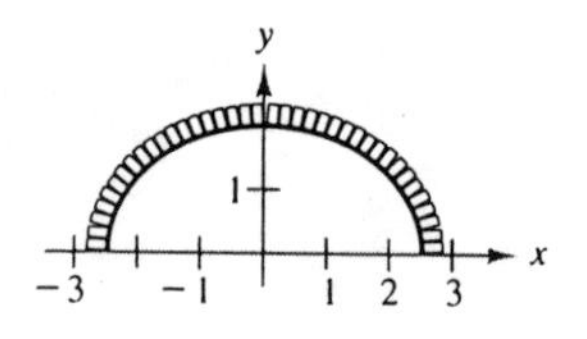

65. Sketch the graph of the ellipse $\frac{x^2}{4} + \frac{y^2}{1} = 1$, making use of the latus recta. (See Exercise 64.)

Solution:

$$\frac{x^2}{4} + \frac{y^2}{1} = 1$$

$a = 2,\ b = 1,\ c = \sqrt{3}$

Points on the ellipse: $(\pm 2,\ 0),\ (0, \pm 1)$

Length of latus recta:

$$\frac{2b^2}{a} = \frac{2(1)^2}{2} = 1$$

Additional points: $\left(-\sqrt{3},\ \pm\frac{1}{2}\right),\ \left(\sqrt{3},\ \pm\frac{1}{2}\right)$

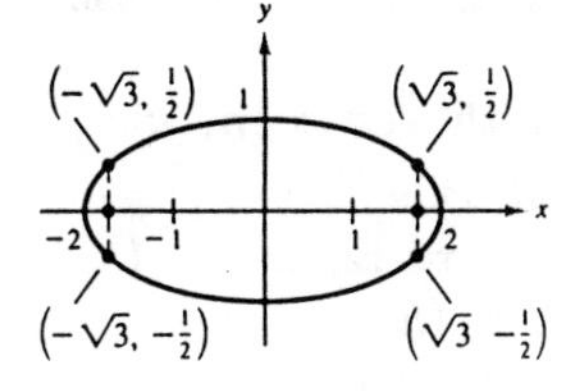

SECTION 5.4

Conic Sections and Translations

You should know the following basic facts about conic sections.

- **Parabola with Vertex (h, k)**

  (a) Vertical Axis

  1. **Standard Equation:** $(x-h)^2 = 4p(y-k)$
  2. **Focus:** $(h, k+p)$
  3. **Directrix:** $y = k - p$

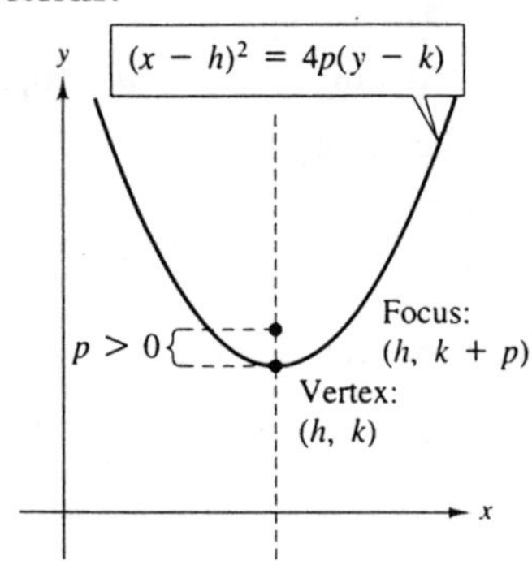

  (b) Horizontal Axis

  1. **Standard Equation:** $(y-k)^2 = 4p(x-h)$
  2. **Focus:** $(h+p, k)$
  3. **Directrix:** $x = h - p$

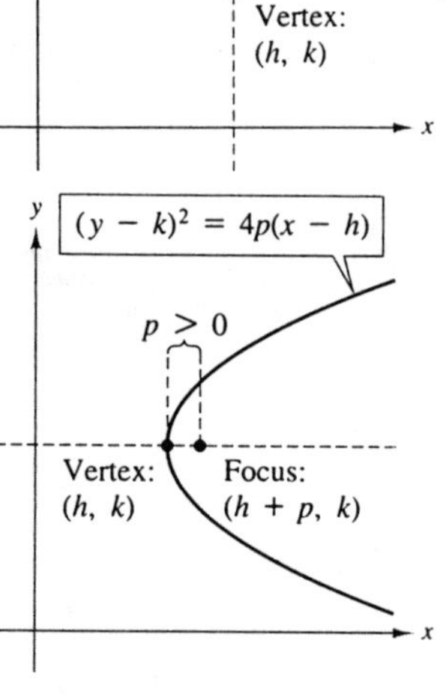

- **Circle with Center (h, k) and Radius r**

  Standard Equation: $(x-h)^2 + (y-k)^2 = r^2$

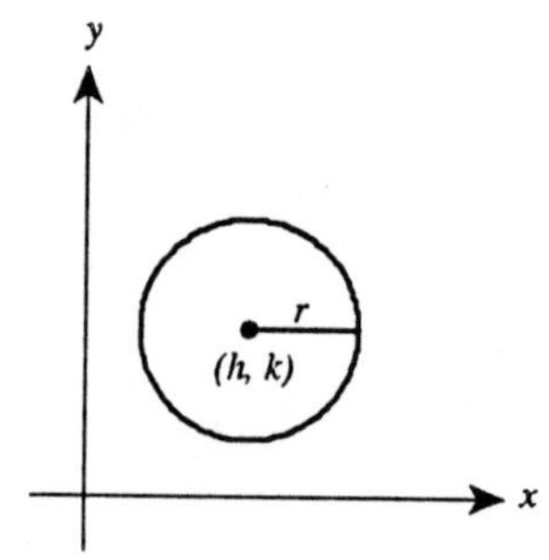

- **Ellipse with Center (h, k)**
  - (a) Horizontal Major Axis:
    1. Standard Equation:
       $$\frac{(x-h)^2}{a^2} + \frac{(y-k)^2}{b^2} = 1$$
    2. Vertices: $(h \pm a,\ k)$
    3. Foci: $(h \pm c,\ k)$
    4. Eccentricity: $e = \dfrac{c}{a}$
    5. $c^2 = a^2 - b^2$

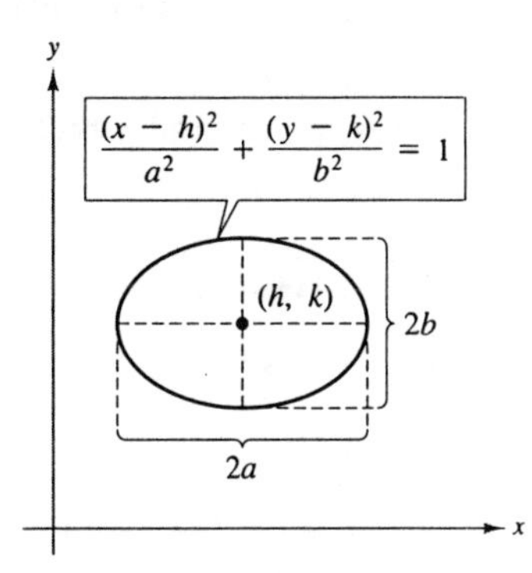

  - (b) Vertical Major Axis:
    1. Standard Equation:
       $$\frac{(x-h)^2}{b^2} + \frac{(y-k)^2}{a^2} = 1$$
    2. Vertices: $(h,\ k \pm a)$
    3. Foci: $(h,\ k \pm c)$
    4. Eccentricity: $e = \dfrac{c}{a}$
    5. $c^2 = a^2 - b^2$

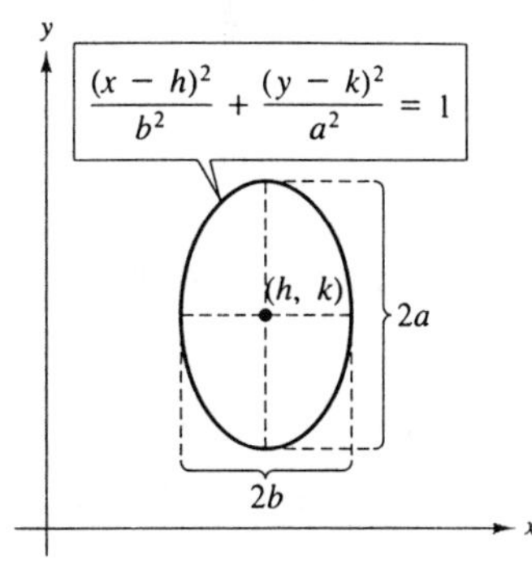

- **Hyperbola with Center (h, k)**
  - (a) Horizontal Transverse Axis:
    1. Standard Equation:
       $$\frac{(x-h)^2}{a^2} - \frac{(y-k)^2}{b^2} = 1$$
    2. Vertices: $(h \pm a,\ k)$
    3. Foci: $(h \pm c,\ k)$
    4. Asymptotes: $y - k = \pm\dfrac{b}{a}(x - h)$
    5. $c^2 = a^2 + b^2$

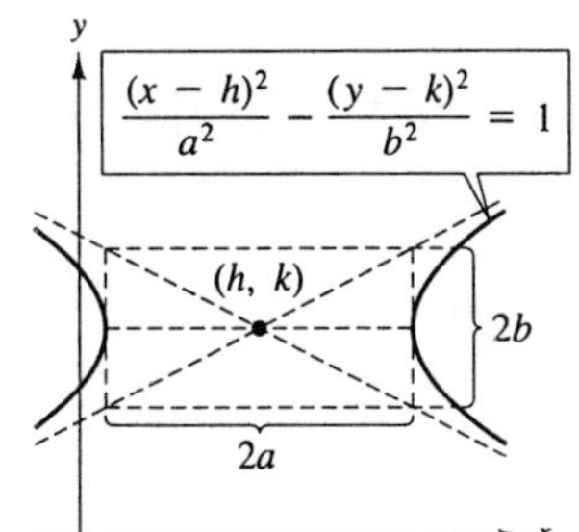

  - (b) Vertical Transverse Axis:
    1. Standard Equation:
       $$\frac{(y-k)^2}{a^2} - \frac{(x-h)^2}{b^2} = 1$$
    2. Vertices: $(h,\ k \pm a)$
    3. Foci: $(h,\ k \pm c)$
    4. Asymptotes: $y - k = \pm\dfrac{a}{b}(x - h)$
    5. $c^2 = a^2 + b^2$

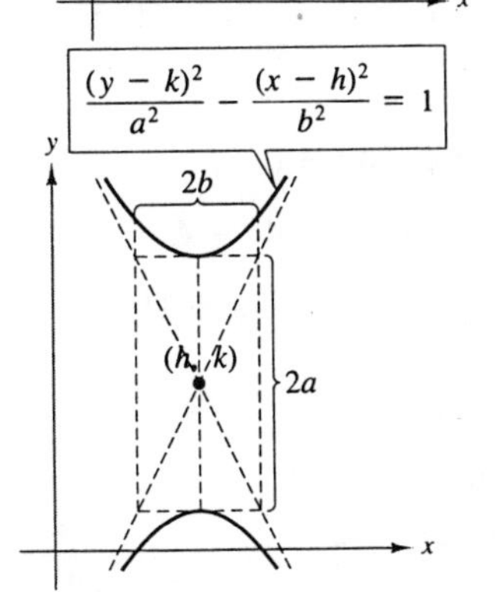

Solutions to Selected Exercises

3. Find the vertex, focus, and directrix of $(y+\frac{1}{2})^2 = 2(x-5)$, and sketch its graph.

Solution:

$$\left(y+\tfrac{1}{2}\right)^2 = 2(x-5)$$

$$\left(y+\tfrac{1}{2}\right)^2 = 4\left(\tfrac{1}{2}\right)(x-5); \quad p = \tfrac{1}{2}$$

Vertex: $(5, -\frac{1}{2})$
Focus: $(5+\frac{1}{2}, -\frac{1}{2}) = (\frac{11}{2}, -\frac{1}{2})$
Directrix: $x = 5 - \frac{1}{2} = \frac{9}{2}$

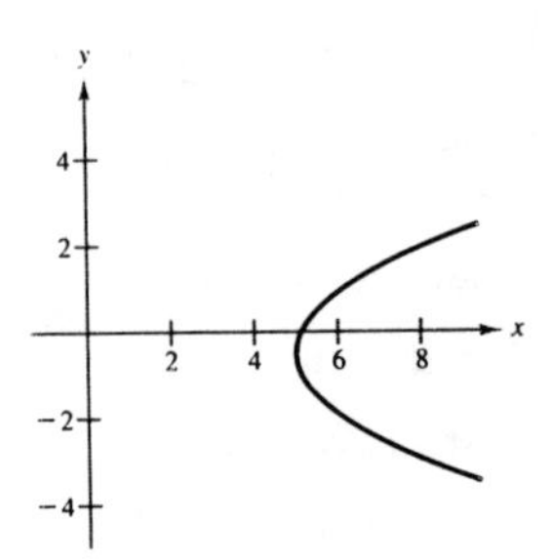

7. Find the vertex, focus, and directrix of $4x - y^2 - 2y - 33 = 0$, and sketch its graph.

Solution:

$$4x - y^2 - 2y - 33 = 0$$

$$4x - 33 + 1 = y^2 + 2y + 1$$

$$4x - 32 = (y+1)^2$$

$$(y+1)^2 = 4(x-8); \quad p = 1$$

Vertex: $(8, -1)$
Focus: $(9, -1)$
Directrix: $x = 7$

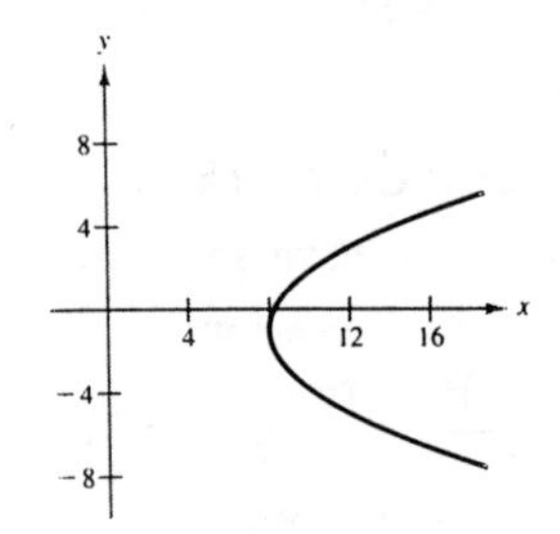

11. Find the vertex, focus, and directrix of $y^2 - 4y - 4x = 0$, and sketch its graph.

Solution:

$$y^2 - 4y - 4x = 0$$

$$y^2 - 4y + 4 = 4x + 4$$

$$(y-2)^2 = 4(x+1); \quad p = 1$$

Vertex: $(-1, 2)$
Focus: $(0, 2)$
Directrix: $x = -2$

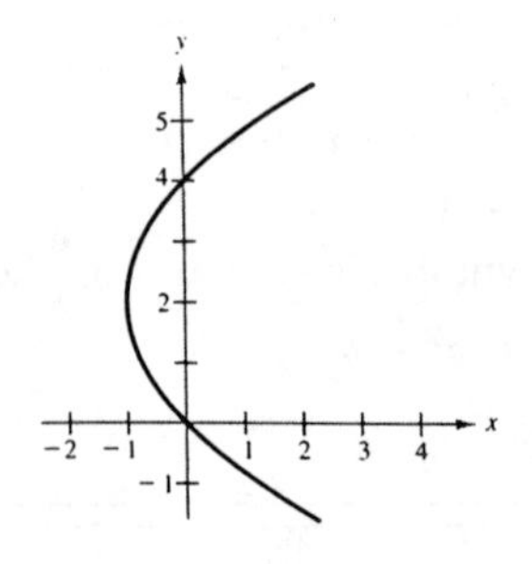

15. Find an equation of the parabola with vertex at $(0, 4)$ and directrix $y = 2$.

Solution:

Vertical axis

$$y = 4 - p = 2; \quad p = 2$$

$$(h, k) = (0, 4)$$

$$(x - 0)^2 = 4(2)(y - 4)$$

$$x^2 = 8(y - 4)$$

19. Find an equation of the parabola shown.

Solution:

Vertex: $(0, 4)$

Opens downward and passes through $(\pm 2, 0)$

$$(x - 0)^2 = 4p(y - 4)$$

$$x^2 = 4p(y - 4)$$

$$4 = 4p(0 - 4) \qquad \text{Since the graph passes through } (\pm 2, 0)$$

$$4 = -16p$$

$$p = -\tfrac{1}{4}$$

$$x^2 = 4(-\tfrac{1}{4})(y - 4)$$

$$x^2 = -(y - 4)$$

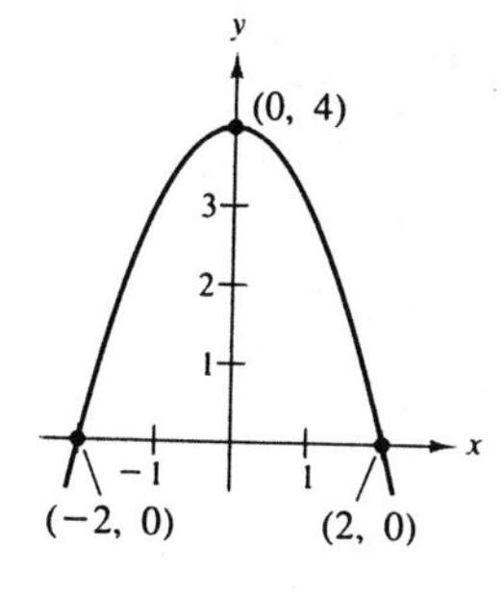

23. Find the center, foci, and vertices of $9x^2+4y^2+36x-24y+36=0$, and sketch its graph.

Solution:

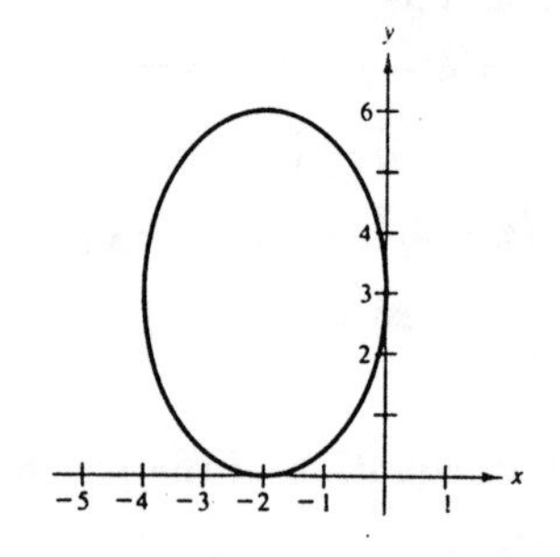

$$\begin{aligned}9x^2+4y^2+36x-24y+36&=0\\9(x^2+4x+4)+4(y^2-6y+9)&=-36+36+36\\9(x+2)^2+4(y-3)^2&=36\\\frac{(x+2)^2}{4}+\frac{(y-3)^2}{9}&=1\end{aligned}$$

$a=3,\ b=2,\ c=\sqrt{9-4}=\sqrt{5}$

Vertical major axis

Center: $(-2,\ 3)$

Foci: $(-2,\ 3\pm\sqrt{5})$

Vertices: $(-2,\ 3\pm3)$

$(-2,\ 6)$ and $(-2,\ 0)$

27. Find the center, foci, and vertices of $12x^2+20y^2-12x+40y-37=0$, and sketch its graph.

Solution:

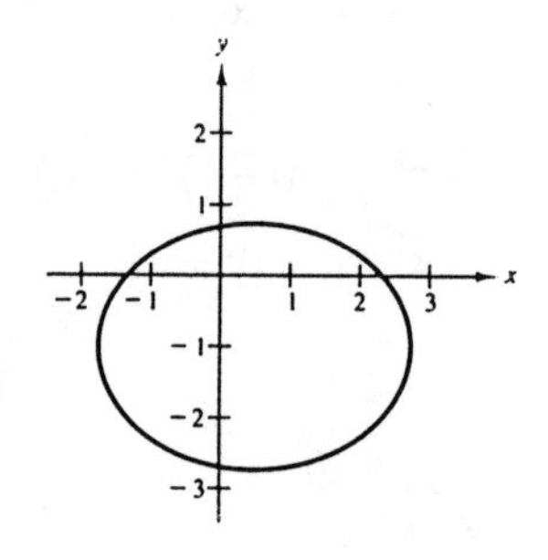

$$\begin{aligned}12x^2+20y^2-12x+40y-37&=0\\12\left(x^2-x+\frac{1}{4}\right)+20(y^2+2y+1)&=37+3+20\\12\left(x-\frac{1}{2}\right)^2+20(y+1)^2&=60\\\frac{(x-1/2)^2}{5}+\frac{(y+1)^2}{3}&=1\end{aligned}$$

$a=\sqrt{5},\ b=\sqrt{3},\ c=\sqrt{2}$

Horizontal major axis

Center: $\left(\frac{1}{2},\ -1\right)$

Foci: $\left(\frac{1}{2}\pm\sqrt{2},\ -1\right)$

Vertices: $\left(\frac{1}{2}\pm\sqrt{5},\ -1\right)$

31. Find an equation of the ellipse with foci at $(0, 0)$ and $(0, 8)$, and a major axis of length 16.

Solution:

Center: $(0, 4)$; $c = 4$

$2a = 16,\ a = 8,\ b = \sqrt{64 - 16} = \sqrt{48}$

Vertical major axis

$$\frac{(x-0)^2}{48} + \frac{(y-4)^2}{64} = 1$$

$$\frac{x^2}{48} + \frac{(y-4)^2}{64} = 1$$

35. Find an equation of the ellipse with center at $(0, 4)$, $a = 2c$, and vertices at $(-4, 4)$ and $(4, 4)$.

Solution:

Center: $(0, 4)$; $a = 2c$

Vertices: $(-4, 4)$, $(4, 4)$; $a = 4$,

$4 = 2c \Rightarrow c = 2;\ b = \sqrt{16 - 4} = \sqrt{12}$

Horizontal major axis

$$\frac{x^2}{16} + \frac{(y-4)^2}{12} = 1$$

39. Find the center, vertices, and foci of $(y + 6)^2 - (x - 2)^2 = 1$, and sketch its graph, using asymptotes as an aid.

Solution:

$(y + 6)^2 - (x - 2)^2 = 1$

$a = b = 1,\ c = \sqrt{1 + 1} = \sqrt{2}$

Center: $(2, -6)$

Vertical transverse axis

Foci: $(2, -6 \pm \sqrt{2})$

Vertices: $(2, -6 \pm 1)$

$(2, -5)$ and $(2, -7)$

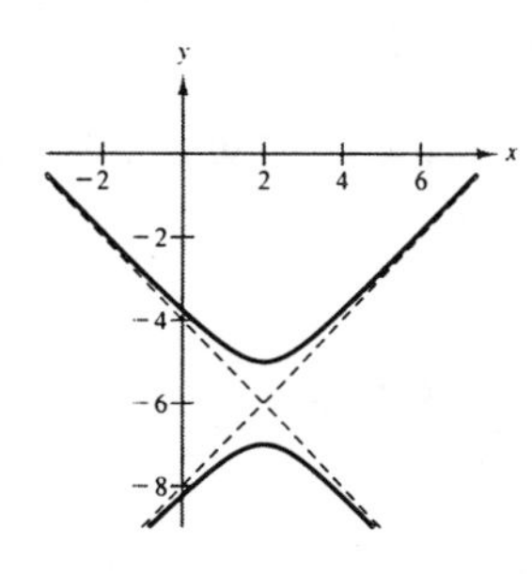

43. Find the center, vertices, and foci of $9y^2 - x^2 + 2x + 54y + 62 = 0$, and sketch its graph, using asymptotes as an aid.

Solution:

$$9y^2 - x^2 + 2x + 54y + 62 = 0$$
$$9(y^2 + 6y + 9) - (x^2 - 2x + 1) = -62 + 81 - 1$$
$$9(y+3)^2 - (x-1)^2 = 18$$
$$\frac{(y+3)^2}{2} - \frac{(x-1)^2}{18} = 1$$

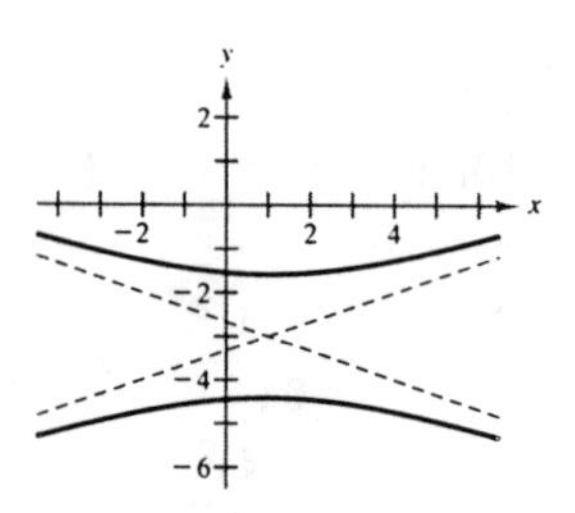

Center: $(1, -3)$

$a = \sqrt{2},\ b = \sqrt{18} = 3\sqrt{2},\ c = \sqrt{20} = 2\sqrt{5}$

Vertical transverse axis

Foci: $\left(1, -3 \pm 2\sqrt{5}\right)$

Vertices: $\left(1, -3 \pm \sqrt{2}\right)$

49. Find an equation for the hyperbola with vertices at $(4, 1)$ and $(4, 9)$, and foci at $(4, 0)$ and $(4, 10)$.

Solution:

Center: $(4, 5)$

Vertical transverse axis

$a = 4,\ c = 5,\ b = \sqrt{25 - 16} = 3$

$$\frac{(y-5)^2}{16} - \frac{(x-4)^2}{9} = 1$$

53. Find an equation for the hyperbola with vertices at $(0, 2)$ and $(6, 2)$ and asymptotes $y = \frac{2}{3}x$ and $y = 4 - \frac{2}{3}x$.

Solution:

Horizontal transverse axis

Center: $(3, 2)$, $a = 3$

Asymptotes: $y - 2 = \pm\frac{b}{3}(x - 3)$

$$y = \pm\frac{b}{3}x \mp b + 2 \quad \Rightarrow \quad \pm\frac{b}{3} = \pm\frac{2}{3}$$

$$b = 2$$

$$\frac{(x-3)^2}{9} - \frac{(y-2)^2}{4} = 1$$

55. Classify the graph of $x^2 + y^2 - 6x + 4y + 9 = 0$ as a circle, a parabola, an ellipse, or a hyperbola.

Solution:

$$x^2 + y^2 - 6x + 4y + 9 = 0$$
$$(x^2 - 6x + 9) + (y^2 + 4y + 4) = -9 + 9 + 4$$
$$(x-3)^2 + (y+2)^2 = 4$$

Circle

59. Classify the graph of $4x^2 + 3y^2 + 8x - 24y + 51 = 0$ as a circle, a parabola, an ellipse, or a hyperbola.

Solution:

$$4x^2 + 3y^2 + 8x - 24y + 51 = 0$$
$$4(x^2 + 2x + 1) + 3(y^2 - 8y + 16) = -51 + 4 + 48$$
$$4(x+1)^2 + 3(y-4)^2 = 1$$
$$\frac{(x+1)^2}{1/4} + \frac{(y-4)^2}{1/3} = 1$$

Ellipse

63. A satellite in a 100-mile high circular orbit around the earth has a velocity of approximately 17,500 miles per hour. If this velocity is multiplied by $\sqrt{2}$, then the satellite will have the minimum velocity necessary to escape the earth's gravity and it will follow a parabolic path with the center of the earth as the focus, as shown in the figure.

(a) Find the escape velocity of the satellite.

(b) Find an equation of its path (assume the radius of the earth is 4000 miles).

Solution:

(a) $V = 17{,}500\sqrt{2}$ mi/hr

$\approx 24{,}750$ mi/hr

(b) $p = -4100$, $(h, k) = (0, 4100)$

$$(x-0)^2 = 4(-4100)(y - 4100)$$
$$x^2 = -16{,}400(y - 4100)$$

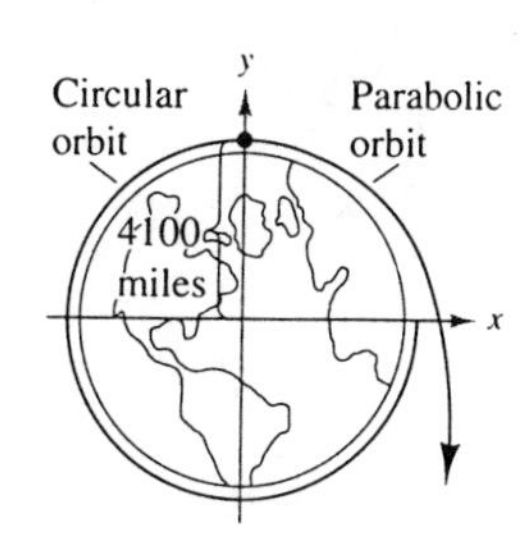

67. The earth moves in an elliptical orbit with the sun at one of the foci (see figure). The length of half of the major axis is 92.957×10^6 miles and the eccentricity is 0.017. Find the least and greatest distances of the earth from the sun.

Solution:

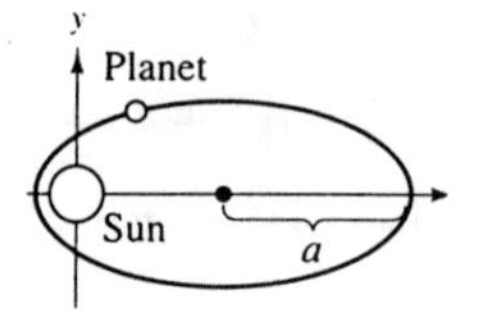

$a = 92.957 \times 10^6$ mi

$e = c/a = 0.017$

$c = 1{,}580{,}269$ mi

Least distance: $a - c = 91{,}376{,}731$ mi

Greatest distance: $a + c = 94{,}537{,}269$ mi

71. Show that the equation of an ellipse can be written as

$$\frac{(x-h)^2}{a^2} + \frac{(y-k)^2}{a^2(1-e^2)} = 1.$$

Note that as e approaches zero, the ellipse approaches a circle of radius a.

Solution:

$$b^2 = a^2 - c^2 = a^2 - (ae)^2 = a^2(1-e^2)$$

Thus, $\dfrac{(x-h)^2}{a^2} + \dfrac{(y-k)^2}{b^2} = 1$ can be written as $\dfrac{(x-h)^2}{a^2} + \dfrac{(y-k)^2}{a^2(1-e^2)} = 1.$

REVIEW EXERCISES FOR CHAPTER 5

Solutions to Selected Exercises

3. Find the domain of the function $g(x) = \dfrac{x^2}{x^2 - 4}$ and identify any horizontal or vertical asymptotes.

Solution:

$$g(x) = \frac{x^2}{x^2 - 4}$$

Domain: All real numbers except ± 2

Vertical asymptotes: $x = -2,\ x = 2$

Horizontal asymptote: $y = 1$ [degree of $p(x)$ = degree of $q(x)$]

9. Sketch the graph of the following rational function. As sketching aids, check for intercepts, symmetry, vertical asymptotes, horizontal asymptotes, and slant asymptotes.

$$p(x) = \frac{x^2}{x^2 + 1}$$

Solution:

$$p(x) = \frac{x^2}{x^2 + 1}$$

Intercept: (0, 0)

y-axis symmetry

Horizontal asymptote: $y = 1$

[degree of $p(x)$ = degree of $q(x)$]

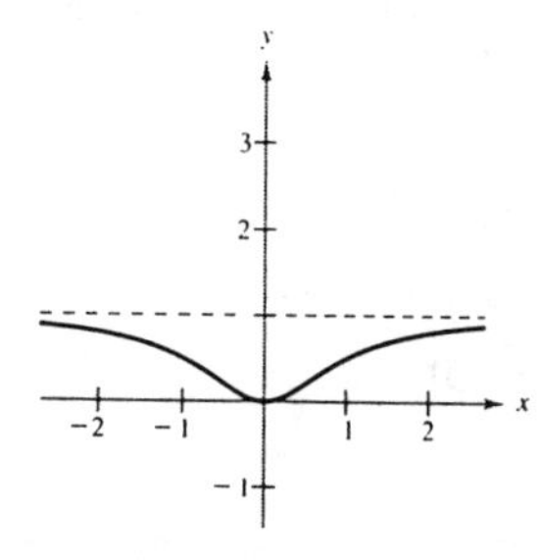

x	± 3	± 2	± 1	0
y	$\frac{9}{10}$	$\frac{4}{5}$	$\frac{1}{2}$	0

13. Sketch the graph of the following rational function. As sketching aids, check for intercepts, symmetry, vertical asymptotes, horizontal asymptotes, and slant asymptotes.

$$f(x) = \frac{-5}{x^2}$$

Solution:

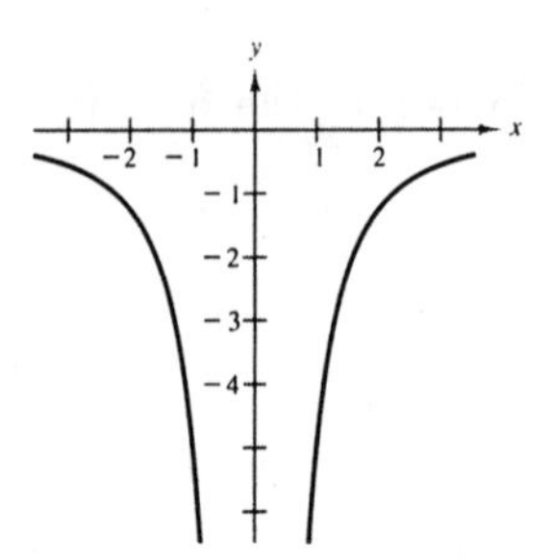

$$f(x) = \frac{-5}{x^2}$$

y-axis symmetry

Vertical asymptote: $x = 0$

Horizontal asymptote: $y = 0$

x	± 3	± 2	± 1
y	$-\frac{5}{9}$	$-\frac{5}{4}$	-5

17. Sketch the graph of the following rational function. As sketching aids, check for intercepts, symmetry, vertical asymptotes, horizontal asymptotes, and slant asymptotes.

$$y = \frac{2x^2}{x^2 - 4}$$

Solution:

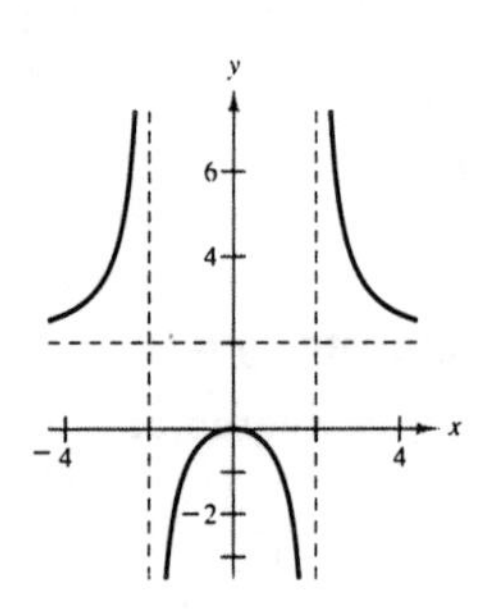

$$y = \frac{2x^2}{x^2 - 4}$$

Intercept: $(0, 0)$

y-axis symmetry

Vertical asymptotes: $x = 2$ and $x = -2$

[degree of $p(x)$ = degree of $q(x)$]

Horizontal asymptote: $y = 2$

x	± 4	± 3	± 1	0
y	$\frac{8}{3}$	$\frac{18}{5}$	$-\frac{2}{3}$	0

23. The cost in millions of dollars for the federal government to seize $p\%$ of an illegal drug as it enters the country is given by

$$C = \frac{528p}{100 - p}, \quad 0 \le p < 100.$$

(a) Find the cost of seizing 25%.

(b) Find the cost of seizing 50%.

(c) Find the cost of seizing 75%.

(d) According to this model, would it be possible to seize 100% of the drug?

Solution:

(a) When $p = 25$, $C = \dfrac{528(25)}{100 - 25} = \$176$ million.

(b) When $p = 50$, $C = \dfrac{528(50)}{100 - 50} = \$528$ million.

(c) When $p = 75$, $C = \dfrac{528(75)}{100 - 75} = \$1584$ million.

(d) As $p \to 100$, $C \to \infty$. No, it is not possible.

27. Write the partial fraction decomposition for $\dfrac{4 - x}{x^2 + 6x + 8}$.

Solution:

$$\frac{4 - x}{x^2 + 6x + 8} = \frac{4 - x}{(x + 2)(x + 4)}$$

$$\frac{4 - x}{(x + 2)(x - 4)} = \frac{A}{x + 2} + \frac{B}{x + 4}$$

$$4 - x = A(x + 4) + B(x + 2) \qquad \text{Basic Equation}$$

Let $x = -2:\quad 6 = 2A \Rightarrow A = 3$

Let $x = -4:\quad 8 = -2B \Rightarrow B = -4$

$$\frac{4 - x}{x^2 + 6x + 8} = \frac{3}{x + 2} - \frac{4}{x + 4}$$

31. Write the partial fraction decomposition for $\dfrac{x^2+2x}{x^3-x^2+x-1}$.

Solution:

$$\frac{x^2+2x}{x^3-x^2+x-1} = \frac{x^2+2x}{(x-1)(x^2+1)}$$

$$\frac{x^2+2x}{(x-1)(x^2+1)} = \frac{A}{x-1}+\frac{Bx+C}{x^2+1}$$ Factor the denominator by grouping

$$x^2+2x = A(x^2+1)+(Bx+C)(x-1)$$ Basic Equation

Let $x=1$: $3=2A,\ A=3/2$

Let $x=0$: $0=A-C,\ C=3/2$

Let $x=2$: $8=5A+2B+C$

$8=(15/2)+2B+(3/2),\ B=-1/2$

$$\frac{3/2}{x-1}+\frac{-(1/2)x+3/2}{x^2+1}=\frac{1}{2}\left(\frac{3}{x-1}-\frac{x-3}{x^2+1}\right)$$

35. Identify the conic $4x-y^2=0$ and sketch its graph.

Solution:

$4x=y^2$

Parabola

Vertex: (0, 0)

Horizontal axis

x	0	1	4
y	0	± 2	± 4

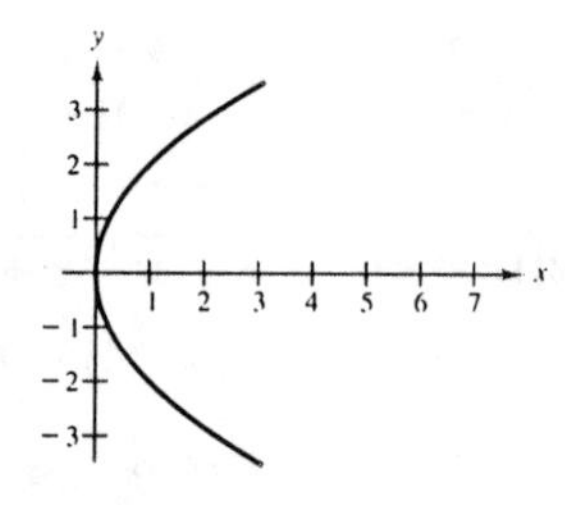

39. Identify $x^2 + y^2 - 2x - 4y + 5 = 0$ and sketch its graph.

Solution:

$$x^2 + y^2 - 2x - 4y + 5 = 0$$
$$(x^2 - 2x + 1) + (y^2 - 4y + 4) = -5 + 1 + 4$$
$$(x - 1)^2 + (y - 2)^2 = 0$$

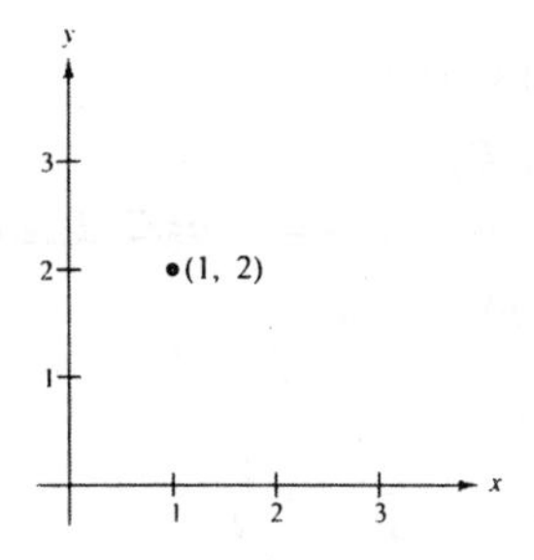

Point: (1, 2)

Note: This is a degenerate conic - a circle of of radius zero.

45. Identify $5y^2 - 4x^2 = 20$ and sketch its graph.

Solution:

$$5y^2 - 4x^2 = 20$$
$$\frac{y^2}{4} - \frac{x^2}{5} = 1$$

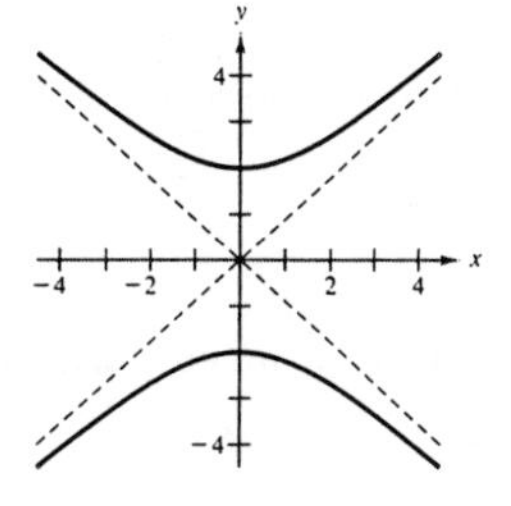

Hyperbola

Vertical transverse axis

Center: (0, 0)

Vertices: $(0, \pm 2)$

47. Find an equation for the parabola with vertex at (4, 2) and focus at (4, 0).

Solution:

Vertical axis, $p = -2$

$$(x - h)^2 = 4p(y - k)$$
$$(x - 4)^2 = 4(-2)(y - 2)$$
$$(x - 4)^2 = -8(y - 2)$$

51. Find an equation for the ellipse with vertices at $(-3, 0)$ and $(7, 0)$ and foci at $(0, 0)$ and $(4, 0)$.

Solution:

Horizontal major axis

Center: $(2, 0)$

$$a = 5, \quad c = 2, \quad b = \sqrt{25 - 4} = \sqrt{21}$$

$$\frac{(x-h)^2}{a^2} + \frac{(y-k)^2}{b^2} = 1$$

$$\frac{(x-2)^2}{25} + \frac{y^2}{21} = 1$$

55. Find an equation of the hyperbola with vertices at $(0, \pm 1)$ and foci at $(0, \pm 3)$.

Solution:

Vertical transverse axis

Center: $(0, 0)$

$$a = 1, \quad c = 3, \quad b = \sqrt{9 - 1} = \sqrt{8}$$

$$\frac{y^2}{1} - \frac{x^2}{8} = 1$$

59. A cross section of a large parabolic antenna (see figure) is given by

$$y = \frac{x^2}{200}, \quad 0 \le x \le 100.$$

The receiving and transmitting equipment is positioned at the focus. Find the coordinates of the focus.

Solution:

$$y = \frac{x^2}{200}, \quad -100 \le x \le 100$$

Vertex: $(0, 0)$

$$x^2 = 200y$$

$$4p = 200$$

$$p = 50$$

Focus: $(0, 50)$

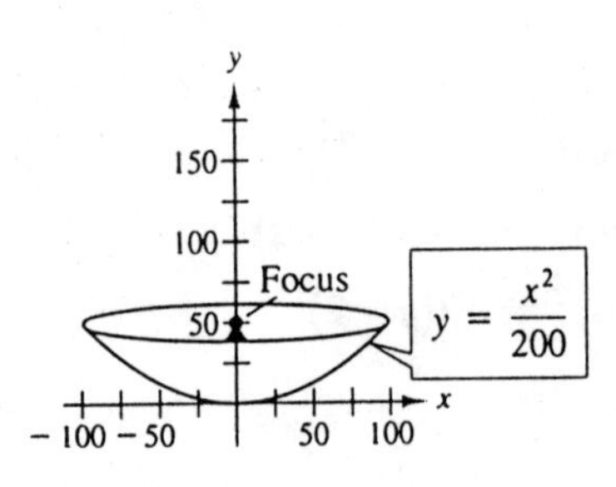

Practice Test for Chapter 5

1. Sketch the graph of $f(x) = \dfrac{x-1}{2x}$ and label all intercepts and asymptotes.

2. Sketch the graph of $f(x) = \dfrac{3x^2-4}{x}$ and label all intercepts and asymptotes.

3. Find all the asymptotes of $f(x) = \dfrac{8x^2-9}{x^2+1}$.

4. Find all the asymptotes of $f(x) = \dfrac{4x^2-2x+7}{x-1}$.

5. Sketch the graph of $f(x) = \dfrac{x-5}{(x-5)^2}$.

For Exercises 6–9, write the partial fraction decomposition for the rational expression.

6. $\dfrac{1-2x}{x^2+x}$

7. $\dfrac{6x}{x^2-x-2}$

8. $\dfrac{6x-17}{(x-3)^2}$

9. $\dfrac{3x^2-x+8}{x^3+2x}$

10. Find the vertex, focus, and directrix of the parabola $x^2 = 20y$.

11. Find the equation of the parabola with vertex (0, 0) and focus (7, 0).

12. Find the center, foci, and vertices of the ellipse $\dfrac{x^2}{144} + \dfrac{y^2}{25} = 1$.

13. Find the equation of the ellipse with foci (±4, 0) and minor axis of length 6.

14. Find the center, vertices, foci, and asymptotes of the hyperbola $\dfrac{y^2}{144} - \dfrac{x^2}{169} = 1$.

15. Find the equation of the hyperbola with vertices (±4, 0) and asymptotes $y = \pm\dfrac{1}{2}x$.

16. Find the equation of the parabola with vertex (6, -1) and focus (6, 3).

17. Find the center, foci, and vertices of the ellipse $16x^2 + 9y^2 - 96x + 36y + 36 = 0$.

18. Find the equation of the ellipse with vertices (-1, 1) and (7, 1) and minor axis of length 2.

19. Find the center, vertices, foci, and asymptotes of the hyperbola $4(x+3)^2 - 9(y-1)^2 = 1$.

20. Find the equation of the hyperbola with vertices (3, 4) and (3, -4) and foci (3, 7) and (3, -7).

CHAPTER 6
Exponential and Logarithmic Functions

SECTION 6.1

Exponential Functions

- You should know that a function of the form $y = a^x$, where $a > 0$, $a \neq 1$, is called an exponential function with base a.
- You should be able to graph exponential functions.
- You should know some properties of exponential functions where $a > 0$ and $a \neq 1$.
  (a) If $a^x = a^y$, then $x = y$.
  (b) If $a^x = b^x$ and $x \neq 0$, then $a = b$.
- You should know formulas for compound interest.
  (a) For n compoundings per year: $A = P\left(1 + \frac{r}{n}\right)^{nt}$.
  (b) For continuous compoundings: $A = Pe^{rt}$.

Solutions to Selected Exercises

3. Use a calculator to evaluate $1000(1.06)^{-5}$. Round your answer to three decimal places.

Solution:

$$1000(1.06)^{-5} \approx 747.258$$

1.06 [y^x] 5 [+/−] [$\times$] 1000 [=]

7. Use a calculator to evaluate $8^{2\pi}$. Round your answer to three decimal places.

Solution:

$$8^{2\pi} \approx 472{,}369.379$$

8 [y^x] [(] 2 [$\times$] π [)] [=]

11. Use a calculator to evaluate e^2. Round your answer to three decimal places.

Solution:

$$e^2 \approx 7.389$$

2 $\boxed{e^x}$ or $\boxed{\text{INV}}$ $\boxed{\ln x}$

15. Match $f(x) = 3^x$ with its graph.

Solution:

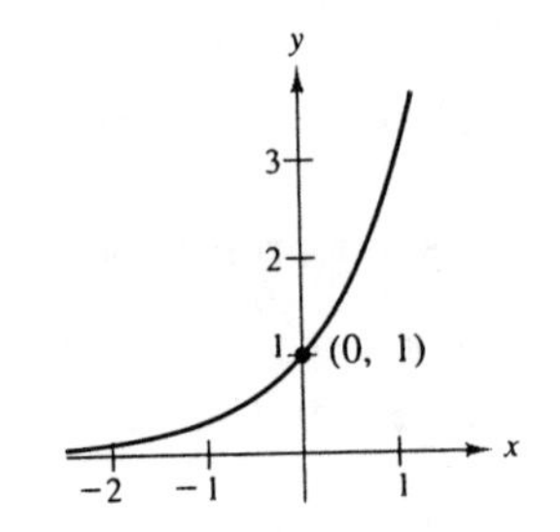

$f(x) = 3^x$

y-intercept: (0, 1)

3^x increases as x increases

Matches graph (g)

19. Match $f(x) = 3^x - 4$ with its graph.

Solution:

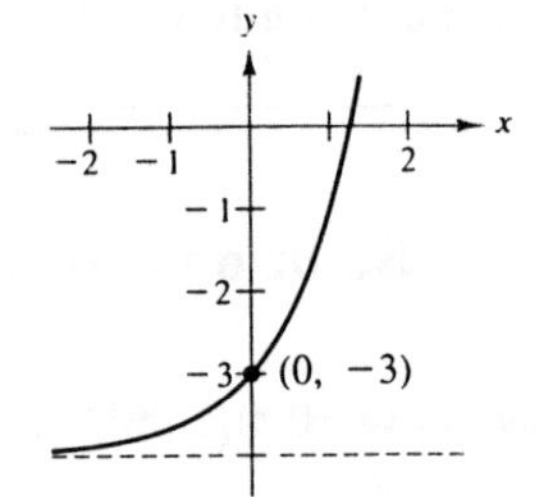

$f(x) = 3^x - 4$

y-intercept: (0, −3)

$3^x - 4$ increases as x increases

Matches graph (d)

23. Sketch the graph of $g(x) = 5^x$.

Solution:

$g(x) = 5^x$

x	−2	−1	0	1	2
$g(x)$	$\frac{1}{25}$	$\frac{1}{5}$	1	5	25

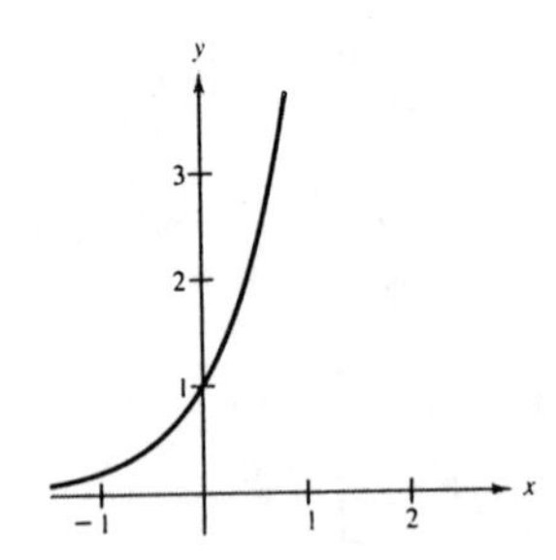

27. Sketch the graph of $h(x) = 5^{x-2}$.

Solution:

$h(x) = 5^{x-2}$

x	-1	0	1	2	3
$h(x)$	$\frac{1}{125}$	$\frac{1}{25}$	$\frac{1}{5}$	1	5

31. Sketch the graph of $f(x) = 3^{x-2} + 1$.

Solution:

$f(x) = 3^{x-2} + 1$

x	-1	0	1	2	3	4
y	$1\frac{1}{27}$	$1\frac{1}{9}$	$1\frac{1}{3}$	2	4	10

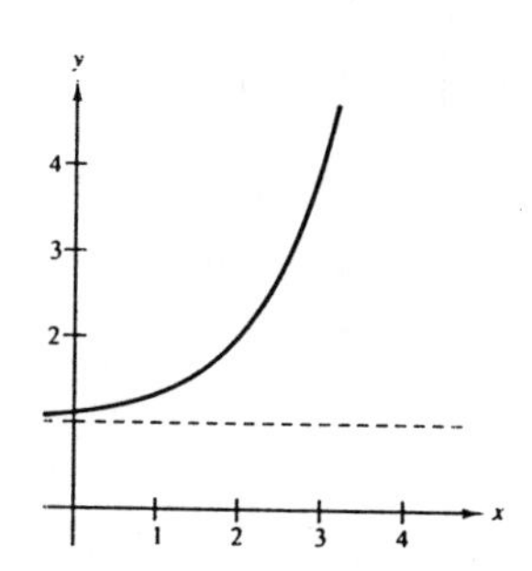

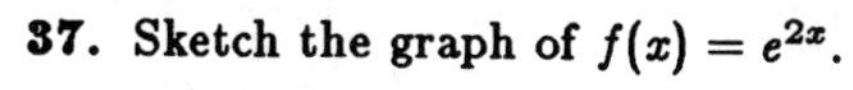

37. Sketch the graph of $f(x) = e^{2x}$.

Solution:

$f(x) = e^{2x}$

x	0	1	2	-1	-2
$f(x)$	1	7.39	54.60	0.135	0.02

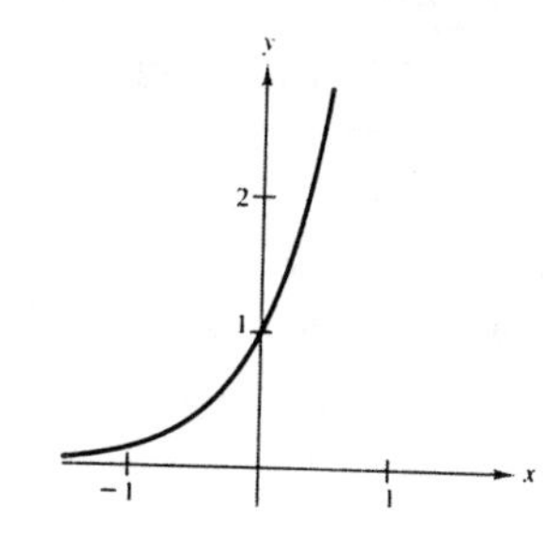

43. Use the table in the textbook to determine the balance A for \$2500 invested at 12% for 20 years and compounded n times per year.

Solution:

$$A = P\left(1 + \frac{r}{n}\right)^{nt}$$

$P = 2500, \quad r = 0.12, \quad t = 20$

–CONTINUED ON NEXT PAGE–

43. –CONTINUED–

When $n = 1$, $A = 2500\left(1 + \frac{0.12}{1}\right)^{(1)(20)} \approx \$24{,}115.73$

When $n = 2$, $A = 2500\left(1 + \frac{0.12}{2}\right)^{(2)(20)} \approx \$25{,}714.29$

When $n = 4$, $A = 2500\left(1 + \frac{0.12}{4}\right)^{(4)(20)} \approx \$26{,}602.23$

When $n = 12$, $A = 2500\left(1 + \frac{0.12}{12}\right)^{(12)(20)} \approx \$27{,}231.38$

When $n = 365$, $A = 2500\left(1 + \frac{0.12}{365}\right)^{(365)(20)} \approx \$27{,}547.07$

For continuous compounding, $A = Pe^{rt}$, $A = 2500e^{(0.12)(20)} \approx \$27{,}557.94$

n	1	2	4	12	365	Continuous compounding
A	\$24,115.73	\$25,714.29	\$26,602.23	\$27,231.38	\$27,547.07	\$27,557.94

45. Use the table in the textbook to determine the amount of money P that should be invested at 9% compounded continuously to produce a balance of \$100,000 in t years.

Solution:

$A = Pe^{rt}$

$100{,}000 = Pe^{0.09t}$

$\frac{100{,}000}{e^{0.09t}} = P$

$P = 100{,}000e^{-0.09t}$

–CONTINUED ON NEXT PAGE–

45. –CONTINUED–

When $t = 1$, $P = 100{,}000e^{-0.09(1)} \approx \$91{,}393.12$

When $t = 10$, $P = 100{,}000e^{-0.09(10)} \approx \$40{,}656.97$

When $t = 20$, $P = 100{,}000e^{-0.09(20)} \approx \$16{,}529.89$

When $t = 30$, $P = 100{,}000e^{-0.09(30)} \approx \$6{,}720.55$

When $t = 40$, $P = 100{,}000e^{-0.09(40)} \approx \$2{,}732.37$

When $t = 50$, $P = 100{,}000e^{-0.09(50)} \approx \$1{,}110.90$

t	1	10	20	30	40	50
P	\$91,393.12	\$40,656.97	\$16,529.89	\$6,720.55	\$2,732.37	\$1,110.90

49. On the day of your grandchild's birth, you deposited \$25,000 in a trust fund that pays 8.75% interest, compounded continuously. Determine the balance in this account on your grandchild's 25th birthday.

Solution:

$$A = 25{,}000e^{(0.0875)(25)} \approx \$222{,}822.57$$

51. The demand equation for a certain product is given by $p = 500 - 0.5e^{0.004x}$. Find the price p for a demand of (a) $x = 1000$ units and (b) $x = 1500$ units.

Solution:

(a) $x = 1000$

$$p = 500 - 0.5e^4 \approx \$472.70$$

(b) $x = 1500$

$$p = 500 - 0.5e^6 \approx \$298.29$$

55. Let Q represent the mass of radium (Ra^{226}) whose half-life is 1620 years. The quantity of radium present after t years is given by

$$Q = 25\left(\frac{1}{2}\right)^{t/1620}.$$

(a) Determine the initial quantity (when $t = 0$).

(b) Determine the quantity present after 1000 years.

(c) Sketch the graph of this function over the interval $t = 0$ to $t = 5000$.

55. —CONTINUED—

Solution:

(a) When $t = 0$, $Q = 25\left(\frac{1}{2}\right)^{0/1620} = 25(1) = 25$ units

(b) When $t = 1000$, $Q = 25\left(\frac{1}{2}\right)^{1000/1620} \approx 16.297$ units

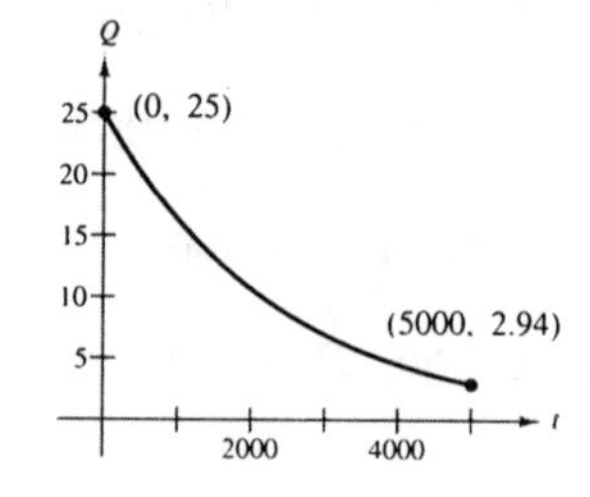

(c)

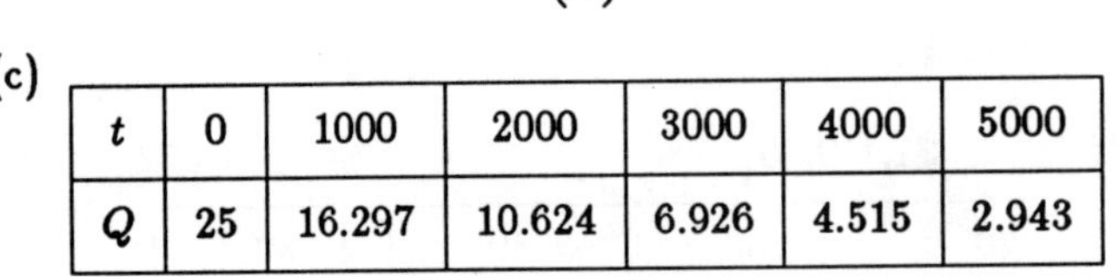

t	0	1000	2000	3000	4000	5000
Q	25	16.297	10.624	6.926	4.515	2.943

59. After t years, the value of a car that cost you \$20,000 is given by

$$V(t) = 20{,}000\left(\frac{3}{4}\right)^t.$$

Sketch a graph of the function and determine the value of the car two years after it was purchased.

Solution:

t	0	1	2
V	\$20,000	\$15,000	\$11,250

t	3	4	5
V	\$8,437.50	\$6,328.13	\$4,746.09

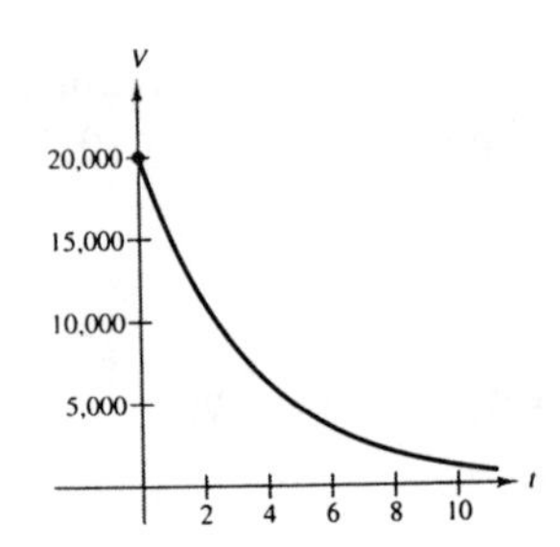

When $t = 2$, $V = \$11{,}250$.

61. Given the exponential function $f(x) = a^x$, show that (a) $f(u+v) = f(u) \cdot f(v)$ and (b) $f(2x) = [f(x)]^2$.

Solution:

(a) $f(u+v) = a^{u+v}$
$= a^u \cdot a^v = f(u) \cdot f(v)$

(b) $f(2x) = a^{2x}$
$= (a^x)^2 = [f(x)]^2$

SECTION 6.2

Logarithmic Functions

- You should know that a function of the form $y = \log_b M$, where $b > 0$, $b \neq 1$, and $M > 0$, is called a logarithm of M to base b.
- You should be able to convert from logarithmic form to exponential form and vice versa.

  $y = \log_b M \iff b^y = M$

- You should know the following properties of logarithms.

  (a) $\log_a 1 = 0$

  (b) $\log_a a = 1$

  (c) $\log_a a^x = x$

- You should know the definition of the natural logarithmic function.

  $\log_e x = \ln x, \quad x > 0$

- You should know the properties of the natural logarithmic function.

  (a) $\ln 1 = 0$

  (b) $\ln e = 1$

  (c) $\ln e^x = x$

- You should be able to graph logarithmic functions.

Solutions to Selected Exercises

5. Evaluate $\log_{16} 4$ without using a calculator.

Solution:

$$\log_{16} 4 = \log_{16} \sqrt{16} = \log_{16} 16^{1/2} = \tfrac{1}{2}$$

9. Evaluate $\log_{10} 0.01$ without using a calculator.

Solution:

$$\log_{10} 0.01 = \log_{10} \tfrac{1}{100} = \log_{10} 10^{-2} = -2$$

13. Evaluate $\ln e^{-2}$ without using a calculator.

Solution:

$$\ln e^{-2} = -2$$

17. Use the definition of a logarithm to write $5^3 = 125$ in logarithmic form.

Solution:

$$5^3 = 125$$
$$\log_5 125 = 3$$

23. Use the definition of a logarithm to write $e^3 = 20.0855\ldots$ in logarithmic form.

Solution:

$$e^3 = 20.0855\ldots$$
$$\log_e 20.0855\ldots = 3$$
$$\ln 20.0855\ldots = 3$$

27. Use a calculator to evaluate $\log_{10} 345$. Round your answer to three decimal places.

Solution:

$$\log_{10} 345 = 2.537819095\ldots \approx 2.538$$

33. Use a calculator to evaluate $\ln(1+\sqrt{3})$. Round your answer to three decimal places.

Solution:

$$\ln(1+\sqrt{3}) = 1.005052539\ldots \approx 1.005$$

37. Demonstrate that $f(x) = e^x$ and $g(x) = \ln x$ are inverses of each other by sketching their graphs on the same coordinate plane.

Solution:

x	-2	-1	0	1	2	3
$f(x)$	0.135	0.368	1	2.718	7.389	20.086
$g(x)$	—	—	—	0	0.693	1.097

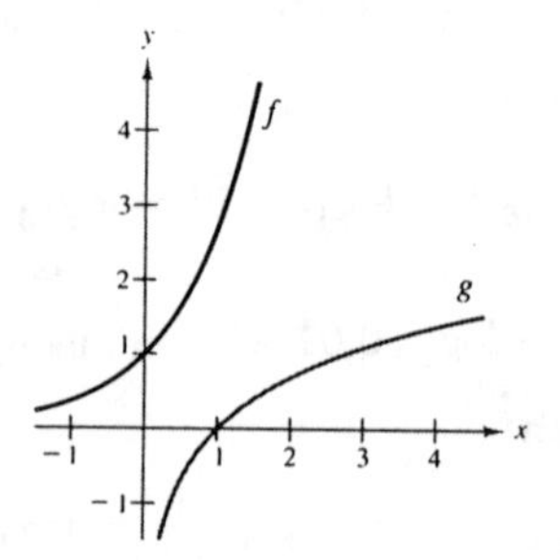

The graph of g is obtained by reflecting the graph of f about the line $y = x$.

41. Use the graph of $y = \ln x$ to match $f(x) = -\ln(x+2)$ to its graph.

Solution:

$f(x) = -\ln(x+2)$

Reflection and horizontal shift two units to the left

Vertical asymptote: $x = -2$

x-intercept: $(-1, 0)$

Matches graph (a)

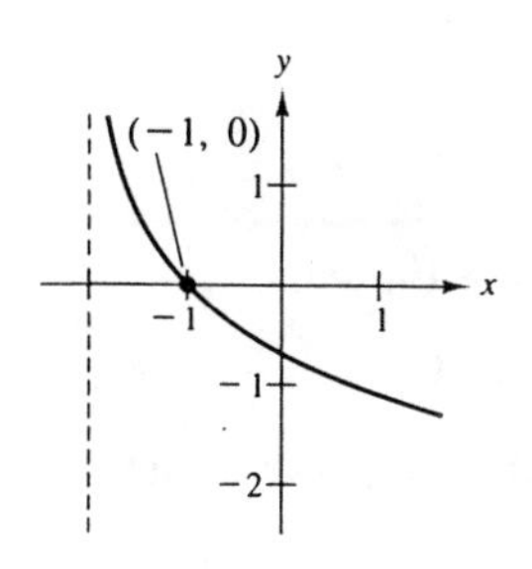

47. Find the domain, vertical asymptote, and x-intercept of $h(x) = \log_4(x-3)$, and sketch its graph.

Solution:

$h(x) = \log_4(x-3)$

Domain: $x - 3 > 0 \Rightarrow x > 3$

The domain is $(3, \infty)$.

Vertical asymptote: $x - 3 = 0 \Rightarrow x = 3$

The vertical asymptote is the line $x = 3$.

x-intercept: $\log_4(x-3) = 0$

$x - 3 = 4^0$

$x - 3 = 1 \Rightarrow x = 4$

The x-intercept is $(4, 0)$.

x	3.5	4	5	7
$h(x)$	−0.5	0	0.5	1

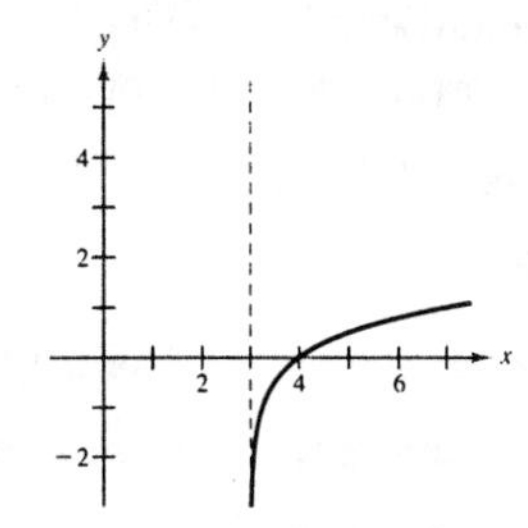

51. Find the domain, vertical asymptote, and x-intercept of $y = \log_{10}\left(\frac{x}{5}\right)$, and sketch its graph.

Solution:

$y = \log_{10}\left(\frac{x}{5}\right)$

Domain: $\frac{x}{5} > 0 \Rightarrow x > 0$

The domain is $(0, \infty)$

Vertical asymptote: $\frac{x}{5} = 0 \Rightarrow x = 0$

The vertical asymptote is the y-axis.

x-intercept: $\log_{10}\left(\frac{x}{5}\right) = 0$

$\frac{x}{5} = 10^0$

$\frac{x}{5} = 1 \Rightarrow x = 5$

The x-intercept is $(5, 0)$.

x	1	2	3	4
y	-0.699	-0.398	-0.222	-0.097

x	5	6	7
y	0	0.079	0.146

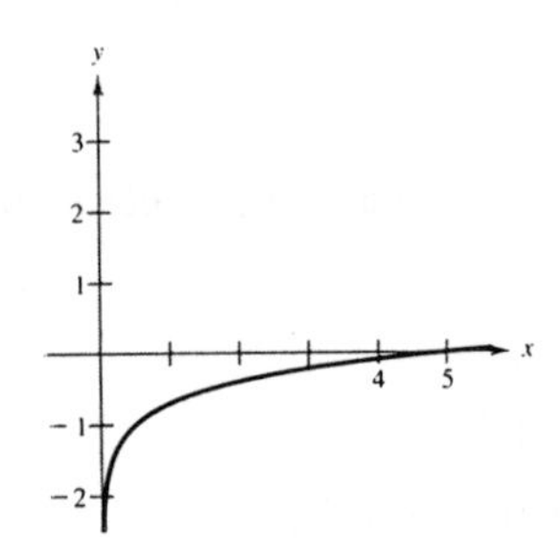

57. Students in a mathematics class were given an exam and then retested monthly with an equivalent exam. The average score for the class was given by the human memory model

$$f(t) = 80 - 17\log_{10}(t+1), \quad 0 \le t \le 12$$

where t is the time in months.

(a) What was the average score on the original exam $(t = 0)$?

(b) What was the average score after four months?

(c) What was the average score after ten months?

Solution:

(a) $f(0) = 80 - 17\log_{10} 1 = 80.0$

(b) $f(4) = 80 - 17\log_{10} 5 \approx 68.1$

(c) $f(10) = 80 - 17\log_{10} 11 \approx 62.3$

61. Use the model

$$y = 80.4 - 11 \ln x$$

which approximates the minimum required ventilation rate in terms of the air space per child in a public school classroom. In the model, x is the air space per child in cubic feet and y is the ventilation rate in cubic feet per minute. Use the model to approximate the required ventilation rate if there are 300 cubic feet of air space per child.

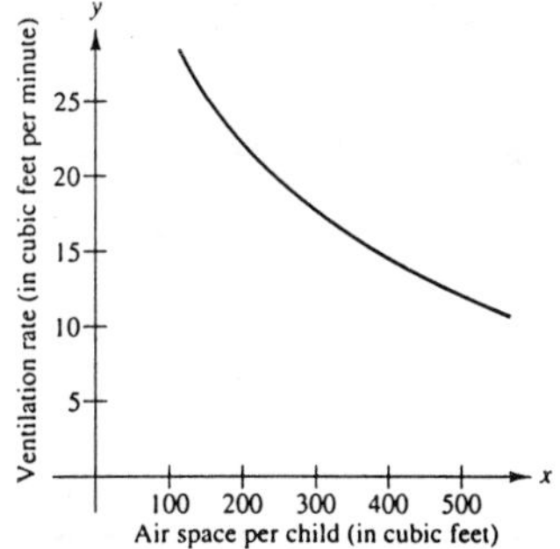

Solution:

$$y = 80.4 - 11 \ln 300 \approx 17.658 \text{ cubic feet per minute}$$

67. The work (in foot-pounds) done in compressing an initial volume of 9 cubic feet at a pressure of 15 pounds per square inch to a volume of 3 cubic feet is

$$W = 19{,}440(\ln 9 - \ln 3).$$

Find W.

Solution:

$$W = 19{,}440(\ln 9 - \ln 3) \approx 21{,}357.023 \text{ foot-pounds}$$

69. (a) Use a calculator to complete the table (shown in the textbook) for the function

$$f(x) = \frac{\ln x}{x}.$$

(b) Use the table in part (a) to determine what $f(x)$ approaches as x increases without bound.

Solution:

(a)

x	1	5	10	10^2	10^4	10^6
$f(x)$	0	0.322	0.230	0.046	0.00092	0.0000138

(b) As $x \to \infty$, $f(x) \to 0$.

SECTION 6.3

Properties of Logarithms

- You should know the following properties of logarithms.

  (a) $\log_a(uv) = \log_a u + \log_a v$

  (b) $\log_a(u/v) = \log_a u - \log_a v$

  (c) $\log_a u^n = n \log_a u$

  (d) $\log_a {}^x = \dfrac{\log_b {}^x}{\log_b {}^a}$

- You should be able to rewrite logarithmic expressions.

Solutions to Selected Exercises

1. Use the change of base formula to write $\log_3 5$ as a multiple of a common logarithm.

Solution:

$$\log_3 5 = \frac{\log_{10} 5}{\log_{10} 3}$$

5. Use the change of base formula to write $\log_3 5$ as a multiple of a natural logarithm.

Solution:

$$\log_3 5 = \frac{\ln 5}{\ln 3}$$

9. Evaluate $\log_3 7$ using the change of base formula. Do the problem twice; once with common logarithms and once with natural logarithms. Round to three decimal places.

Solution:

$$\log_3 7 = \frac{\log_{10} 7}{\log_{10} 3} \approx 1.771$$

$$\log_3 7 = \frac{\ln 7}{\ln 3} \approx 1.771$$

13. Evaluate $\log_9(0.4)$ using the change of base formula. Do the problem twice; once with common logarithms and once with natural logarithms. Round to three decimal places.

Solution:

$$\log_9(0.4) = \frac{\log_{10} 0.4}{\log_{10} 9} \approx -0.417$$

$$\log_9(0.4) = \frac{\ln 0.4}{\ln 9} \approx -0.417$$

17. Use the properties of logarithms to write $\log_{10} 5x$ as a sum, difference and/or multiple of logarithms.

Solution:

$$\log_{10} 5x = \log_{10} 5 + \log_{10} x$$

21. Use the properties of logarithms to write $\log_8 x^4$ as a sum, difference, and/or multiple of logarithms.

Solution:

$$\log_8 x^4 = 4 \log_8 x$$

25. Use the properties of logarithms to write $\ln xyz$ as a sum, difference, and/or multiple of logarithms.

Solution:

$$\ln xyz = \ln[x(yz)] = \ln x + \ln yz = \ln x + \ln y + \ln z$$

29. Use the properties of logarithms to write $\ln z(z-1)^2$ as a sum, difference and/or multiple of logarithms.

Solution:

$$\ln z(z-1)^2 = \ln z + \ln(z-1)^2 = \ln z + 2\ln(z-1)$$

33. Use the properties of logarithms to write the following expression as a sum, difference, and/or multiple of logarithms.

$$\ln \frac{x^4\sqrt{y}}{z^5}$$

Solution:

$$\ln \frac{x^4\sqrt{y}}{z^5} = \ln x^4\sqrt{y} - \ln z^5 = \ln x^4 + \ln\sqrt{y} - \ln z^5 = 4\ln x + \tfrac{1}{2}\ln y - 5\ln z$$

35. Use the properties of logarithms to write the following expression as a sum, difference, and/or multiple of logarithms.

$$\log_b \frac{x^2}{y^2 z^3}$$

Solution:

$$\log_b \frac{x^2}{y^2 z^3} = \log_b x^2 - \log_b y^2 z^3 = \log_b x^2 - [\log_b y^2 + \log_b z^3] = 2\log_b x - 2\log_b y - 3\log_b z$$

39. Write $\log_4 z - \log_4 y$ as the logarithm of a single quantity.

Solution:

$$\log_4 z - \log_4 y = \log_4 \frac{z}{y}$$

45. Write $\ln x - 3\ln(x+1)$ as the logarithm of a single quantity.

Solution:

$$\ln x - 3\ln(x+1) = \ln x - \ln(x+1)^3 = \ln \frac{x}{(x+1)^3}$$

49. Write $\ln x - 2[\ln(x+2) + \ln(x-2)]$ as the logarithm of a single quantity.

Solution:

$$\begin{aligned} \ln x - 2[\ln(x+2) + \ln(x-2)] &= \ln x - 2\ln(x+2)(x-2) \\ &= \ln x - 2\ln(x^2-4) \\ &= \ln x - \ln(x^2-4)^2 \\ &= \ln \frac{x}{(x^2-4)^2} \end{aligned}$$

53. Write $\frac{1}{3}[\ln y + 2\ln(y+4)] - \ln(y-1)$ as the logarithm of a single quantity.

Solution:

$$\begin{aligned} \frac{1}{3}[\ln y + 2\ln(y+4)] - \ln(y-1) &= \frac{1}{3}[\ln y + \ln(y+4)^2] - \ln(y-1) \\ &= \frac{1}{3}\ln[y(y+4)^2] - \ln(y-1) \\ &= \ln \sqrt[3]{y(y+4)^2} - \ln(y-1) \\ &= \ln \frac{\sqrt[3]{y(y+4)^2}}{y-1} \end{aligned}$$

57. Approximate $\log_b 6$ using the properties of logarithms, given $\log_b 2 \approx 0.3562$ and $\log_b 3 \approx 0.5646$.

Solution:

$$\log_b 6 = \log_b(2 \cdot 3) = \log_b 2 + \log_b 3 \approx 0.3562 + 0.5646 = 0.9208$$

63. Approximate $\log_b \sqrt{2}$ using the properties of logarithms, given $\log_b 2 \approx 0.3562$.

Solution:

$$\log_b \sqrt{2} = \log_b(2^{1/2}) = \tfrac{1}{2}\log_b 2 \approx \tfrac{1}{2}(0.3562) = 0.1781$$

65. Approximate $\log_b \frac{1}{4}$ using the properties of logarithms, given $\log_b 2 \approx 0.3562$.

Solution:

$$\log_b \tfrac{1}{4} = \log_b 1 - \log_b 4 = 0 - \log_b 2^2 = -2\log_b 2 \approx -2(0.3562) = -0.7124$$

69. Approximate the following using the properties of logarithms, given $\log_b 2 \approx 0.3562$ and $\log_b 3 \approx 0.5646$.

$$\log_b\left[\frac{(4.5)^3}{\sqrt{3}}\right]$$

Solution:

$$\begin{aligned}\log_b\left[\frac{(4.5)^3}{\sqrt{3}}\right] &= 3\log_b 4.5 - \frac{1}{2}\log_b 3\\ &= 3\log_b\frac{9}{2} - \frac{1}{2}\log_b 3\\ &= 3[\log_b 9 - \log_b 2] - \frac{1}{2}\log_b 3\\ &= 3[2\log_b 3 - \log_b 2] - \frac{1}{2}\log_b 3\\ &\approx 3[2(0.5646) - 0.3562] - \frac{1}{2}(0.5646) = 2.0367\end{aligned}$$

73. Find the exact value of $\log_4 16^{1.2}$.

Solution:

$$\log_4 16^{1.2} = 1.2\log_4 16 = 1.2\log_4 4^2 = (1.2)(2)\log_4 4 = (2.4)(1) = 2.4$$

77. Use the properties of logarithms to simplify $\log_4 8$.

Solution:

$$\log_4 8 = \log_4 2^3 = 3\log_4 2 = 3\log_4\sqrt{4} = 3\log_4 4^{1/2} = 3\left(\tfrac{1}{2}\right)\log_4 4 = \tfrac{3}{2}$$

81. Use the properties of logarithms to simplify $\log_5 \frac{1}{250}$.

Solution:

$$\begin{aligned}\log_5 \tfrac{1}{250} &= \log_5 1 - \log_5 250 = 0 - \log_5(125 \cdot 2)\\ &= -\log_5(5^3 \cdot 2) = -[\log_5 5^3 + \log_5 2]\\ &= -[3\log_5 5 + \log_5 2] = -3 - \log_5 2\end{aligned}$$

85. The relationship between the number of decibels β and the intensity of a sound I in watts per meter squared is given by

$$\beta = 10\log_{10}\left(\frac{I}{10^{-16}}\right).$$

Use properties of logarithms to write the formula in simpler form, and determine the number of decibels of a sound with an intensity of 10^{-10} watts per meter squared.

Solution:

$$\beta = 10\log_{10}\left(\frac{I}{10^{-16}}\right) = 10\left[\log_{10} I - \log_{10} 10^{-16}\right]$$

$$= 10[\log_{10} I + 16] = 160 + 10\log_{10} I$$

When $I = 10^{-10}$, we have

$$\beta = 160 + 10\log_{10} 10^{-10} = 160 + 10(-10) = 60 \text{ decibels.}$$

87. Prove that $\log_b \frac{u}{v} = \log_b u - \log_b v$.

Solution:

Let $x = \log_b u$ and $y = \log_b v$, then $b^x = u$ and $b^y = v$.

$$\frac{u}{v} = \frac{b^x}{b^y} = b^{x-y}$$

$$\log_b\left(\frac{u}{v}\right) = \log_b(b^{x-y}) = x - y = \log_b u - \log_b v$$

SECTION 6.4

Solving Exponential and Logarithmic Equations

- **You should be able to solve exponential and logarithmic equations.**
- **To solve an exponential equation, take the logarithm of both sides.**
- **To solve a logarithmic equation, rewrite it in exponential form.**

Solutions to Selected Exercises

5. Solve $\left(\frac{3}{4}\right)^x = \frac{27}{64}$ for x without a calculator.

Solution:

$$\left(\frac{3}{4}\right)^x = \frac{27}{64}$$
$$\left(\frac{3}{4}\right)^x = \left(\frac{3}{4}\right)^3$$
$$x = 3$$

9. Solve $\log_{10} x = -1$ for x without a calculator.

Solution:

$$\log_{10} x = -1$$
$$x = 10^{-1} = \frac{1}{10}$$

13. Apply the inverse properties of $\ln x$ and e^x to simplify $e^{\ln(5x+2)}$.

Solution:

$$e^{\ln(5x+2)} = 5x + 2$$

17. Solve $e^x = 10$. Round to three decimal places.

Solution:

$$e^x = 10$$
$$x = \ln 10 \approx 2.303$$

21. Solve $e^x - 5 = 10$. Round to three decimal places.

Solution:

$$e^x - 5 = 10$$
$$e^x = 15$$
$$x = \ln 15 \approx 2.708$$

27. Solve $500e^{-x} = 300$. Round to three decimal places.

Solution:

$$500e^{-x} = 300$$
$$e^{-x} = \frac{3}{5}$$
$$-x = \ln \frac{3}{5}$$
$$x = -\ln \frac{3}{5} = \ln \frac{5}{3} \approx 0.511$$

31. Solve $e^{2x} - 4e^x - 5 = 0$. Round to three decimal places.

Solution:

$$e^{2x} - 4e^x - 5 = 0$$
$$(e^x + 1)(e^x - 5) = 0$$

$$e^x + 1 = 0 \quad \text{or} \quad e^x - 5 = 0$$
$$e^x = -1 \qquad\qquad e^x = 5$$
$$\text{No solution} \qquad\qquad x = \ln 5 \approx 1.609$$

33. Solve $3(1 + e^{2x}) = 4$. Round to three decimal places.

Solution:

$$3(1 + e^{2x}) = 4$$
$$1 + e^{2x} = \frac{4}{3}$$
$$e^{2x} = \frac{1}{3}$$
$$2x = \ln \frac{1}{3}$$
$$x = \frac{1}{2} \ln \frac{1}{3} \approx -0.549$$

37. Solve $10^x = 42$. Round to three decimal places.

Solution:

$$10^x = 42$$
$$x = \log_{10} 42 \approx 1.623$$

41. Solve $5^{-t/2} = 0.20$. Round to three decimal places.

Solution:

$$5^{-t/2} = 0.20$$
$$5^{-t/2} = \frac{1}{5}$$
$$5^{-t/2} = 5^{-1}$$
$$-\frac{t}{2} = -1$$
$$t = 2$$

45. Solve $3\left(5^{x-1}\right) = 21$. Round to three decimal places.

Solution:

$$3\left(5^{x-1}\right) = 21$$
$$5^{x-1} = 7$$
$$\ln 5^{x-1} = \ln 7$$
$$(x-1)\ln 5 = \ln 7$$
$$x - 1 = \frac{\ln 7}{\ln 5}$$
$$x = 1 + \frac{\ln 7}{\ln 5} \approx 2.209$$

49. Solve $\left(1+\frac{0.10}{12}\right)^{12t} = 2$. Round to three decimal places.

Solution:

$$\left(1+\frac{0.10}{12}\right)^{12t} = 2$$

$$\ln\left(1+\frac{0.10}{12}\right)^{12t} = \ln 2$$

$$12t\ln\left(1+\frac{0.10}{12}\right) = \ln 2$$

$$t = \frac{\ln 2}{12\ln\left(1+\frac{0.10}{12}\right)} \approx 6.960$$

53. Solve $\ln 2x = 2.4$. Round to three decimal places.

Solution:

$$\ln 2x = 2.4$$

$$2x = e^{2.4}$$

$$x = \frac{e^{2.4}}{2} \approx 5.512$$

57. Solve $\ln\sqrt{x+2} = 1$. Round to three decimal places.

Solution:

$$\ln\sqrt{x+2} = 1$$

$$\sqrt{x+2} = e^1$$

$$x+2 = e^2$$

$$x = -2 + e^2 \approx 5.389$$

61. Solve $\log_{10}(z-3) = 2$. Round to three decimal places.

Solution:

$$\log_{10}(z-3) = 2$$

$$z - 3 = 10^2$$

$$z = 10^2 + 3 = 103$$

63. Solve $\log_{10}(x+4) - \log_{10} x = \log_{10}(x+2)$. Round to three decimal places.

Solution:

$$\begin{aligned}
\log_{10}(x+4) - \log_{10} x &= \log_{10}(x+2) \\
\log_{10}\left(\frac{x+4}{x}\right) &= \log_{10}(x+2) \\
\frac{x+4}{x} &= x+2 \\
x+4 &= x^2+2x \\
0 &= x^2+x-4 \\
x &= \frac{-1 \pm \sqrt{17}}{2} = -\frac{1}{2} \pm \frac{\sqrt{17}}{2} \quad \text{Quadratic Formula}
\end{aligned}$$

Choosing the positive value of x (the negative value is extraneous), we have

$$-\frac{1}{2} + \frac{\sqrt{17}}{2} \approx 1.562.$$

67. Solve $\ln(x+5) = \ln(x-1) - \ln(x+1)$. Round to three decimal places.

Solution:

$$\begin{aligned}
\ln(x+5) &= \ln(x-1) - \ln(x+1) \\
\ln(x+5) &= \ln\left(\frac{x-1}{x+1}\right) \\
x+5 &= \frac{x-1}{x+1} \\
(x+5)(x+1) &= x-1 \\
x^2+6x+5 &= x-1 \\
x^2+5x+6 &= 0 \\
(x+2)(x+3) &= 0 \\
x = -2 \text{ or } x &= -3
\end{aligned}$$

Both of these solutions are extraneous, so the equation has no solution.

71. Find the time required for a $1000 investment to double at an interest rate of $r = 0.085$ compounded continuously.

Solution:

$$A = Pe^{rt}$$
$$2000 = 1000e^{0.085t}$$
$$2 = e^{0.085t}$$
$$\ln 2 = 0.085t$$
$$\frac{\ln 2}{0.085} = t$$
$$t \approx 8.2 \text{ years}$$

75. The demand equation for a certain product is given by $p = 500 - 0.5(e^{0.004x})$. Find the demand x for a price of (a) $p = \$350$ and (b) $p = \$300$.

Solution:

(a)
$$350 = 500 - 0.5(e^{0.004x})$$
$$-150 = -0.5(e^{0.004x})$$
$$300 = e^{0.004x}$$
$$0.004x = \ln 300$$
$$x = \frac{\ln 300}{0.004} \approx 1426 \text{ units}$$

(b)
$$300 = 500 - 0.5(e^{0.004x})$$
$$-200 = -0.5(e^{0.004x})$$
$$400 = e^{0.004x}$$
$$0.004x = \ln 400$$
$$x = \frac{\ln 400}{0.004} \approx 1498 \text{ units}$$

SECTION 6.5

Exponential and Logarithmic Applications

- **You should be able to solve compound interest problems.**

  (a) Compound interest formulas:

  1. $A = P\left(1 + \frac{r}{n}\right)^{nt}$
  2. $A = Pe^{rt}$

  (b) Doubling time:

  1. $t = \frac{\ln 2}{n \ln[1 + (r/n)]}$, **$n$ compoundings per year**
  2. $t = \frac{\ln 2}{r}$, **continuous compounding**

  (c) Effective yield:

  1. **Effective yield** $= \left(1 + \frac{r}{n}\right)^{n} - 1$, **$n$ compoundings per year**
  2. **Effective yield** $= e^{r} - 1$, **continuous compounding**

- **You should be able to solve growth and decay problems.**

  $Q(t) = Ce^{kt}$

  (a) If $k > 0$, the population grows.

  (b) If $k < 0$, the population decays.

  (c) Ratio of Carbon 14 to Carbon 12 is $R(t) = \frac{1}{10^{12}}e^{-t/8223}$

- **You should be able to solve logistics model problems.**

  $y = \frac{a}{1 + be^{-(x-c)/d}}$

- **You should be able to solve intensity model problems.**

  $S = k \log_{10} \frac{I}{I_0}$

Solutions to Selected Exercises

5. Five hundred dollars is deposited into an account with continuously compounded interest. If the balance is \$1292.85 after 10 years, find the annual percentage rate, the effective yield, and the time to double.

Solution:

$$P = 500, \;\; A = 1292.85, \;\; t = 10$$

$$A = Pe^{rt}$$

$$1292.85 = 500e^{10r}$$

$$\frac{1292.85}{500} = e^{10r}$$

$$10r = \ln\left(\frac{1292.85}{500}\right)$$

$$r = \frac{1}{10}\ln\left(\frac{1292.85}{100}\right) \approx 0.095 = 9.5\%$$

Effective yield $= e^{0.095} - 1 \approx 0.09966 \approx 9.97\%$

Time to double: $1000 = 500e^{0.095t}$

$$2 = e^{0.095t}$$

$$0.095t = \ln 2$$

$$t = \frac{\ln 2}{0.095} \approx 7.30 \text{ years}$$

9. Five thousand dollars is deposited into an account with continuously compounded interest. If the effective yield is 8.33%, find the annual percentage rate, the time to double, and the amount after 10 years.

Solution:

$P = 5000$

Effective yield $= 8.33\%$

$0.0833 = e^r - 1$

$$r = \ln 1.0833 \approx 0.0800 = 8\%$$

Time to double: $10{,}000 = 5000e^{0.08t}$

$$t = \frac{\ln 2}{0.08} \approx 8.66 \text{ years}$$

After 10 years: $A = 5000e^{0.08(10)} = \$11{,}127.70$

13. Determine the time necessary for \$1000 to double if it is invested at 11% compounded (a) annually, (b) monthly, (c) daily, and (d) continuously.

Solution:

$$P = 1000, \; r = 11\%$$

(a) $n = 1$

$$t = \frac{\ln 2}{\ln(1+0.11)} \approx 6.642 \text{ years}$$

(b) $n = 12$

$$t = \frac{\ln 2}{12 \ln\left(1 + \frac{0.11}{12}\right)} \approx 6.330 \text{ years}$$

(c) $n = 365$

$$t = \frac{\ln 2}{365 \ln\left(1 + \frac{0.11}{365}\right)} \approx 6.302 \text{ years}$$

(d) Continuously

$$t = \frac{\ln 2}{0.11} \approx 6.301 \text{ years}$$

17. The half-life of the isotope Ra^{226} is 1620 years. If the initial quantity is 10 grams, how much will remain after 1000 years, and after 10,000 years?

Solution:

$$Q(t) = Ce^{kt}$$

$$Q = 10 \quad \text{when } t = 0 \Rightarrow 10 = Ce^{\circ} \Rightarrow 10 = C$$

$$Q(t) = 10e^{kt}$$

$$Q = 5 \quad \text{when } t = 1620$$

$$5 = 10e^{1620k}$$

$$k = \tfrac{1}{1620} \ln\left(\tfrac{1}{2}\right)$$

$$Q(t) = 10e^{[\ln(1/2)/1620]t}$$

When $t = 1000$, $Q(t) = 10e^{[\ln(1/2)/1620](1000)} \approx 6.52$ grams.

When $t = 10{,}000$, $Q(t) = 10e^{[\ln(1/2)/1620](10000)} \approx 0.14$ gram.

21. The half-life of the isotope Pu^{230} is 24,360 years. If 2.1 grams remain after 1000 years, what is the initial quantity and how much will remain after 10,000 years?

Solution:

$$y = Ce^{[\ln(1/2)/24360]t}$$

$$2.1 = Ce^{[\ln(1/2)/24360](1000)}$$

$$C \approx 2.16$$

The initial quantity is 2.16 grams.

When $t = 10{,}000$, $y = 2.16e^{[\ln(1/2)/24360](10000)} \approx 1.63$ grams.

23. Find the constant k such that the exponential function $y = Ce^{kt}$ passes through the points $(0, 1)$ and $(4, 10)$.

Solution:

$$y = Ce^{kt}$$

Since the graph passes through the point $(0, 1)$, we have

$$1 = Ce^{k(0)},$$

$$1 = C.$$

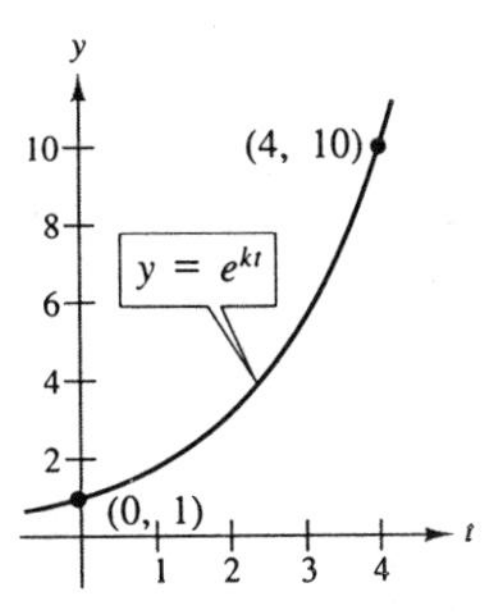

$$y = e^{kt}$$

Since the graph passes through the point $(4, 10)$, we have

$$10 = e^{4k},$$

$$4k = \ln 10$$

$$k = \frac{\ln 10}{4} \approx 0.5756.$$

27. The population P of a city is given by

$$P = 105{,}300e^{0.015t}$$

where t is the time in years with $t = 0$ corresponding to 1990. According to this model, in what year will the city have a population of 150,000?

Solution:

$$150{,}000 = 105{,}300e^{0.015t}$$

$$0.015t = \ln\left(\frac{150{,}000}{105{,}300}\right)$$

$$t = \frac{1}{0.015}\ln\left(\frac{150{,}000}{105{,}300}\right) \approx 23.588 \text{ years}$$

$$1990 + 23 = 2013$$

The city will have a population of 150,000 in the year 2013.

31. The population of Dhaka, Bangladesh was 4.22 million in 1990, and its projected population for the year 2000 is 6.49 million. (*Source:* U.S. Bureau of Census) Find the exponential growth model $y = Ce^{kt}$ for the population growth of Dhaka by letting $t = 0$ correspond to 1990. Use the model to predict the population of the city in 2010.

Solution:

$$P = Ce^{kt}$$

$$4.22 = Ce^{k(0)} \Rightarrow C = 4.22$$

$$6.49 = 4.22e^{10k}$$

$$\frac{6.49}{4.22} = e^{10k}$$

$$k = \frac{1}{10}\ln\left(\frac{6.49}{4.22}\right) \approx 0.0430$$

$$P = 4.22e^{0.0430t}$$

When $t = 20$: $P \approx 9.97$ million

35. The half-life of radioactive radium (Ra^{226}) is 1620 years. What percentage of a present amount of radioactive radium will remain after 100 years?

Solution:

$$Q(t) = Ce^{kt}$$

$$\tfrac{1}{2}C = Ce^{k(1620)}$$

$$\tfrac{1}{2} = e^{1620k}$$

$$k = \tfrac{1}{1620}\ln(\tfrac{1}{2}) \approx -0.0004$$

$$Q(t) = Ce^{-0.0004t}$$

When $t = 100$: $Q(100) \approx 0.961C = 96.1\%$ of C.

39. The sales S (in thousands of units) of a new product after it has been on the market t years are given by

$$S(t) = 100(1 - e^{kt}).$$

(a) Find S as a function of t if 15,000 units have been sold after one year.

(b) How many units will be sold after five years?

Solution:

(a) $S(t) = 100(1 - e^{kt}),\quad S = 15$ when $t = 1$

$$15 = 100(1 - e^{k})$$

$$0.15 = 1 - e^{k}$$

$$e^{k} = 0.85$$

$$k = \ln 0.85 \approx -0.1625$$

$$S(t) = 100[1 - e^{-0.1625t}]$$

(b) $S(5) = 100[1 - e^{-0.1625t}] \approx 55.625$ thousands of units $= 55{,}625$ units

41. A certain lake was stocked with 500 fish and the fish population increased according to the logistics curve

$$p(t) = \frac{10{,}000}{1 + 19e^{-t/5}}$$

where t is measured in months (see figure).

(a) Estimate the fish population after 5 months.

(b) After how many months will the fish population be 2000?

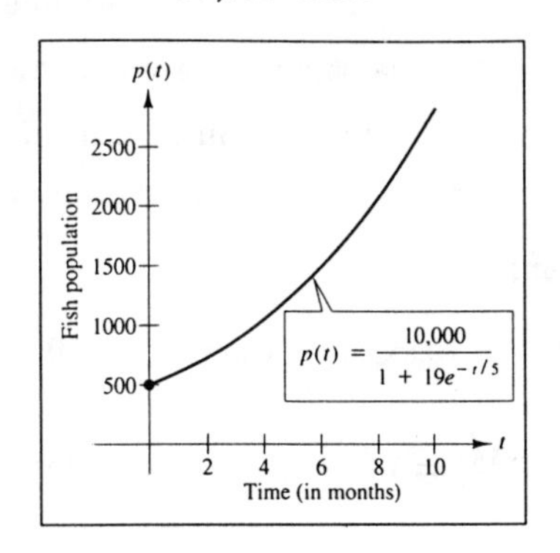

Solution:

$$p(t) = \frac{10{,}000}{1 + 19e^{-t/5}}$$

(a) $p(5) = \dfrac{10{,}000}{1 + 19e^{-1}} \approx 1252$ fish

(b)

$$2000 = \frac{10{,}000}{1 + 19e^{-t/5}}$$

$$1 + 19e^{-t/5} = 5$$

$$e^{-t/5} = \frac{4}{19}$$

$$-\frac{t}{5} = \ln\left(\frac{4}{19}\right)$$

$$t = -5\ln\left(\frac{4}{19}\right) \approx 7.8 \text{ months}$$

45. Find the magnitude R of an earthquake of intensity I (let $I_0 = 1$).

(a) $I = 80{,}500{,}000$ (b) $I = 48{,}275{,}000$

Solution:

$$R = \log_{10} \frac{I}{I_0} = \log_{10} I \text{ since } I_0 = 1$$

(a) $R = \log_{10} 80{,}500{,}000 \approx 7.9$

(b) $R = \log_{10} 48{,}275{,}000 \approx 7.7$

47. The intensity level β, in decibels, of a sound wave is defined by

$$\beta(I) = 10 \log_{10} \frac{I}{I_0}$$

where I_0 is an intensity of 10^{-16} watts per square centimeter, corresponding roughly to the faintest sound that can be heard. Determine $\beta(I)$ for the following conditions.

(a) $I = 10^{-14}$ watts per square centimeter (whisper)

(b) $I = 10^{-9}$ watts per square centimeter (busy street corner)

(c) $I = 10^{-6.5}$ watts per square centimeter (air hammer)

(d) $I = 10^{-4}$ watts per square centimeter (threshold of pain)

Solution:

$$\beta(I) = 10 \log_{10} \frac{I}{I_0} \text{ where } I_0 = 10^{-16} \text{ watt/cm}^2$$

(a) $\beta(10^{-14}) = 10 \log_{10} \frac{10^{-14}}{10^{-16}} = 10 \log_{10} 10^2 = 20$ decibels

(b) $\beta(10^{-9}) = 10 \log_{10} \frac{10^{-9}}{10^{-16}} = 10 \log_{10} 10^7 = 70$ decibels

(c) $\beta(10^{-6.5}) = 10 \log_{10} \frac{10^{-6.5}}{10^{-16}} = 10 \log_{10} 10^{9.5} = 95$ decibels

(d) $\beta(10^{-4}) = 10 \log_{10} \frac{10^{-4}}{10^{-16}} = 10 \log_{10} 10^{12} = 120$ decibels

51. Use the acidity model $\text{pH} = -\log_{10}[\text{H}^+]$, where acidity (pH) is a measure of the hydrogen ion concentration $[\text{H}^+]$ (measured in moles of hydrogen per liter) of a solution. Find the pH if $[\text{H}^+] = 2.3 \times 10^{-5}$.

Solution:

$$\text{pH} = -\log_{10}[\text{H}^+] = -\log_{10}[2.3 \times 10^{-5}] \approx 4.64$$

57. At 8:30 A.M. a coroner was called to the home of a person who had died during the night. In order to estimate the time of death, the coroner took the person's temperature twice. At 9:00 A.M. the temperature was 85.7° and at 9:30 A.M. the temperature was 82.8°. From these two temperatures the coroner was able to determine that the time elapsed since death and the body temperature were related by the formula

$$t = -2.5 \ln \frac{T - 70}{98.6 - 70}$$

where t is the time in hours that has elapsed since the person died and T is the temperature (in degrees Fahrenheit) of the person's body at 9:00 A.M. Assume that the person had a normal body temperature of 98.6° at death, and that the room temperature was a constant 70°. (This formula is derived from a general cooling principle called Newton's Law of Cooling.) Use this formula to estimate the time of death of the person.

Solution:

$$t = -2.5 \ln \left(\frac{T - 70}{98.6 - 70}\right)$$

At 9:00 A.M. we have: $t = -2.5 \ln \left(\frac{85.7 - 70}{98.6 - 70}\right) \approx 1.5$ hours

From this we can conclude that the person died at 7:30 A.M.

REVIEW EXERCISES FOR CHAPTER 6

Solutions to Selected Exercises

5. Match $f(x) = \log_2 x$ with the sketch of its graph.

Solution:

$f(x) = \log_2 x$

Vertical asymptote: $x = 0$

Intercept: $(1, 0)$

Matches (c)

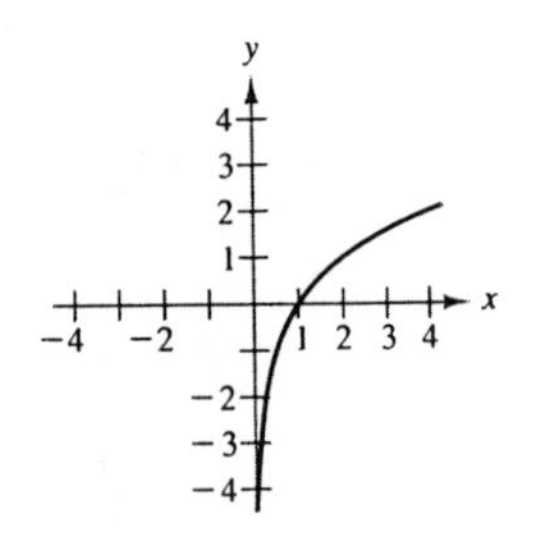

9. Sketch the graph of $g(x) = 6^{-x}$.

Solution:

$g(x) = 6^{-x} = \left(\frac{1}{6}\right)^x$

x	0	1	−1
$g(x)$	1	$\frac{1}{6}$	6

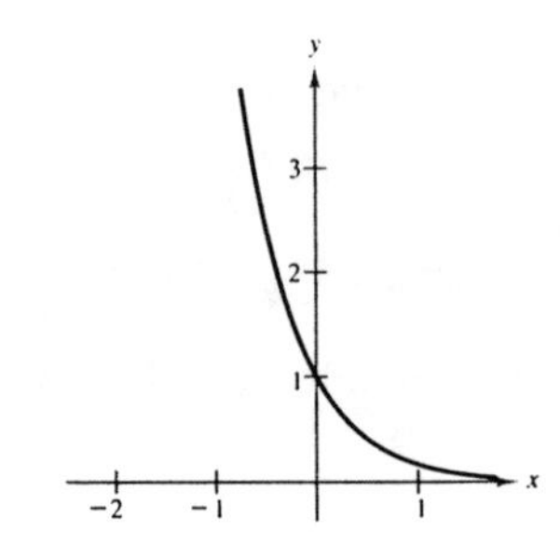

15. Use the table in the textbook to determine the balance A for \$3500 invested at 10.5% for 10 years and compounded n times per year.

Solution:

$n = 1: \quad A = 3500(1 + 0.105)^{10} \approx \$9{,}499.28$

$n = 2: \quad A = 3500\left(1 + \frac{0.105}{2}\right)^{20} \approx \$9{,}738.91$

$n = 4: \quad A = 3500\left(1 + \frac{0.105}{4}\right)^{40} \approx \$9{,}867.22$

$n = 12: \quad A = 3500\left(1 + \frac{0.105}{12}\right)^{120} \approx \$9{,}956.20$

$n = 365: \quad A = 3500\left(1 + \frac{0.105}{365}\right)^{3650} \approx \$10{,}000.27$

Continuous: $A = 3500e^{(0.105)(10)} \approx \$10{,}001.78$

n	1	2	4	12	365	Continuous compounding
A	\$9,499.28	\$9,738.91	\$9,867.22	\$9,956.20	\$10,000.27	\$10,001.78

17. Use the table in the textbook to determine the amount of money P that should be invested at 8% compounded continuously to produce a final balance of \$200,000 in t years.

Solution:

$$A = Pe^{rt}$$
$$P = Ae^{-rt} = 200{,}000e^{-0.08t}$$

t	1	10	20	30	40	50
P	\$184,623.27	\$89,865.79	\$40,379.30	\$18,143.59	\$8,152.44	\$3,663.13

21. A solution of a certain drug contained 500 units per milliliter when prepared. It was analyzed after 40 days and found to contain 300 units per milliliter. Assuming that the rate of decomposition is proportional to the amount present, the equation giving the amount A after t days is

$$A = 500e^{-0.013t}.$$

Use this model to find A when $t = 60$.

Solution:

$$A = 500e^{-0.013(60)} \approx 229.2 \text{ units per milliliter}$$

23. A certain automobile gets 28 miles per gallon of gasoline for speeds up to 50 miles per hour. Over 50 miles per hour, the number of miles per gallon drops at the rate of 12% for each 10 miles per hour. If s is the speed and y is the number of miles per gallon, then

$$y = 28e^{0.6-0.012s}, \quad s \geq 50.$$

Use this function to complete the table shown in the textbook.

Solution:

When $s = 50$, $y = 28e^{0.6-0.012(50)} = 28$ miles per gallon

When $s = 55$, $y = 28e^{0.6-0.012(55)} \approx 26.4$ miles per gallon

When $s = 60$, $y = 28e^{0.6-0.012(60)} \approx 24.8$ miles per gallon

When $s = 65$, $y = 28e^{0.6-0.012(65)} \approx 23.4$ miles per gallon

When $s = 70$, $y = 28e^{0.6-0.012(70)} \approx 22.0$ miles per gallon

Speed	50	55	60	65	70
Miles per gallon	28	26.4	24.8	23.4	22.0

27. Sketch the graph of $f(x) = \ln x + 3$.

Solution:

$f(x) = \ln x + 3$

Domain: $(0, \infty)$

x	1	2	3	$\frac{1}{2}$	$\frac{1}{4}$
$f(x)$	3	3.69	4.10	2.31	1.61

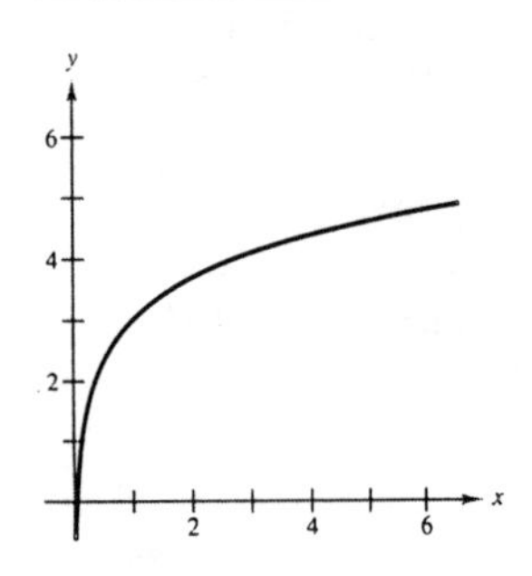

29. Sketch the graph of $h(x) = \ln(e^{x-1})$.

Solution:

$$h(x) = \ln(e^{x-1}) = (x-1)\ln e = x - 1$$

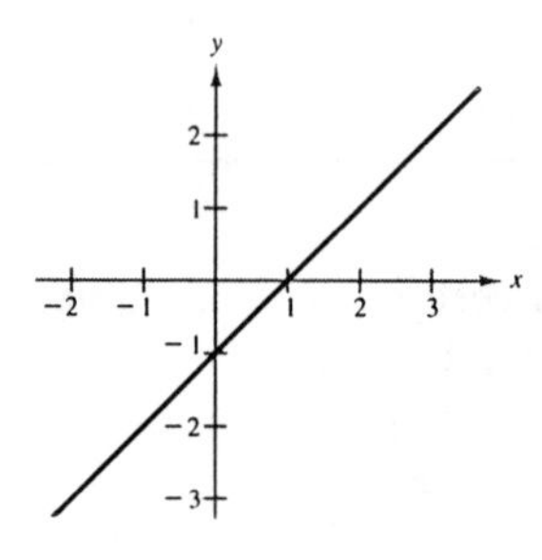

33. Evaluate $\log_{10} 1000$ without using a calculator.

Solution:

$$\log_{10} 1000 = \log_{10} 10^3 = 3$$

37. Evaluate $\ln e^7$ without using a calculator.

Solution:

$$\ln e^7 = 7 \ln e = 7(1) = 7$$

41. Evaluate $\log_4 9$ using the change of base formula. Do the problem twice; once with common logarithms and once with natural logarithms. Round to three decimal places.

Solution:

$$\log_4 9 = \frac{\log_{10} 9}{\log_{10} 4} \approx 1.585$$

$$\log_4 9 = \frac{\ln 9}{\ln 4} \approx 1.585$$

47. Use the properties of logarithms to write the following expression as a sum, difference, and/or multiple of logarithms.

$$\log_{10} \frac{5\sqrt{y}}{x^2}$$

Solution:

$$\begin{aligned}\log_{10} \frac{5\sqrt{y}}{x^2} &= \log_{10} 5\sqrt{y} - \log_{10} x^2 \\ &= \log_{10} 5 + \log_{10} \sqrt{y} - \log_{10} x^2 \\ &= \log_{10} 5 + \frac{1}{2} \log_{10} y - 2 \log_{10} x\end{aligned}$$

49. Use the properties of logarithms to write the following expression as a sum, difference, and/or multiple of logarithms.

$$\ln[(x^2+1)(x-1)]$$

Solution:

$$\ln[(x^2+1)(x-1)] = \ln(x^2+1) + \ln(x-1)$$

53. Write $\frac{1}{2} \ln|2x-1| - 2\ln|x+1|$ as the logarithm of a single quantity.

Solution:

$$\frac{1}{2} \ln|2x-1| - 2\ln|x+1| = \ln\sqrt{|2x-1|} - \ln(x+1)^2 = \ln \frac{\sqrt{|2x-1|}}{(x+1)^2}$$

55. Write $\ln 3 + \frac{1}{3}\ln(4-x^2) - \ln x$ as the logarithm of a single quantity.

Solution:

$$\ln 3 + \frac{1}{3}\ln(4-x^2) - \ln x = \ln 3 + \ln \sqrt[3]{4-x^2} - \ln x = \ln\left(3\sqrt[3]{4-x^2}\right) - \ln x = \ln \frac{3\sqrt[3]{4-x^2}}{x}$$

59. Determine whether the equation $\ln(x+y) = \ln x + \ln y$ is true or false.

Solution:

False, since $\ln x + \ln y = \ln(xy)$

63. Approximate $\log_b \sqrt{3}$ using the properties of logarithms given $\log_b 3 \approx 0.5646$.

Solution:

$$\log_b \sqrt{3} = \tfrac{1}{2} \log_b 3 \approx \tfrac{1}{2}(0.5646) = 0.2823$$

67. Solve $e^x = 12$. Round to three decimal places.

Solution:

$$e^x = 12$$
$$x = \ln 12 \approx 2.485$$

71. Solve $e^{2x} - 7e^x + 10 = 0$. Round to three decimal places.

Solution:

$$e^{2x} - 7e^x + 10 = 0$$
$$(e^x - 2)(e^x - 5) = 0$$

$e^x = 2$ or $e^x = 5$

$x = \ln 2$ $\quad$ $x = \ln 5$

$x \approx 0.693$ $\quad$ $x \approx 1.609$

75. Solve $\ln x - \ln 3 = 2$. Round to three decimal places.

Solution:

$$\ln x - \ln 3 = 2$$
$$\ln\left(\frac{x}{3}\right) = 2$$
$$\frac{x}{3} = e^2$$
$$x = 3e^2 \approx 22.167$$

79. Find the exponential function $y = Ce^{kt}$ that passes through the points $(0, 4)$ and $(5, \frac{1}{2})$.

Solution:

Since the graph passes through the point $(0, 4)$, we have

$$4 = Ce^{k(0)},$$
$$4 = C(1) \text{ so } y = 4e^{kt}.$$

Since the graph passes through the point $(5, \frac{1}{2})$, we have

$$\tfrac{1}{2} = 4e^{5k},$$
$$\tfrac{1}{8} = e^{5k}$$
$$5k = \ln \tfrac{1}{8}$$
$$k \approx -0.4159.$$

Thus, $y = 4e^{-0.4159t}$.

81. The demand equation for a certain product is given by

$$p = 500 - 0.5e^{0.004x}.$$

Find the demand x for a price of (a) $p = \$450$ and (b) $p = \$400$.

Solution:

(a) $p = 450$

$$450 = 500 - 0.5e^{0.004x}$$
$$0.5e^{0.004x} = 50$$
$$e^{0.004x} = 100$$
$$0.004x = \ln 100$$
$$x \approx 1151 \text{ units}$$

(b) $p = 400$

$$400 = 500 - 0.5e^{0.004x}$$
$$0.5e^{0.004x} = 100$$
$$e^{0.004x} = 200$$
$$0.004x = \ln 200$$
$$x \approx 1325 \text{ units}$$

Practice Test for Chapter 6

1. Solve for x: $x^{3/5} = 8$.

2. Solve for x: $3^{x-1} = \frac{1}{81}$.

3. Graph $f(x) = 2^{-x}$.

4. Graph $g(x) = e^x + 1$.

5. If \$5000 is invested at 9% interest, find the amount after three years if the interest is compounded

(a) monthly (b) quarterly (c) continuously.

6. Write the equation in logarithmic form: $7^{-2} = \frac{1}{49}$.

7. Solve for x: $x - 4 = \log_2 \frac{1}{64}$.

8. Given $\log_b 2 = 0.3562$ and $\log_b 5 = 0.8271$, evaluate $\log_b \sqrt[4]{8/25}$.

9. Write $5 \ln x - \frac{1}{2} \ln y + 6 \ln z$ as a single logarithm.

10. Using your calculator and the change of base formula, evaluate $\log_9 28$.

11. Use your calculator to solve for $N : \log_{10} N = 0.6646$.

12. Graph $y = \log_4 x$.

13. Determine the domain of $f(x) = \log_3(x^2 - 9)$.

14. Graph $y = \ln(x - 2)$.

15. True or False: $\dfrac{\ln x}{\ln y} = \ln(x - y)$.

16. Solve for x: $5^x = 41$.

17. Solve for x: $x - x^2 = \log_5 \frac{1}{25}$.

18. Solve for x: $\log_2 x + \log_2(x - 3) = 2$.

19. Solve for x: $\dfrac{e^x + e^{-x}}{3} = 4$.

20. Six thousand dollars is deposited into a fund at an annual percentage rate of 13%. Find the time required for the investment to double if the interest is compounded continuously.

CHAPTER 7

Systems of Equations and Inequalities

SECTION 7.1

Systems of Equations

- **You should be able to solve systems of equations by the method of substitution.**
  1. **Solve one of the equations for one of the variables.**
  2. **Substitute this expression into the other equation and solve.**
  3. **Back substitute into the first equation to find the value of the other variable.**
- **You should be able to find solutions graphically. (See Example 5 in the textbook.)**

Solutions to Selected Exercises

5. Solve the following system by the method of substitution.

$$x + 3y = 15$$
$$x^2 + y^2 = 25$$

Solution:

$$x + 3y = 15 \Rightarrow x = 15 - 3y$$

$$\begin{aligned} x^2 + y^2 &= 25 \\ (15 - 3y)^2 + y^2 &= 25 \\ 225 - 90y + 10y^2 &= 25 \\ 10y^2 - 90y + 200 &= 0 \\ 10(y^2 - 9y + 20) &= 0 \\ 10(y-4)(y-5) &= 0 \end{aligned}$$

$y = 4$ or $y = 5$

$x = 3$ $\quad$ $x = 0$

Solutions: $(3, 4)$, $(0, 5)$

9. Solve the following system by the method of substitution.

$$x - 3y = -4$$
$$x^2 - y^3 = 0$$

Solution:

$$x - 3y = -4 \Rightarrow x = 3y - 4$$
$$x^2 - y^3 = 0$$
$$(3y-4)^2 - y^3 = 0$$
$$9y^2 - 24y + 16 - y^3 = 0$$
$$y^3 - 9y^2 + 24y - 16 = 0 \quad \text{Multiply both sides by } -1.$$
$$(y-1)(y-4)^2 = 0 \quad \text{See Section 4.5.}$$

$y = 1$ or $y = 4$

$x = -1$ $\quad$ $x = 8$

Solutions: $(-1, 1)$, $(8, 4)$

13. Solve the following system by the method of substitution.

$$2x - y + 2 = 0$$
$$4x + y - 5 = 0$$

Solution:

$$2x - y + 2 = 0 \Rightarrow y = 2x + 2$$
$$4x + y - 5 = 0$$
$$4x + (2x + 2) - 5 = 0$$
$$6x = 3$$
$$x = \tfrac{1}{2}$$
$$y = 3$$

Solution: $(\frac{1}{2}, 3)$

17. Solve the following system by the method of substitution.

$$\begin{aligned} \tfrac{1}{5}x + \tfrac{1}{2}y &= 8 \\ x + \quad y &= 20 \end{aligned}$$

Solution:

$$\tfrac{1}{5}x + \tfrac{1}{2}y = 8 \Rightarrow 2x + 5y = 80$$

$$x + y = 20 \Rightarrow y = 20 - x$$

$$\begin{aligned} 2x + 5(20 - x) &= 80 \\ -3x &= -20 \\ x &= \tfrac{20}{3} \\ y &= \tfrac{40}{3} \end{aligned}$$

Solution: $\left(\tfrac{20}{3}, \tfrac{40}{3}\right)$

23. Solve the following system by the method of substitution.

$$\begin{aligned} 3x - 7y + 6 &= 0 \\ x^2 - y^2 &= 4 \end{aligned}$$

Solution:

$$3x - 7y + 6 = 0 \Rightarrow x = \frac{7y - 6}{3}$$

$$\begin{aligned} x^2 - y^2 &= 4 \\ \left(\frac{7y-6}{3}\right)^2 - y^2 &= 4 \\ \frac{49y^2 - 84y + 36}{9} - y^2 &= 4 \\ 49y^2 - 84y + 36 - 9y^2 &= 36 \\ 40y^2 - 84y &= 0 \\ 4y(10y - 21) &= 0 \end{aligned}$$

$$y = 0 \qquad \text{or} \qquad y = \frac{21}{10}$$

$$x = -2 \qquad\qquad x = \frac{29}{10}$$

Solutions: $(-2, 0)$, $\left(\dfrac{29}{10}, \dfrac{21}{10}\right)$

27. Solve the following system by the method of substitution.

$$y = x^4 - 2x^2 + 1$$
$$y = 1 - x^2$$

Solution:

$$x^4 - 2x^2 + 1 = 1 - x^2$$
$$x^4 - x^2 = 0$$
$$x^2(x^2 - 1) = 0$$
$$x^2(x + 1)(x - 1) = 0$$

$x = 0$ or $x = -1$ or $x = 1$

$y = 1$ $\quad y = 0$ $\quad y = 0$

Solutions: $(0, 1)$, $(\pm 1, 0)$

29. Solve the following system by the method of substitution.

$$xy - 1 = 0$$
$$2x - 4y + 7 = 0$$

Solution:

$$xy - 1 = 0$$
$$2x - 4y + 7 = 0 \Rightarrow x = \frac{4y - 7}{2}$$
$$\left(\frac{4y - 7}{2}\right) y - 1 = 0$$
$$4y^2 - 7y - 2 = 0$$
$$(4y + 1)(y - 2) = 0$$

$y = -\dfrac{1}{4}$ or $y = 2$

$x = -4$ $\quad x = \dfrac{1}{2}$

Solutions: $\left(-4, -\dfrac{1}{4}\right)$, $\left(\dfrac{1}{2}, 2\right)$

33. Find all points of intersection of the graphs of the given pair of equations. [*Hint:* A graphical approach, as demonstrated in Example 5, may be helpful.]

$$2x - y + 3 = 0$$
$$x^2 + y^2 - 4x = 0$$

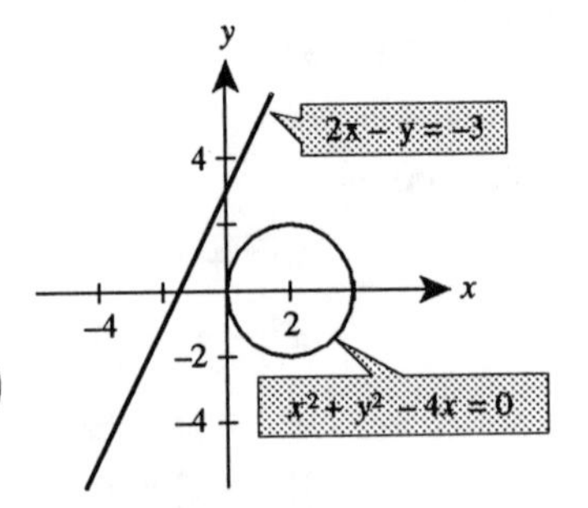

Solution:

$2x - y + 3 = 0 \Rightarrow y = 2x + 3$ (Line)

$x^2 + y^2 - 4x = 0 \Rightarrow (x - 2)^2 + y^2 = 4$ (Circle)

No points of intersection

37. Find all points of intersection of the graphs of the given pair of equations. [*Hint:* A graphical approach, as demonstrated in Example 5, may be helpful.]

$$y = e^x$$
$$x - y + 1 = 0$$

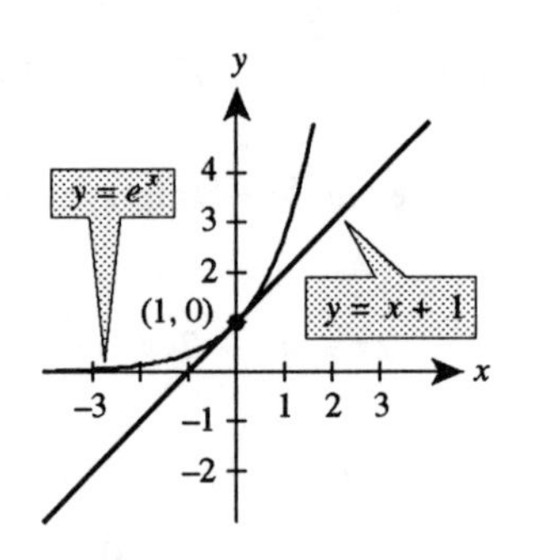

Solution:

$y = e^x$ (Exponential curve)

$x - y + 1 = 0 \Rightarrow y = x + 1$ (Line)

The graphs intersect at the point $(0,\ 1)$.

41. Find all points of intersection of the graphs of the given pair of equations. [*Hint:* A graphical approach, as demonstrated in Example 5, may be helpful.]

$$x^2 + y^2 = 169$$
$$x^2 - 8y = 104$$

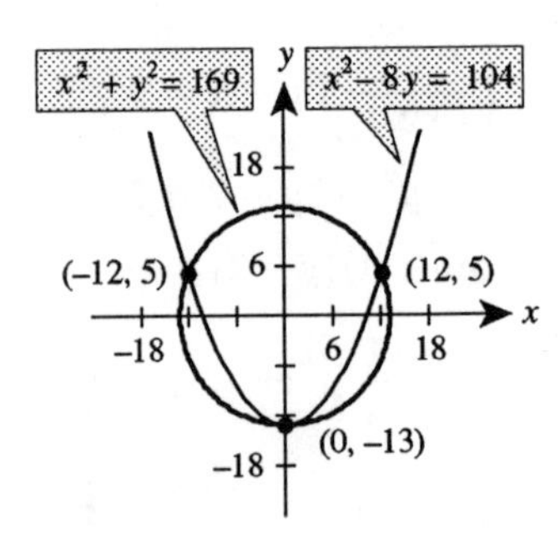

Solution:

$x^2 + y^2 = 169 \Rightarrow x^2 = 8(y + 13)$ (Circle)

$x^2 - 8y = 104$ (Parabola)

The points of intersection are $(0,\ -13)$ and $(\pm 12,\ 5)$.

43. Find the sales necessary to break even ($R = C$) for $C = 8650x + 250{,}000$ and $R = 9950x$. (Round your answer to the nearest whole unit.)

Solution:

$$R = C$$
$$9950x = 8650x + 250{,}000$$
$$1300x = 250{,}000$$
$$x \approx 192 \text{ units}$$

47. Suppose you are setting up a small business and have invested \$16,000 to produce an item that will sell for \$5.95. If each unit can be produced for \$3.45, how many units must be sold to break even?

Solution:

Let $x =$ the number of units

$$C = 3.45x + 16{,}000$$
$$R = 5.95x$$

To break even: $R = C$

$$5.95x = 3.45x + 16,000$$
$$2.5x = 16{,}000$$
$$x = 6400 \text{ units}$$

51. Suppose you are offered two different jobs selling dental supplies. One company offers a straight commission of 6% of sales. The other company offers a salary of \$250 per week *plus* 3% of the sales. How much would you have to sell in a week in order to make the straight commission offer better?

Solution:

Let $x =$ total sales

Job #1: $y = 0.06x$

Job #2: $y = 250 + 0.03x$

You break even when

$$0.06x = 250 + 0.03x$$
$$0.03x = 250$$
$$x \approx \$8333.33.$$

If you sell more than \$8333.33 in a week, the straight commission offer is better.

55. What are the dimensions of a rectangular tract of land if its perimeter is 40 miles and its area is 96 square miles?

Solution:

Let l = the length of the rectangle and w = the width of the rectangle.

Perimeter: $2l + 2w = 40 \Rightarrow w = 20 - l$

Area: $lw = 96 \Rightarrow l(20 - l) = 96$

$$20l - l^2 = 96$$

$$0 = l^2 - 20l + 96$$

$$0 = (l - 8)(l - 12)$$

$l = 8$ or $l = 12$

$w = 12$ $\quad$ $w = 8$

The dimensions are 12 miles by 8 miles.

SECTION 7.2

Systems of Linear Equations in Two Variables

- You should be able to solve a linear system by the method of elimination.
- You should know that for a system of two linear equations, one of the following is true.
  (a) There are infinitely many solutions; the lines are identical.
  (b) There is no solution; the lines are parallel.
  (c) There is one solution; the lines intersect at one point.

Solutions to Selected Exercises

5. Solve the linear system by elimination. Identify and label each line with the appropriate equation.

$$x - y = 1$$
$$-2x + 2y = 5$$

Solution:

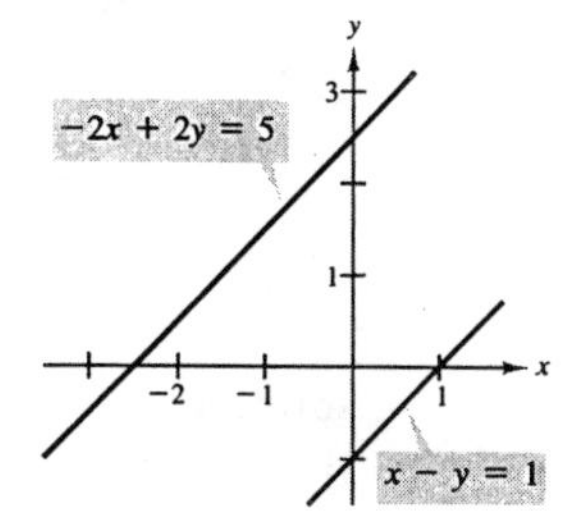

$$\begin{aligned} x - y = 1 &\Rightarrow \quad 2x - 2y = 2 \\ -2x + 2y = 5 &\Rightarrow \underline{-2x + 2y = 5} \\ & \qquad\qquad 0 = 7 \end{aligned}$$

Inconsistent; no solution

9. Solve the linear system by elimination. Identify and label each line with the appropriate equation.

$$9x - 3y = -1$$
$$3x + 6y = -5$$

Solution:

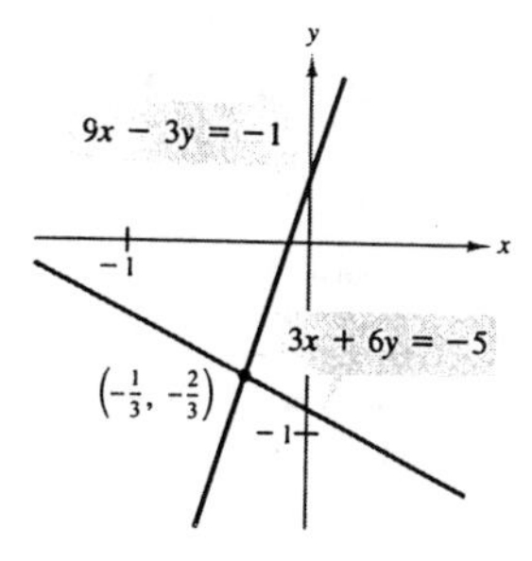

$$\begin{aligned} 9x - 3y = -1 &\Rightarrow 18x - 6y = -2 \\ 3x + 6y = -5 &\Rightarrow \underline{\ 3x + 6y = -5} \\ & \qquad 21x \qquad = -7 \\ & \qquad\qquad x = -\tfrac{1}{3} \\ & \qquad\qquad y = -\tfrac{2}{3} \end{aligned}$$

Consistent; one solution

Solution: $\left(-\frac{1}{3}, -\frac{2}{3}\right)$

13. Solve the system by elimination.

$$\begin{aligned} 2x + 3y &= 18 \\ 5x - y &= 11 \end{aligned}$$

Solution:

$$\begin{aligned} 2x + 3y = 18 &\Rightarrow 2x + 3y = 18 \\ 5x - y = 11 &\Rightarrow \underline{15x - 3y = 33} \\ & 17x = 51 \\ & x = 3 \\ & y = 4 \end{aligned}$$

Consistent; one solution
Solution: (3, 4)

17. Solve the system by elimination.

$$\begin{aligned} 2u + v &= 120 \\ u + 2v &= 120 \end{aligned}$$

Solution:

$$\begin{aligned} 2u + v = 120 &\Rightarrow -4u - 2v = -240 \\ u + 2v = 120 &\Rightarrow \underline{u + 2v = 120} \\ & -3u = -120 \\ & u = 40 \\ & v = 40 \end{aligned}$$

Consistent; one solution
Solution: (40, 40)

23. Solve the system by elimination.

$$\begin{aligned} \frac{x+3}{4} + \frac{y-1}{3} &= 1 \\ 2x - y &= 12 \end{aligned}$$

Solution:

$$\begin{aligned} \frac{x+3}{4} + \frac{y-1}{3} = 1 \Rightarrow 3(x+3) + 4(y-1) = 12 &\Rightarrow 3x + 4y = 7 \\ 2x - y = 12 &\Rightarrow \underline{8x - 4y = 48} \\ & 11x = 55 \\ & x = 5 \\ & y = -2 \end{aligned}$$

Consistent; one solution
Solution: (5, −2)

25. Solve the system by elimination.

$$\begin{aligned} 2.5x - 3y &= 1.5 \\ 10x - 12y &= 6 \end{aligned}$$

Solution:

$$\begin{aligned} 2.5x - 3y = 1.5 &\Rightarrow 25x - 30y = 15 \Rightarrow 5x - 6y = 3 \\ 10x - 12y = 6 &\Rightarrow 5x - 6y = 3 \Rightarrow -5x + 6y = -3 \\ & \qquad\qquad\qquad\qquad\qquad 0 = 0 \end{aligned}$$

Consistent; infinite solutions of the form $\left(a, \frac{5}{6}a - \frac{1}{2}\right)$

29. Solve the system by elimination.

$$\begin{aligned} 4b + 3m &= 3 \\ 3b + 11m &= 13 \end{aligned}$$

Solution:

$$\begin{aligned} 4b + 3m = 3 &\Rightarrow 44b + 33m = 33 \\ 3b + 11m = 13 &\Rightarrow -9b - 33m = -39 \\ & \quad\; 35b = -6 \\ & \quad\; b = -\tfrac{6}{35} \\ & \quad\; m = \tfrac{43}{35} \end{aligned}$$

Consistent; one solution

Solution: $\left(-\frac{6}{35}, \frac{43}{35}\right)$

33. An airplane flying into a headwind travels the 1800-mile flying distance between two cities in 3 hours and 36 minutes. On the return flight, the distance is traveled in 3 hours. Find the ground speed of the plane and the speed of the wind, assuming that both remain constant.

Solution:

Use (rate)(time) = distance.

Let g = ground speed and w = wind speed.

$$\begin{aligned} (g - w)\left(3 + \tfrac{36}{60}\right) = 1800 &\Rightarrow g - w = 500 \\ (g + w)(3) = 1800 &\Rightarrow g + w = 600 \\ & \quad\; 2g = 1100 \\ & \quad\; g = 550 \\ & \quad\; w = 50 \end{aligned}$$

The ground speed of the plane is 550 mph and the speed of the wind is 50 mph.

35. Ten gallons of a 30% acid solution are obtained by mixing a 20% solution with a 50% solution. How much of each must be used?

Solution:

Let $x =$ amount of 20% solution and $y =$ amount of 50% solution.

$$\begin{aligned} x + y &= 10 \Rightarrow -2x - 2y = -20 \\ 0.2x + 0.5y = 0.3(10) &\Rightarrow \underline{2x + 5y = 30} \\ 3y &= 10 \\ y &= \tfrac{10}{3} \\ x &= \tfrac{20}{3} \end{aligned}$$

Solution: $\frac{20}{3}$ gallons of 20% solution, $\frac{10}{3}$ gallons of 50% solution

39. Five hundred tickets were sold for a certain performance of a play. The tickets for adults and children sold for \$7.50 and \$4.00, respectively, and the receipts for the performance were \$3312.50. How many of each kind of ticket were sold?

Solution:

Let $x =$ number of adult tickets and $y =$ number of child tickets. Then

$$\begin{aligned} x + y = 500 &\Rightarrow -4x - 4y = -2000 \\ \text{and } 7.50x + 4.00y = 3312.50 &\Rightarrow \underline{7.5x + 4y = 3312.50} \\ 3.5x &= 1312.50 \\ x &= 375 \\ y &= 125 \end{aligned}$$

Thus, 375 adult tickets and 125 child tickets were sold.

43. Find the point of equilibrium.

Demand: $p = 300 - x$

Supply: $p = 100 + x$

Solution:

$$\begin{aligned} \text{Supply} &= \text{Demand} \\ 100 + x &= 300 - x \\ 2x &= 200 \\ x &= 100 \\ p &= 200 \end{aligned}$$

The point of equilibrium occurs when $x = 100$ units and the price per unit is \$200 (assuming that the price is in dollars).

47. On a 300 mile trip two people do the driving. One person drives three times as far as the other. Find the distance that each person drives.

Solution:

Let $x =$ distance of the first driver and $y =$ distance of the second driver. Then

$$\begin{aligned} \text{Then } x + y = 300 &\Rightarrow 3x + 3y = 900 \\ \text{and } x = 3y &\Rightarrow \underline{x - 3y = 0} \\ & 4x = 900 \\ & x = 225 \\ & y = 75 \end{aligned}$$

One person drives 225 miles and the other person drives 75 miles.

51. Find the least squares regression line $y = ax + b$. To find the line, solve the following system for a and b.

$$\begin{aligned} 7b + 21a &= 35.1 \\ 21b + 91a &= 114.2 \end{aligned}$$

Solution:

$$\begin{aligned} 7b + 21a = 35.1 &\Rightarrow -21b - 63a = -105.3 \\ 21b + 91a = 114.2 &\Rightarrow \underline{21b + 91a = 114.2} \\ 28a &= 8.9 \\ a &= \frac{8.9}{28} = \frac{89}{280} \\ b &= \frac{1137}{280} \end{aligned}$$

Thus, $y = \frac{89}{280}x + \frac{1137}{280} \approx 0.318x + 4.061$.

55. Find the least squares regression line, $y = ax+b$ for the points $(0, 4)$, $(1, 3)$, $(1, 1)$, $(2, 0)$. The *least squares regression line*, $y = ax+b$, for the points (x_1, y_1), (x_2, y_2), $\ldots$, (x_n, y_n) is obtained by solving the following system of linear equations for a and b.

$$\begin{aligned} nb + \left(\sum_{i=1}^{n} x_i\right) a &= \sum_{i=1}^{n} y_i \\ \left(\sum_{i=1}^{n} x_i\right) b + \left(\sum_{i=1}^{n} x_i^2\right) a &= \sum_{i=1}^{n} x_i y_i \end{aligned}$$

55. –CONTINUED–

Solution:

$$n = 4$$

$$\sum_{i=1}^{4} x_i = 4$$

$$\sum_{i=1}^{4} x_i^2 = 6$$

$$\sum_{i=1}^{4} y_i = 8$$

$$\sum_{i=1}^{4} x_i y_i = 4$$

$$\begin{aligned} 4b + 4a &= \ \ 8 \\ -(4b + 6a &= \ \ 4) \\ \hline -2a &= \ \ 4 \\ a &= -2 \\ b &= \ \ 4 \end{aligned}$$

$$y = -2x + 4$$

59. Find a system of linear equations having the solution $(3, \frac{5}{2})$. (The answer is not unique.)

Solution:

Since $x = 3$ and $y = \frac{5}{2}$, one possible system of linear equations is

$$\begin{aligned} x + 2y &= \ \ 8 \\ x - 2y &= -2. \end{aligned}$$

SECTION 7.3

Linear Systems in More Than Two Variables

- **You should know the operations that lead to equivalent systems of equations:**
  - **(a) Interchange any two equations.**
  - **(b) Multiply all terms of an equation by a nonzero constant.**
  - **(c) Replace an equation by the sum of itself and a constant multiple of any other equation in the system.**
- **You should be able to use the method of elimination.**

Solutions to Selected Exercises

3. Solve the following system of linear equations.

$$\begin{aligned} 4x + y - 3z &= 11 \\ 2x - 3y + 2z &= 9 \\ x + y + z &= -3 \end{aligned}$$

Solution:

$$\begin{array}{r} 4x + y - 3z = 11 \\ \underline{-4x + 6y - 4z = -18} \\ 7y - 7z = -7 \\ y - z = -1 \end{array} \qquad \begin{array}{r} 2x - 3y + 2z = 9 \\ \underline{-2x - 2y - 2z = 6} \\ -5y = 15 \end{array}$$

$y = -3,\ z = -2,\ x = 2$

Solution: $(2, -3, -2)$

7. Solve the following system of linear equations.

$$\begin{aligned} 3x - 2y + 4z &= 1 \\ x + y - 2z &= 3 \\ 2x - 3y + 6z &= 8 \end{aligned}$$

Solution:

$$\begin{array}{r} 3x - 2y + 4z = 1 \\ \underline{-3x - 3y + 6z = -9} \\ -5y + 10z = -8 \end{array} \qquad \begin{array}{r} -2x - 2y + 4z = -6 \\ \underline{2x - 3y + 6z = 8} \\ -5y + 10z = 2 \end{array} \qquad \begin{array}{r} -5y + 10z = -8 \\ \underline{5y - 10z = -2} \\ 0 = -10 \end{array}$$

Inconsistent; no solution

11. Solve the following system of linear equations.

$$\begin{aligned} x + 2y - 7z &= -4 \\ 2x + y + z &= 13 \\ 3x + 9y - 36z &= -33 \end{aligned}$$

Solution:

$$\begin{array}{r} 2x + 4y - 14z = -8 \\ -2x - y - z = -13 \\ \hline 3y - 15z = -21 \\ y - 5z = -7 \end{array} \qquad \begin{array}{r} 3x + 6y - 21z = -12 \\ -3x - 9y + 36z = 33 \\ \hline -3y + 15z = 21 \\ y - 5z = -7 \end{array} \qquad \begin{array}{r} y - 5z = -7 \\ -y + 5z = 7 \\ \hline 0 = 0 \end{array}$$

Consistent; infinite solutions

Let $z = a$. Then

$$\begin{aligned} y &= 5a - 7 \\ x &= -4 - 2(5a - 7) + 7a = -3a + 10. \end{aligned}$$

Solution: $(-3a + 10,\ 5a - 7,\ a)$, a is any real number.

15. Solve the following system of linear equations.

$$\begin{aligned} x - 2y + 5z &= 2 \\ 3x + 2y - z &= -2 \end{aligned}$$

Solution:

$$\begin{array}{r} x - 2y + 5z = 2 \\ 3x + 2y - z = -2 \\ \hline 4x \quad + 4z = 0 \end{array} \Rightarrow x = -z$$

Let $z = a$. Then

$$\begin{aligned} x &= -a \\ y &= \tfrac{1}{2}[-a + 5a - 2] = 2a - 1. \end{aligned}$$

Solution: $(-a,\ 2a - 1,\ a)$, a is any real number.

21. Solve the following system of linear equations.

$$\begin{aligned} x \quad\;\; + 4z &= 1 \\ x + y + 10z &= 10 \\ 2x - y + 2z &= -5 \end{aligned}$$

Solution:

$$\begin{array}{r} -x \quad\;\; - 4z = -1 \\ x + y + 10z = 10 \\ \hline y + 6z = 9 \end{array} \qquad \begin{array}{r} -2x \quad\;\; - 8z = -2 \\ 2x - y + 2z = -5 \\ \hline -y - 6z = -7 \end{array} \qquad \begin{array}{r} y + 6z = 9 \\ -y - 6z = -7 \\ \hline 0 = 2 \end{array}$$

Inconsistent; no solution

23. Solve the following system of linear equations.

$$\begin{aligned} 4x + 3y + 17z &= 0 \\ 5x + 4y + 22z &= 0 \\ 4x + 2y + 19z &= 0 \end{aligned}$$

Solution:

$$\begin{array}{r} 20x + 15y + 85z = 0 \\ -20x - 16y - 88z = 0 \\ \hline -y - 3z = 0 \end{array} \qquad \begin{array}{r} 4x + 3y + 17z = 0 \\ -4x - 2y - 19z = 0 \\ \hline y - 2z = 0 \\ -y - 3z = 0 \\ \hline -5z = 0 \end{array}$$

$$z = 0,\ y = 0,\ x = 0$$

Solution: $(0, 0, 0)$

27. Find the equation of the parabola $y = ax^2 + bx + c$ that passes through the points $(0, -4)$, $(1, 1)$, and $(2, 10)$.

Solution:

$$\begin{aligned} -4 &= a(0)^2 + b(0) + c \Rightarrow -4 = c \\ 1 &= a(1)^2 + b(1) + c \Rightarrow 1 = a + b + c \\ 10 &= a(2)^2 + b(2) + c \Rightarrow 10 = 4a + 2b + c \end{aligned}$$

$$\begin{aligned} 1 &= a + b - 4 \Rightarrow a + b = 5 \Rightarrow -a - b = -5 \\ 10 &= 4a + 2b - 4 \Rightarrow 4a + 2b = 14 \Rightarrow 2a + b = 7 \end{aligned}$$

$$a = 2$$

$$b = 3$$

Thus, $y = 2x^2 + 3x - 4$.

31. Find the equation of the circle $x^2+y^2+Dx+Ey+F=0$ that passes through the points (0, 0), (2, −2), and (4, 0).

Solution:

$$\begin{aligned}
(0)^2 + (0)^2 + D(0) + E(0) + F = 0 &\Rightarrow \quad F = 0\\
(2)^2 + (-2)^2 + D(2) + E(-2) + F = 0 &\Rightarrow 2D - 2E + F = -8\\
(4)^2 + (0)^2 + D(4) + E(0) + F = 0 &\Rightarrow 4D \quad + F = -16
\end{aligned}$$

$$\begin{aligned}
2D - 2E + 0 &= -8 \Rightarrow D - E = -4\\
4D \quad + 0 &= -16 \Rightarrow D \quad = -4 \Rightarrow E = 0
\end{aligned}$$

Thus, $x^2 + y^2 - 4x + 0y + 0 = 0$

$$x^2 + y^2 - 4x = 0.$$

35. Find a, v_0, and s_0 in the position equation $s = \frac{1}{2}at^2 + v_0t + s_0$.

At $t = 1$ second, $s = 128$ feet.
At $t = 2$ seconds, $s = 80$ feet.
At $t = 3$ seconds, $s = 0$ feet.

Solution:

$s = \frac{1}{2}at^2 + v_0t + s_0$

(1, 128), (2, 80), (3, 0)

$$\begin{aligned}
128 &= \tfrac{1}{2}a + v_0 + s_0 \Rightarrow a + 2v_0 + 2s_0 = 256\\
80 &= 2a + 2v_0 + s_0 \Rightarrow 2a + 2v_0 + s_0 = 80\\
0 &= \tfrac{9}{2}a + 3v_0 + s_0 \Rightarrow 9a + 6v_0 + 2s_0 = 0
\end{aligned}$$

$$\begin{aligned}
2a + 4v_0 + 4s_0 &= 512\\
-2a - 2v_0 - s_0 &= -80\\
\hline
2v_0 + 3s_0 &= 432
\end{aligned}
\qquad
\begin{aligned}
18a + 18v_0 + 9s_0 &= 720\\
-18a - 12v_0 - 4s_0 &= 0\\
\hline
6v_0 + 5s_0 &= 720\\
-6v_0 - 9s_0 &= -1296\\
\hline
-4s_0 &= -576\\
s_0 &= 144\\
v_0 &= 0\\
a &= -32
\end{aligned}$$

Thus, $s = \frac{1}{2}(-32)t^2 + (0)t + 144$

$$= -16t^2 + 144.$$

39. An inheritance of \$16,000 was divided among three investments yielding a total of \$900 in interest per year. The interest rates for the three investments were 5%, 6%, and 7%. Find the amount placed in each investment if the 5% and 6% investments were \$3000 and \$2000 less than the 7% investment, respectively.

Solution:

Let x = amount at 5%, y = amount at 6%, and z = amount at 7%.

$$x + y + z = 16{,}000$$

$$0.05x + 0.06y + 0.07z = 990$$

$$x = z - 3000$$

$$y = z - 2000$$

$$(z - 3000) + (z - 2000) + z = 16{,}000$$

$$3z = 21{,}000$$

$$z = \$7000 \text{ at } 7\%$$

$$y = \$5000 \text{ at } 6\%$$

$$x = \$4000 \text{ at } 5\%$$

Check: $0.05(4000) + 0.06(5000) + 0.07(7000) = 990$

43. Consider an investor with a portfolio totaling \$500,000 that is to be allocated among the following types of investments: (1) certificates of deposit, (2) municipal bonds, (3) blue-chip stocks, and (4) growth or speculative stocks. How much should be allocated to each type of investment?

The certificates of deposit pay 10% annually, and the municipal bonds pay 8% annually. Over a five-year period, the investor expects the blue-chip stocks to return to 12% annually, and expects the growth stocks to return to 13% annually. The investor wants a combined annual return of 10% and also wants to have only one-fourth of the portfolio invested in stocks.

Solution:

Let x = amount in certificates of deposit, y = amount in municipal bonds, z = amount in blue-chip stocks, and w = amount in growth or speculative stocks.

$$x + y + z + w = 500{,}000$$

$$0.10x + 0.08y + 0.12z + 0.13w = 0.10(500{,}000)$$

$$z + w = 0.25(500{,}000)$$

43. –CONTINUED–

Using elimination, we find **infinitely** many solutions:

$w = a$

$z = 125{,}000 - a$

$y = 125{,}000 + \frac{1}{2}a$

$x = 250{,}000 - \frac{1}{2}a$

One **possible** solution is to let $a = 50{,}000$.

Certificates of deposit:	$225,000
Municipal bonds:	$150,000
Blue-chip stocks:	$75,000
Growth or speculative stocks:	$50,000

47. A small company that manufactures products A and B has an order for 15 units of product A and 16 units of product B. The company has trucks of three different sizes that can haul the products, as shown in the following table. How many trucks of each size are needed to deliver the order? (Give *two* possible solutions.)

	Product	
Truck	A	B
Large	6	3
Medium	4	4
Small	0	3

Solution:

Possible solutions:

(1) 4 medium trucks

(2) 2 large trucks, 1 medium truck, 2 small trucks

(3) 3 large trucks, 1 medium truck, 1 small truck

(4) 3 large trucks, 3 small trucks

51. Use a system of linear equations to decompose the following rational fraction into partial fractions. (See Example 9 in this section.)

$$\frac{1}{x^3 - x} = \frac{A}{x} + \frac{B}{x-1} + \frac{C}{x+1}$$

Solution:

$$\frac{1}{x^3 - x} = \frac{A}{x} + \frac{B}{x-1} + \frac{C}{x+1}$$

$$1 = A(x+1)(x-1) + Bx(x+1) + Cx(x-1)$$

$$1 = Ax^2 - A + Bx^2 + Bx + Cx^2 - Cx$$

$$1 = (A+B+C)x^2 + (B-C)x - A$$

By equating coefficients, we have
$$\begin{aligned} 0 &= A+B+C \\ 0 &= B-C \\ 1 &= -A \end{aligned} \quad \Rightarrow \quad \begin{aligned} A &= -1 \\ B + C &= 1 \\ B - C &= 0 \\ \hline 2B &= 1 \Rightarrow B = \frac{1}{2},\ C = \frac{1}{2}. \end{aligned}$$

$$\frac{A}{x} + \frac{B}{x-1} + \frac{C}{x+1} = \frac{-1}{x} + \frac{1/2}{x-1} + \frac{1/2}{x+1} = \frac{1}{2}\left(-\frac{2}{x} + \frac{1}{x-1} + \frac{1}{x+1}\right)$$

57. Find the least squares regression parabola using the points (0, 0), (2, 2), (3, 6), (4, 12). *The least squares regression parabola*, $y = ax^2 + bx + c$, for the points $(x_1, y_1), (x_2, y_2), \ldots, (x_n, y_n)$ is obtained by solving the following system of linear equations for a, b, and c.

$$\begin{aligned} nc + \left(\sum_{i=1}^{n} x_i\right)b + \left(\sum_{i=1}^{n} x_i^2\right)a &= \sum_{i=1}^{n} y_i \\ \left(\sum_{i=1}^{n} x_i\right)c + \left(\sum_{i=1}^{n} x_i^2\right)b + \left(\sum_{i=1}^{n} x_i^3\right)a &= \sum_{i=1}^{n} x_i y_i \\ \left(\sum_{i=1}^{n} x_i^2\right)c + \left(\sum_{i=1}^{n} x_i^3\right)b + \left(\sum_{i=1}^{n} x_i^4\right)a &= \sum_{i=1}^{n} x_i^2 y_i \end{aligned}$$

Solution:

$n = 4$

$\sum x_i = 9$ $\sum y_i = 20$

$\sum x_i^2 = 29$ $\sum x_i^3 = 99$

$\sum x_i^4 = 353$ $\sum x_i y_i = 70$

$\sum x_i^2 y_i = 254$

$$\begin{aligned} 353a + 99b + 29c &= 254 \\ 99a + 29b + 9c &= 70 \\ 29a + 9b + 4c &= 20 \end{aligned}$$

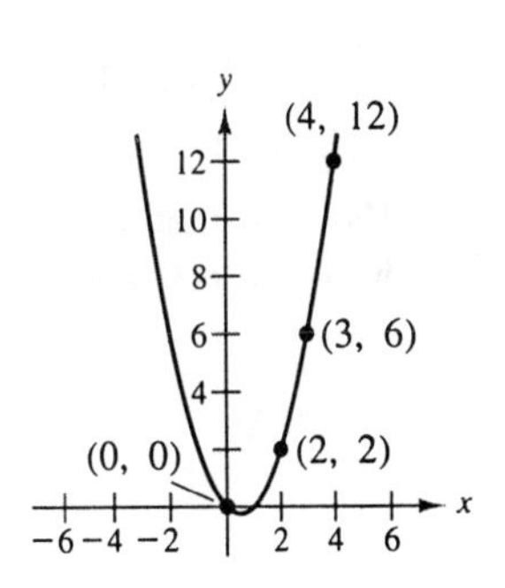

By solving, we get $a = 1$, $b = -1$, $c = 0$. Thus, $y = x^2 - x$.

SECTION 7.4
Systems of Inequalities

- You should be able to sketch the graph of an inequality in two variables:
  (a) Replace the inequality with an equal sign and graph the equation. Use a dashed line for $<$ or $>$, a solid line for $\leq$ or $\geq$.
  (b) Test a point in each region formed by the graph. If the point satisfies the inequality, shade the whole region.

Solutions to Selected Exercises

5. Match $x^2 + y^2 < 4$ with its graph.

Solution:

Since $x^2 + y^2 = 4$ is a circle with center (0, 0) and radius $r = 2$, it matches graph a.

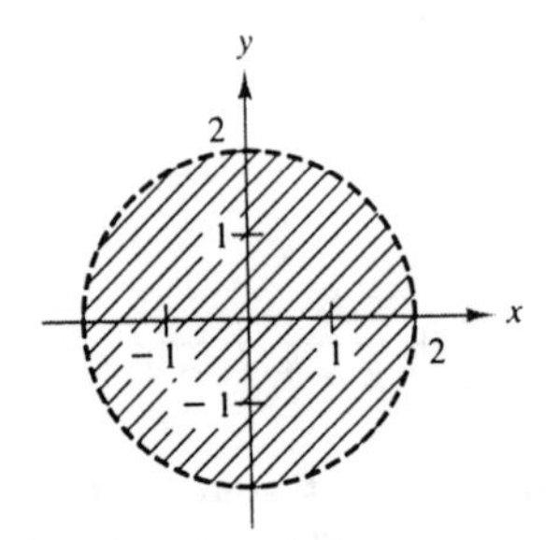

9. Sketch the graph of $x \geq 2$.

Solution:

Using a solid line, sketch the graph of the vertical line $x = 2$. Test point (3, 0). Shade the half-plane to the right of $x = 2$.

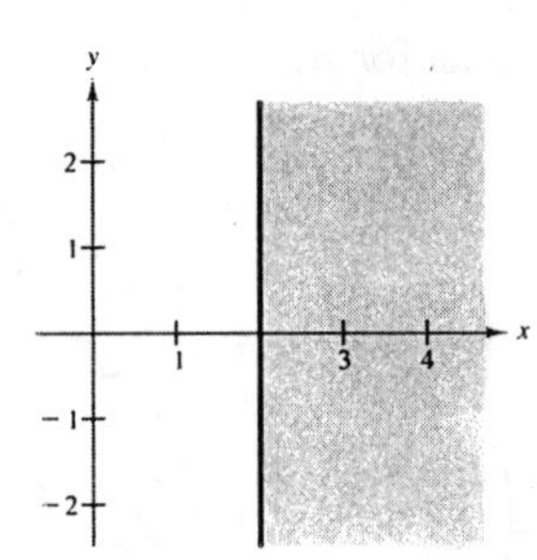

13. Sketch the graph of $y < 2 - x$.

Solution:

Using a dashed line, graph $x + y = 2$, and then shade the half-plane below the line. [Use (0, 0) as a test point.]

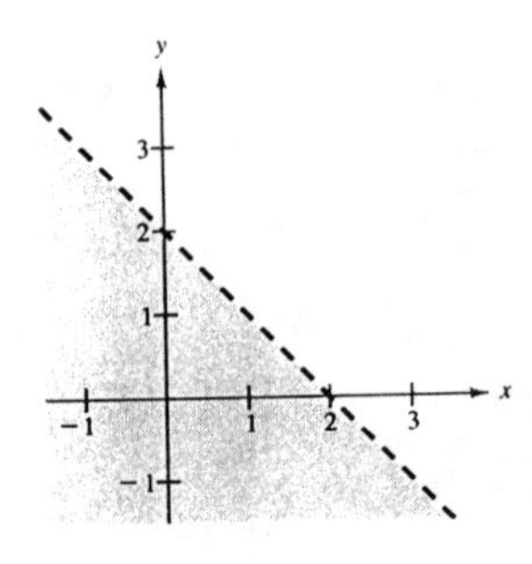

17. Sketch the graph of $(x+1)^2 + (y-2)^2 < 9$.

Solution:

Using a dashed line, sketch the circle $(x+1)^2 + (y-2)^2 = 9$.

Center: $(-1,\ 2)$
Radius: 3

Test Point: $(0,\ 0)$. Shade the inside of the circle.

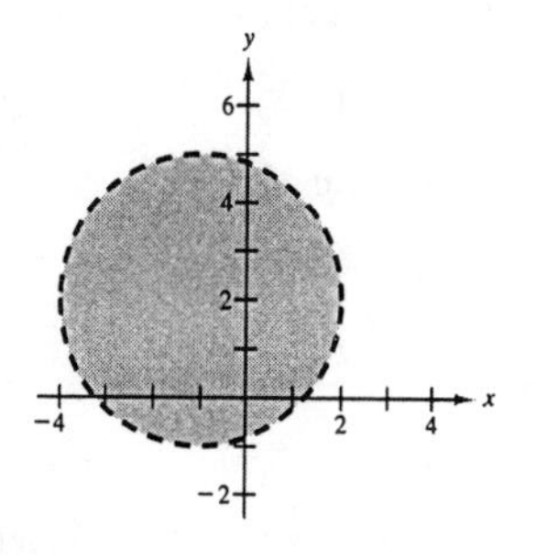

21. Sketch the graph of the solution of the system of inequalities.

$$x + y \le 1$$
$$-x + y \le 1$$
$$y \ge 0$$

Solution:

First, find the points of intersection of each pair of equations.

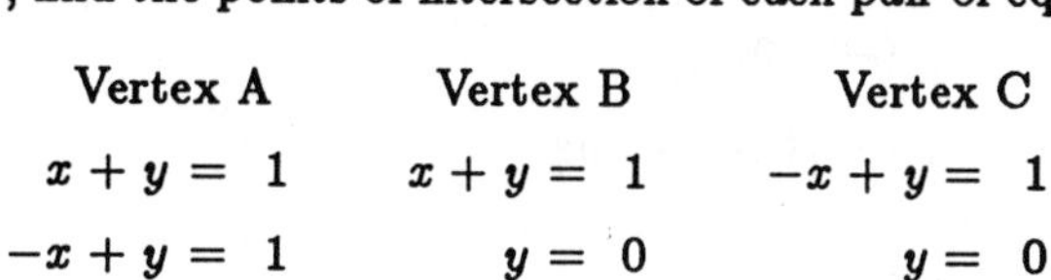

Vertex A	Vertex B	Vertex C
$x + y = 1$	$x + y = 1$	$-x + y = 1$
$-x + y = 1$	$y = 0$	$y = 0$
$(0,\ 1)$	$(1,\ 0)$	$(-1,\ 0)$

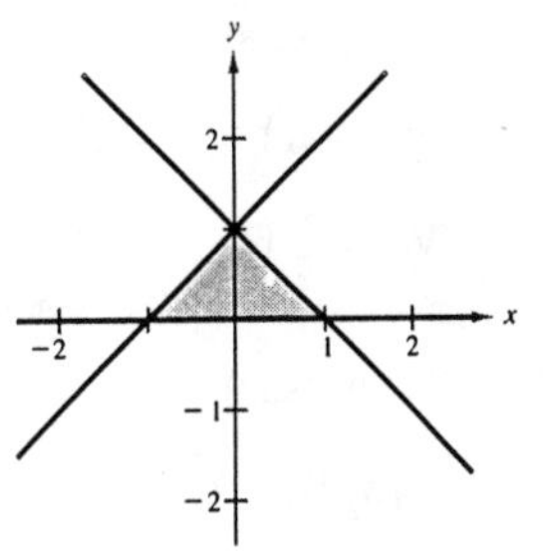

25. Sketch the graph of the solution of the system of inequalities.

$$-3x + 2y < 6$$
$$x + 4y > -2$$
$$2x + y < 3$$

Solution:

First, find the points of intersection of each pair of equations.

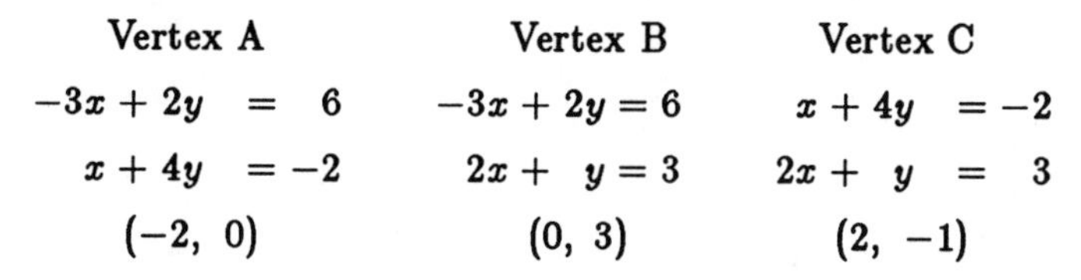

Vertex A	Vertex B	Vertex C
$-3x + 2y = 6$	$-3x + 2y = 6$	$x + 4y = -2$
$x + 4y = -2$	$2x + y = 3$	$2x + y = 3$
$(-2,\ 0)$	$(0,\ 3)$	$(2,\ -1)$

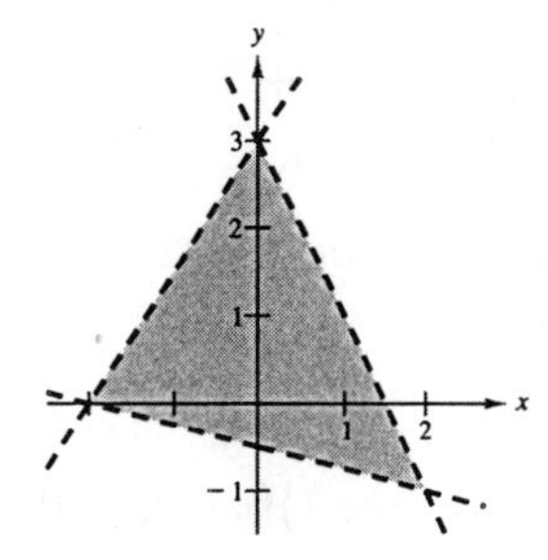

29. Sketch the graph of the solution of the system of inequalities.

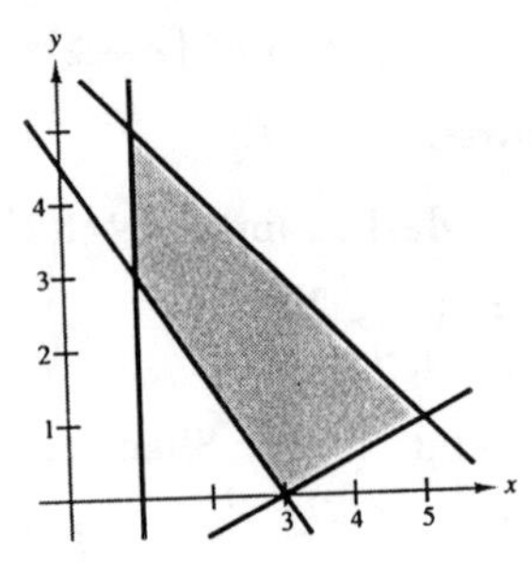

$$x \geq 1$$
$$x - 2y \leq 3$$
$$3x + 2y \geq 9$$
$$x + y \leq 6$$

Solution:

First, find the points of intersection of each pair of equations.

Vertex A	Vertex B	Vertex C
$x = 1$	$x = 1$	$x = 1$
$x - 2y = 3$	$3x + 2y = 9$	$x + y = 6$
$(1, -1)$	$(1, 3)$	$(1, 5)$

Vertex D	Vertex E	Vertex F
$x - 2y = 3$	$x - 2y = 3$	$3x + 2y = 9$
$3x + 2y = 9$	$x + y = 6$	$x + y = 6$
$(3, 0)$	$(5, 1)$	$(-3, 9)$

By shading each inequality, we find that the vertices of the region are $(1, 5)$, $(1, 3)$, $(3, 0)$, and $(5, 1)$.

33. Sketch the graph of the solution of the system of inequalities.

$$x > y^2$$
$$x < y + 2$$

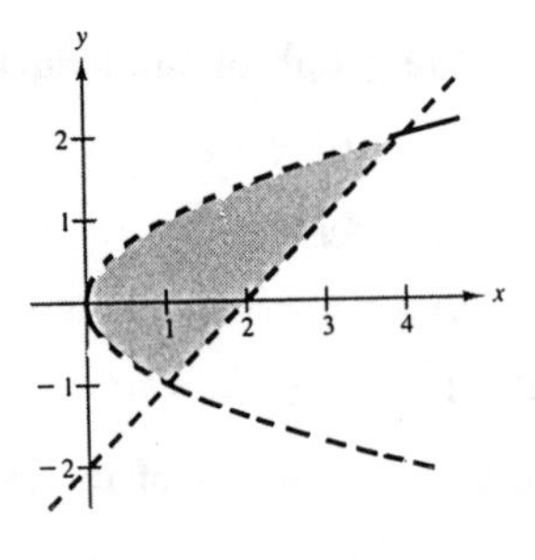

Solution:

Points of intersection:

$$y^2 = y + 2$$
$$y^2 - y - 2 = 0$$
$$(y + 1)(y - 2) = 0$$
$$y = -1,\ 2$$
$$(1, -1),\ (4, 2)$$

37. Sketch the graph of the solution of the system of inequalities.

$$y < x^3 - 2x + 1$$
$$y > -2x$$
$$x \le 1$$

$y = x^3 - 2x + 1$

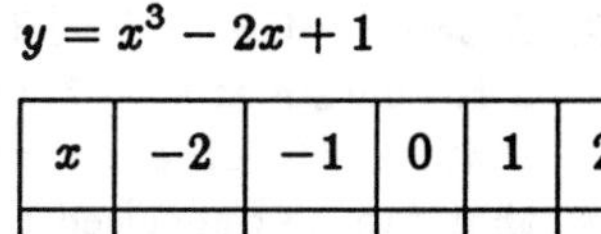

x	-2	-1	0	1	2
y	-3	2	1	0	5

Solution:

Points of intersection:

$$x^3 - 2x + 1 = -2x \qquad x = 1 \qquad x = 1$$
$$x^3 + 1 = 0 \qquad y = x^3 - 2x + 1 \qquad y = -2x$$
$$x = -1 \qquad (1,\ 0) \qquad (1,\ -2)$$
$$(-1,\ 2)$$

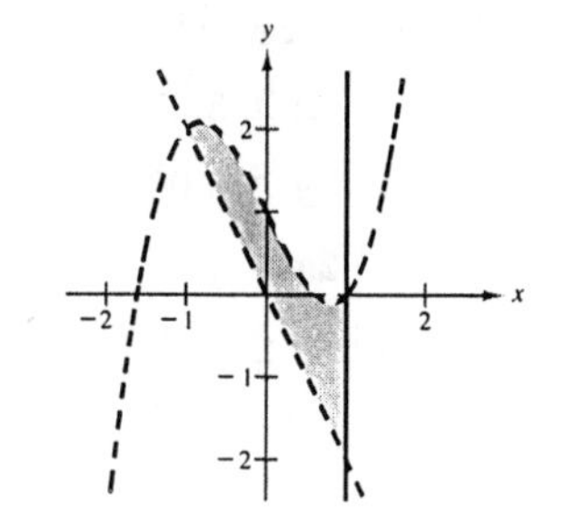

41. Derive a set of inequalities to describe the rectangular region with vertices at (2, 1), (5, 1), (5, 7), and (2, 7).

Solution:

$$x \ge 2$$
$$x \le 5$$
$$y \ge 1$$
$$y \le 7$$

Thus, $2 \le x \le 5,\ 1 \le y \le 7$.

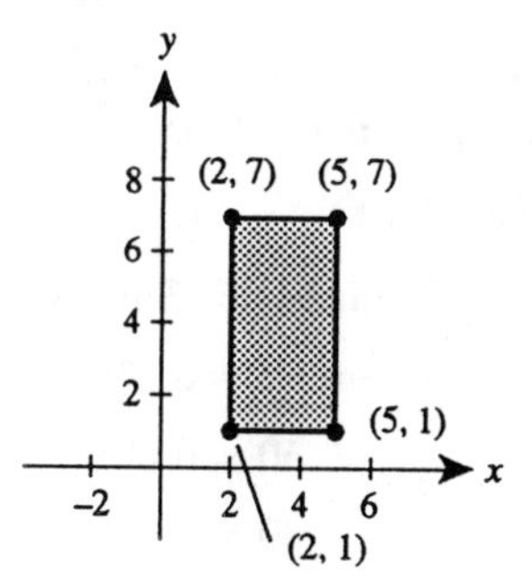

47. A furniture company can sell all the tables and chairs it produces. Each table requires 1 hour in the assembly center and $1\frac{1}{3}$ hours in the finishing center. Each chair requires $1\frac{1}{2}$ hours in the assembly center and $1\frac{1}{2}$ hours in the finishing center. The company's assembly center is available 12 hours per day, and its finishing center is available 15 hours per day. If x is the number of tables produced per day and y is the number of chairs, find a system of inequalities describing all possible production levels. Sketch the graph of the system.

Solution:

Assembly center constraint: $x + \frac{3}{2}y \le 12$
Finishing center constraint: $\frac{4}{3}x + \frac{3}{2}y \le 15$
Point of intersection: (9, 2)
Physical constraints: $x \ge 0$ and $y \ge 0$

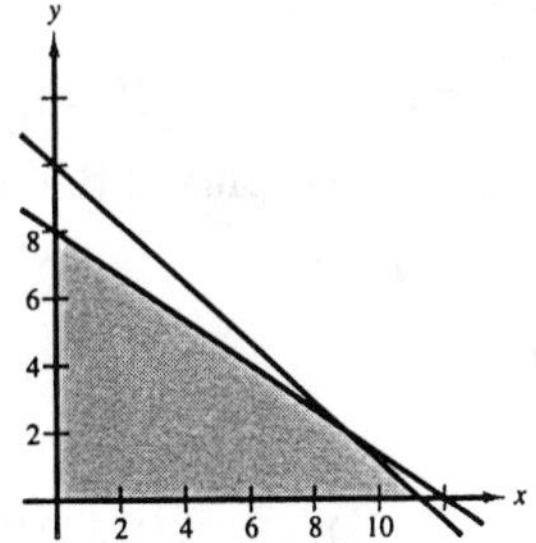

49. A person plans to invest \$20,000 in two different interest bearing accounts. Each account is to contain at least \$5000. Moreover, one account should have at least twice the amount that is in the other account. Find a system of inequalities to describe the various amounts that can be deposited in each account, and sketch the graph of the system.

Solution:

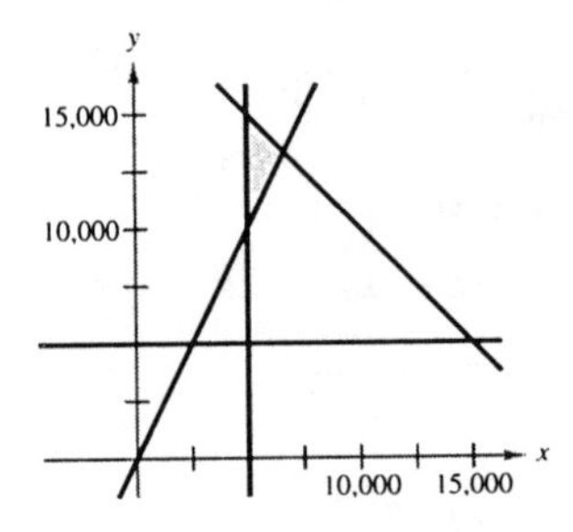

Account constraints:

$$x \geq 5{,}000$$

$$y \geq 5{,}000$$

$$2x \leq y$$

$$x + y \leq 20{,}000$$

53. Find the consumer surplus and producer surplus for the given pair of equations.

Demand: $p = 50 - 0.5x$
Supply: $p = 0.125x$

Solution:

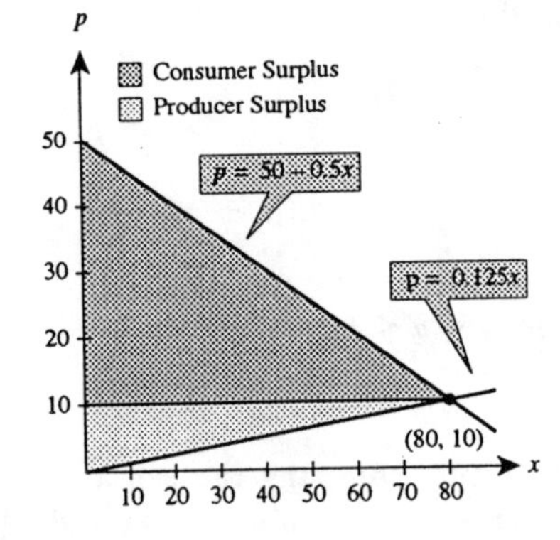

$$\text{Demand} = \text{Supply}$$

$$50 - 0.5x = 0.125x$$

$$50 = 0.625x$$

$$80 = x$$

$$10 = p$$

Point of equilibrium: (80, 10)

The consumer surplus is the area of the triangle bounded by

$$p \leq 50 - 0.5x$$

$$p \geq 10$$

$$x \geq 0.$$

Consumer surplus $= \frac{1}{2}(\text{base})(\text{height}) = \frac{1}{2}(80)(40) = \$1600$

The producer surplus is the area of the triangle bounded by

$$p \geq 0.125x$$

$$p \leq 10$$

$$x \geq 0.$$

Producer surplus $= \frac{1}{2}(\text{base})(\text{height}) = \frac{1}{2}(80)(10) = \$400$

57. Find the consumer surplus and producer surplus for the given pair of equations.

Demand: $p = 140 - 0.00002x$
Supply: $p = 80 + 0.00001x$

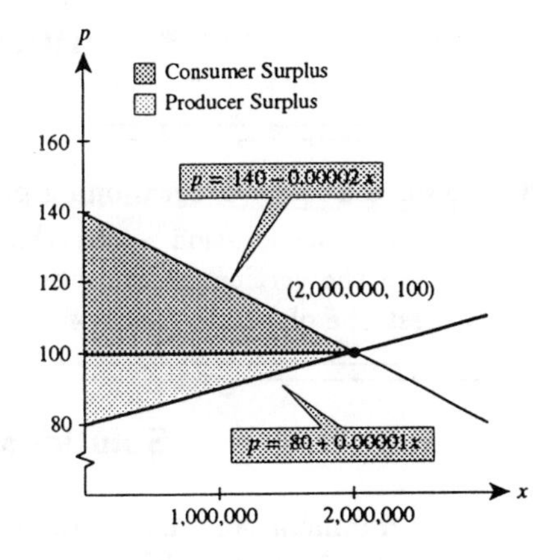

Solution:

$$\begin{aligned} \text{Demand} &= \text{Supply} \\ 140 - 0.00002x &= 80 + 0.00001x \\ 60 &= 0.00003x \\ 2{,}000{,}000 &= x \\ 100 &= p \end{aligned}$$

Point of equilibrium: (2,000,000, 100)

The consumer surplus is the area of the triangle bounded by

$$p \le 140 - 0.00002x$$
$$p \ge 100$$
$$x \ge 0.$$

Consumer surplus $= \frac{1}{2}(\text{base})(\text{height}) = \frac{1}{2}(2{,}000{,}000)(40) = \$40{,}000{,}000$ or \$40 million

The producer surplus is the area of the triangle bounded by

$$p \ge 80 + 0.00001x$$
$$p \le 100$$
$$x \ge 0.$$

Producer surplus $= \frac{1}{2}(\text{base})(\text{height}) = \frac{1}{2}(2{,}000{,}000)(20) = \$20{,}000{,}000$ or \$20 million

SECTION 7.5
Linear Programming

- **To solve a linear programming problem:**
  1. **Sketch the solution set for the system of constraints.**
  2. **Find the vertices of the region.**
  3. **Test the objective function at each of the vertices.**

Solutions to Selected Exercises

5. Find the minimum and maximum values of the objective function, $z = 3x + 2y$, subject to the following constraints.

$$x \geq 0$$
$$y \geq 0$$
$$x + 3y \leq 15$$
$$4x + y \leq 16$$

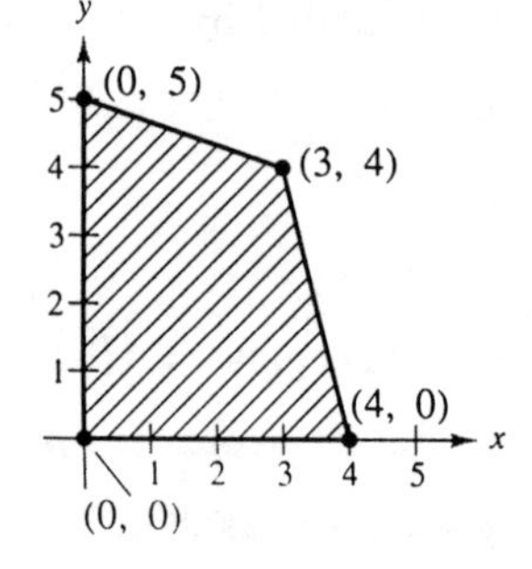

Solution:

Minimum and maximum values occur at the vertices of the constrained region.

Vertex	Value of $z = 3x + 2y$
(0, 0)	$z = 0$, minimum value
(4, 0)	$z = 12$
(0, 5)	$z = 10$
(3, 4)	$z = 17$, maximum value

11. Find the minimum and maximum values of the objective function, $z = 25x + 30y$, subject to the following constraints.

$$0 \leq x \leq 60$$
$$0 \leq y \leq 45$$
$$5x + 6y \leq 420$$

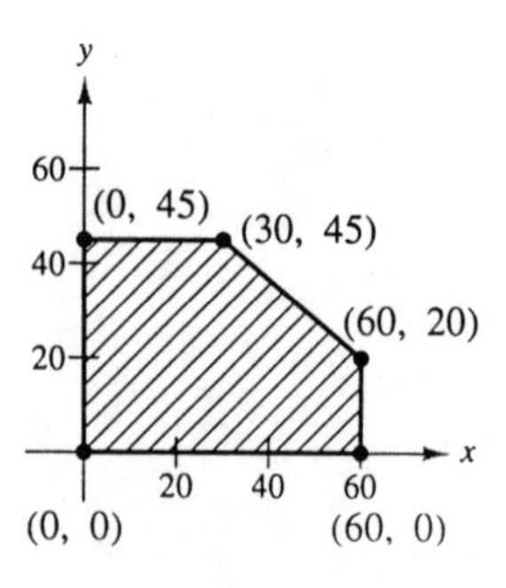

11. –CONTINUED–

Solution:

Vertex	Value of $z = 25x + 30y$
(0, 0)	$z = 0$, minimum value
(60, 0)	$z = 1500$
(60, 20)	$z = 2100$, maximum value
(30, 45)	$z = 2100$, maximum value
(0, 45)	$z = 1350$

Maximum value of 2100 occurs at any point on the line segment connecting (60, 20) and (30, 45).

15. Sketch the region determined by the constraints. Then find the minimum and maximum values of the objective function, $z = 9x + 24y$, subject to the following constraints.

$$x \geq 0$$
$$y \geq 0$$
$$2x + 5y \leq 10$$

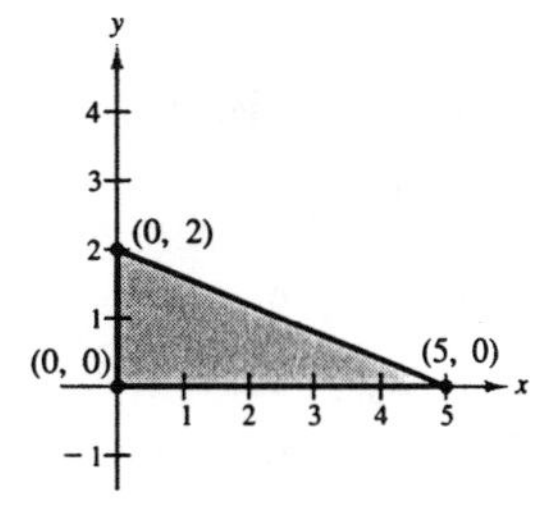

Solution:

Vertex	Value of $z = 9x + 24y$
(0, 0)	$z = 0$, minimum value
(0, 2)	$z = 48$, maximum value
(5, 0)	$z = 45$

21. Sketch the region determined by the constraints. Then find the minimum and maximum values of the objective function, $z = 4x + y$, subject to the following constraints.

$$x \geq 0$$
$$y \geq 0$$
$$x + 2y \leq 40$$
$$x + y \geq 30$$
$$2x + 3y \geq 72$$

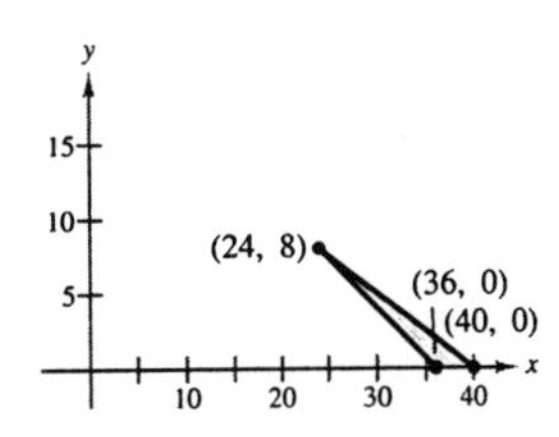

Solution:

Vertex	Value of $z = 4x + y$
(36, 0)	$z = 144$
(40, 0)	$z = 160$, maximum value
(24, 8)	$z = 104$, minimum value

25. Maximize the objective function, $z = 2x + y$, subject to the following constraints.

$$3x + y \leq 15$$
$$4x + 3y \leq 30$$
$$x \geq 0$$
$$y \geq 0$$

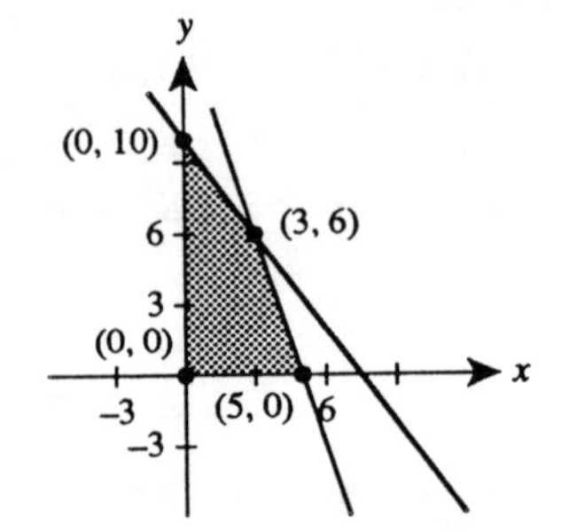

Solution:

Vertex	Value of $z = 2x + y$
(0, 0)	$z = 0$
(0, 10)	$z = 10$
(3, 6)	$z = 12$, maximum value
(5, 0)	$z = 10$

29. Maximize the objective function, $z = x + 5y$, subject to the following constraints.

$$x + 4y \leq 20$$
$$x + y \leq 8$$
$$3x + 2y \leq 21$$
$$x \geq 0$$
$$y \geq 0.$$

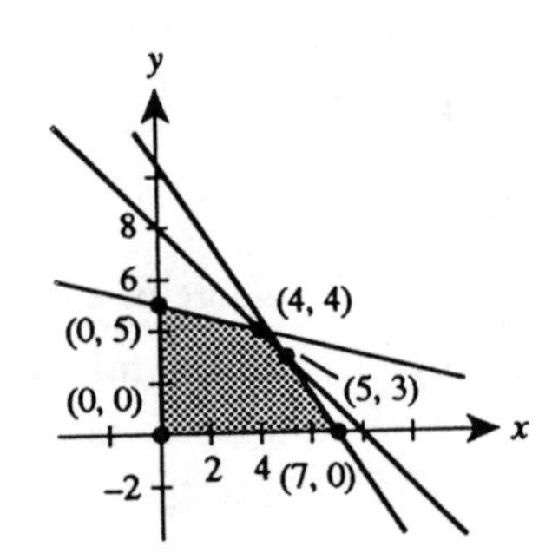

Solution:

Vertex	Value of $z = x + 5y$
(0, 0)	$z = 0$
(0, 5)	$z = 25$, maximum value
(4, 4)	$z = 24$
(5, 3)	$z = 20$
(7, 0)	$z = 7$

33. A merchant plans to sell two models of home computers at costs of \$250 and \$400, respectively. The \$250 model yields a profit of \$45 and the \$400 model yields a profit of \$50. The merchant estimates that the total monthly demand will not exceed 250 units. Find the number of units of each model that should be stocked in order to maximize profit. Assume that the merchant does not want to invest more than \$70,000 in computer inventory.

Solution:

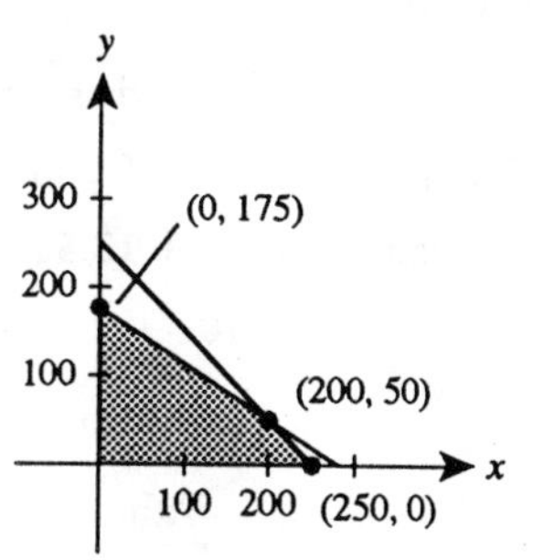

Objective function: Maximize $P = 45x + 50y$.

Constraints:

$$x \geq 0$$
$$y \geq 0$$
$$x + y \leq 250$$
$$250x + 400y \leq 70{,}000$$

Testing the vertices shows that the profit is maximized when $x = 200$ units and $y = 50$ units.

35. A farmer mixes two brands of cattle feed. Brand X costs \$25 per bag and contains 2 units of nutritional element A, 2 units of element B, and 2 units of element C. Brand Y costs \$20 per bag and contains 1 unit of nutritional element A, 9 units of element B, and 3 units of element C. Find the number of bags of each brand that should be mixed to produce a mixture having a minimum cost per bag. The minimum requirements of nutrients A, B, and C are 12 units, 36 units, and 24 units, respectively.

Solution:

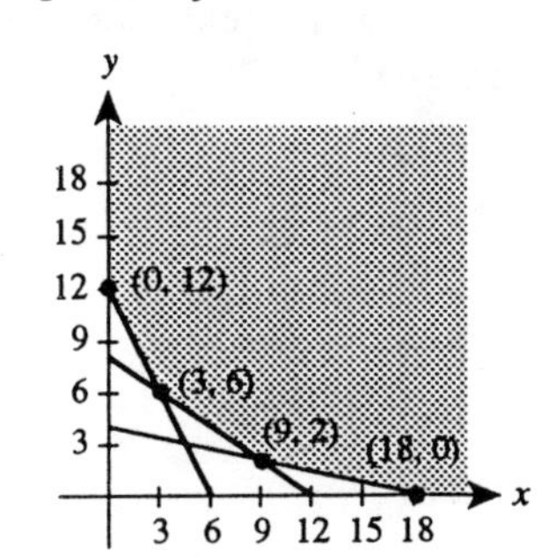

Objective function: Minimize $z = 25x + 20y$.

Constraints:

$$2x + y \geq 12$$
$$2x + 9y \geq 36$$
$$2x + 3y \geq 24$$
$$x \geq 0, \ y \geq 0$$

Vertex	Value of $z = 25x + 20y$
(0, 12)	$z = 240$
(3, 6)	$z = 195$, minimum value
(9, 2)	$z = 265$
(18, 0)	$z = 450$

To minimize cost, use three bags of Brand X and six bags of Brand Y.

39. An accounting firm has 900 hours of staff time and 100 hours of reviewing time available each week. The firm charges $2000 for an audit and $300 for a tax return. Each audit requires 100 hours of staff time and 10 hours of review time. Each tax return requires 12.5 hours of staff time and 2.5 hours of review time. What number of audits and tax returns will yield the maximum revenue?

Solution:

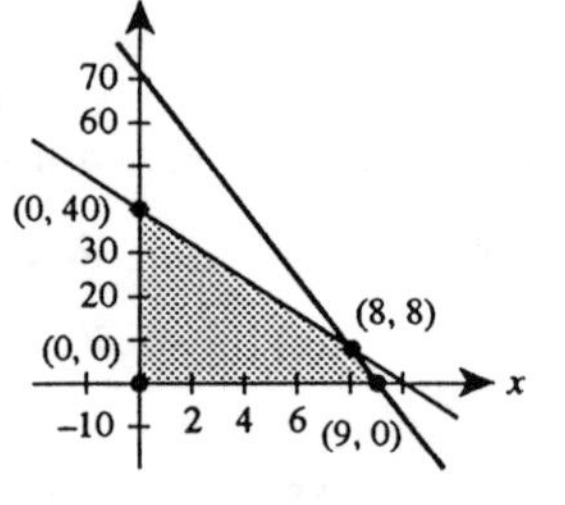

$x =$ the number of audits, and $y =$ the number of tax returns.

Objective function: Maximize $R = 2000x + 300y$.

Constraints: $100x + 12.5y \leq 900$

$10x + 2.5y \leq 100$

$x \geq 0, \ y \geq 0$

Vertex	Value of $R = 2000x + 300y$
(0, 0)	$R = 2000(0) + 300(0) = 0$
(0, 40)	$R = 2000(0) + 300(40) = 12{,}000$
(8, 8)	$R = 2000(8) + 300(8) = 18{,}400$, maximum value
(9, 0)	$R = 2000(9) + 300(0) = 18{,}000$

The revenue will be maximum if the firm does 8 audits and 8 tax returns each week.

43. Sketch a graph of the solution region for the given linear programming problem and describe its unusual characteristic. (The objective function, $z = -x + 2y$, is to be maximized.)

Constraints: $x \geq 0$

$y \geq 0$

$x \leq 10$

$x + y \leq 7$

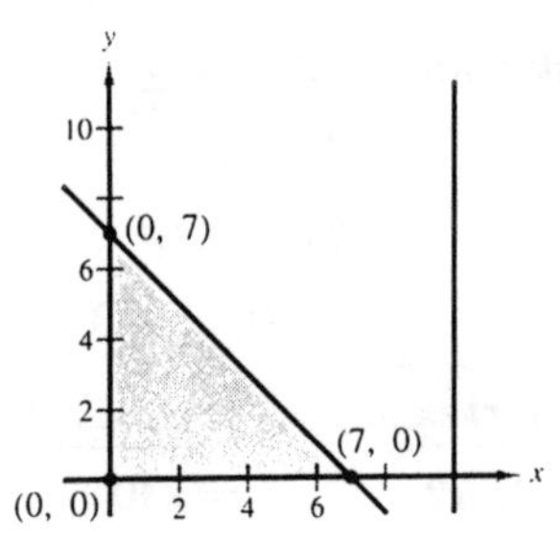

Solution:

The constraint $x \leq 10$ is extraneous.

Vertex	Value of $z = -x + 2y$
(0, 0)	$z = 0$
(7, 0)	$z = -7$
(0, 7)	$z = 14$, maximum value

47. Determine the values of t such that the objective function, $z = 3x + ty$, has a maximum value at the indicated vertex.

Constraints: $x \geq 0$

$y \geq 0$

$x + 3y \leq 15$

$4x + y \leq 16$

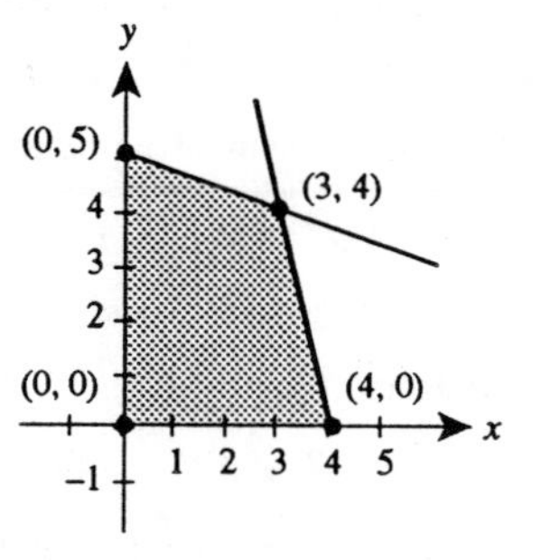

(a) (0, 5) (b) (3, 4)

Solution:

Vertex	Value of $z = 3x + ty$
(0, 0)	$z = 0$
(0, 5)	$z = 5t$
(3, 4)	$z = 9 + 4t$
(4, 0)	$z = 12$

(a) For the maximum value to be at (0, 5), $z = 5t$ must be greater than $z = 9 + 4t$ and $z = 12$.

$5t > 9 + 4t$ and $5t > 12$

$t > 9$ $\qquad t > \frac{12}{5}$

Thus, $t > 9$.

(b) For the maximum value to be at (3, 4), $z = 9 + 4t$ must be greater than $z = 5t$ and $z = 12$.

$9 + 4t > 5t$ and $9 + 4t > 12$

$9 > t$ $\qquad t > 3$

$t > \frac{3}{4}$

Thus, $\frac{3}{4} < t < 9$.

REVIEW EXERCISES FOR CHAPTER 7

Solutions to Selected Exercises

3. Solve the system of equations by the method of substitution.

$$x^2 - y^2 = 9$$
$$x - y = 1$$

Solution:

$$x^2 - y^2 = 9$$
$$x - y = 1 \Rightarrow x = y + 1$$
$$(y+1)^2 - y^2 = 9$$
$$2y + 1 = 9$$
$$y = 4$$
$$x = 5$$

Solution: $(5, 4)$

7. Solve the system of equations by the method of substitution.

$$y^2 - 2y + x = 0$$
$$x + y = 0$$

Solution:

$$y^2 - 2y + x = 0$$
$$x + y = 0 \Rightarrow x = -y$$
$$y^2 - 2y - y = 0$$
$$y(y-3) = 0$$
$$y = 0 \quad \text{or} \quad y = 3$$
$$x = 0 \qquad\qquad x = -3$$

Solutions: $(0, 0)$, $(-3, 3)$

13. Solve the system of equations by elimination.

$$0.2x + 0.3y = 0.14$$
$$0.4x + 0.5y = 0.20$$

Solution:

$$\begin{aligned} 0.2x + 0.3y = 0.14 \Rightarrow 20x + 30y = 14 \Rightarrow \quad 20x + 30y &= \ \ 14 \\ 0.4x + 0.5y = 0.20 \Rightarrow \ \ 4x + \ \ 5y = \ \ 2 \Rightarrow -20x - 25y &= -10 \\ \hline 5y &= \ \ 4 \\ y &= \ \ \tfrac{4}{5} \\ x &= -\tfrac{1}{2} \end{aligned}$$

Solution: $\left(-\frac{1}{2},\ \frac{4}{5}\right)$ or $(-0.5,\ 0.8)$

19. One hundred gallons of a 60% acid solution are obtained by mixing a 75% solution with a 50% solution. How many gallons of each must be used to obtain the desired mixture?

Solution:

Let $x =$ the amount of 75% solution, and $y =$ amount of 50% solution.

$$\begin{aligned} x + y = 100 &\Rightarrow y = 100 - x \\ 0.75x + 0.50y &= 0.60(100) \\ 0.75x + 0.50(100 - x) &= 60 \\ 0.75x + 50 - 0.50x &= 60 \\ 0.25x &= 10 \\ x &= 40 \\ y &= 100 - x = 60 \end{aligned}$$

Answer: 40 gallons of 75% solution, 60 gallons of 60% solution

23. Find the point of equilibrium.

Demand: $p = 37 - 0.0002x$
Supply: $p = 22 + 0.00001x$

Solution:

$$\begin{aligned} \text{Demand} &= \text{Supply} \\ 37 - 0.0002x &= 22 + 0.00001x \\ 15 &= 0.00021x \\ x &\approx 71{,}429 \text{ units} \\ p &= \$22.71 \end{aligned}$$

Point of equilibrium: (71,429, \$22.71)

25. Solve the system of equations.

$$\begin{aligned} x + 2y + 6z &= 4 \\ -3x + 2y - z &= -4 \\ 4x \qquad + 2z &= 16 \end{aligned}$$

Solution:

$$\begin{array}{r} 3x + 6y + 18z = 12 \\ -3x + 2y - z = -4 \\ \hline 8y + 17z = 8 \end{array} \qquad \begin{array}{r} 2x + 4y + 12z = 8 \\ -2x \qquad - z = -8 \\ \hline 4y + 11z = 0 \end{array} \qquad \begin{array}{r} 8y + 17z = 8 \\ -8y - 22z = 0 \\ \hline -5z = 8 \end{array}$$

$$\begin{aligned} z &= -\tfrac{8}{5} = -1.6 \\ y &= \tfrac{1}{8}[8 - 17(-1.6)] = 4.4 \\ x &= \tfrac{1}{2}[8 - (-1.6)] = 4.8 \end{aligned}$$

Solution: (4.8, 4.4, −1.6)

29. Solve the system of equations.

$$2x + 5y - 19z = 34$$
$$3x + 8y - 31z = 54$$

Solution:

$$\begin{aligned} 2x + 5y - 19z = 34 &\Rightarrow \quad 6x + 15y - 57z = 102 \\ 3x + 8y - 31z = 54 &\Rightarrow \underline{-6x - 16y + 62z = -108} \\ &\qquad\qquad -y + 5z = -6 \end{aligned}$$

Let $z = a$. Then,

$$y = 5a + 6$$
$$x = \tfrac{1}{2}[34 - 5(5a + 6) + 19a] = -3a + 2.$$

Solution: $(-3a + 2,\ 5a + 6,\ a)$ where a is any real number.

31. Find the equation of the parabola $y = ax^2 + bx + c$ that passes through the points $(0, -6)$, $(1, -3)$, and $(2, 4)$.

Solution:

$$y = ax^2 + bx + c$$

$$\begin{aligned} &\text{At } (0, -6): -6 = c \\ &\text{At } (1, -3): -3 = a + b + c \Rightarrow a + b = 3 \Rightarrow -a - b = -3 \\ &\text{At } (2, 4): \quad 4 = 4a + 2b + c \Rightarrow 4a + 2b = 10 \Rightarrow \underline{2a + b = 5} \\ &\qquad\qquad\qquad\qquad\qquad\qquad\qquad\qquad a = 2 \\ &\qquad\qquad\qquad\qquad\qquad\qquad\qquad\qquad b = 1 \end{aligned}$$

Thus, $y = 2x^2 + x - 6$.

35. A mixture of 6 gallons of chemical A, 8 gallons of chemical B, and 13 gallons of chemical C is required to kill a certain destructive crop insect. Commercial spray X contains 1, 2, and 2 parts, respectively, of these chemicals. Commercial spray Y contains only chemical C. Commercial spray Z contains chemicals A, B, and C in equal amounts. How much of each type of commerical spray is needed to get the desired mixture?

Solution:

From the following chart we obtain our system of equations.

	A	B	C
Mixture X	$\frac{1}{5}$	$\frac{2}{5}$	$\frac{2}{5}$
Mixture Y	0	0	1
Mixture Z	$\frac{1}{3}$	$\frac{1}{3}$	$\frac{1}{3}$
Desired Mixture	$\frac{6}{27}$	$\frac{8}{27}$	$\frac{13}{27}$

$$\left.\begin{aligned}\tfrac{1}{5}x + \tfrac{1}{3}z &= \tfrac{6}{27}\\ \tfrac{2}{5}x + \tfrac{1}{3}z &= \tfrac{8}{27}\end{aligned}\right\} x = \tfrac{10}{27},\ z = \tfrac{12}{27}$$

$$\tfrac{2}{5}x + y + \tfrac{1}{3}z = \tfrac{13}{27} \Rightarrow y = \tfrac{5}{27}$$

To obtain the desired mixture, use 10 gallons of X, 5 gallons of Y, and 12 gallons of Z.

37. Find the least squares regression line $y = ax + b$ for the points $(x_1, y_1), (x_2, y_2), \ldots, (x_n, y_n)$. To find the line, solve the following system of linear equations for a and b.

$$5b + 10a = 17.8$$
$$10b + 30a = 45.7$$

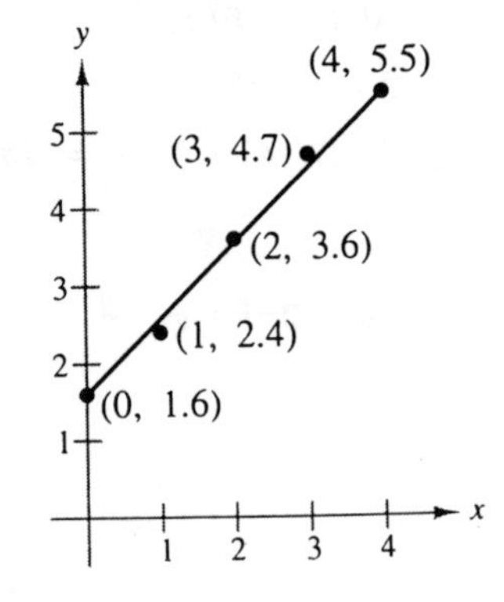

Solution:

$$\begin{aligned}5b + 10a = 17.8 &\Rightarrow -10b - 20a = -35.6\\ 10b + 30a = 45.7 &\Rightarrow \underline{\quad 10b + 30a = \quad 45.7}\\ &\qquad\qquad 10a = \quad 10.1\\ &\qquad\qquad\quad a = \quad 1.01\\ &\qquad\qquad\quad b = \quad 1.54\end{aligned}$$

Least squares regression line: $y = 1.01x + 1.54$

41. Sketch the graph of the solution set of the system of inequalities.

$$3x + 2y \geq 24$$
$$x + 2y \geq 12$$
$$2 \leq x \leq 15$$
$$y \leq 15$$

Solution:

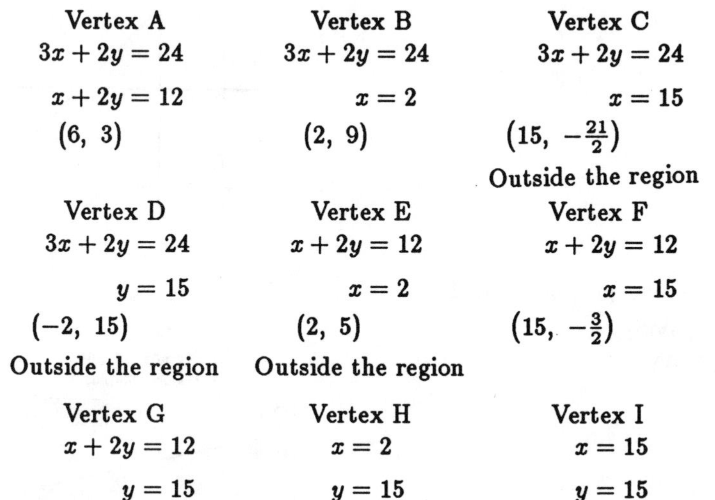

Vertex A
$3x + 2y = 24$
$x + 2y = 12$
$(6,\ 3)$

Vertex B
$3x + 2y = 24$
$x = 2$
$(2,\ 9)$

Vertex C
$3x + 2y = 24$
$x = 15$
$(15,\ -\frac{21}{2})$
Outside the region

Vertex D
$3x + 2y = 24$
$y = 15$
$(-2,\ 15)$
Outside the region

Vertex E
$x + 2y = 12$
$x = 2$
$(2,\ 5)$
Outside the region

Vertex F
$x + 2y = 12$
$x = 15$
$(15,\ -\frac{3}{2})$

Vertex G
$x + 2y = 12$
$y = 15$
$(-18,\ 15)$
Outside the region

Vertex H
$x = 2$
$y = 15$
$(2,\ 15)$

Vertex I
$x = 15$
$y = 15$
$(15,\ 15)$

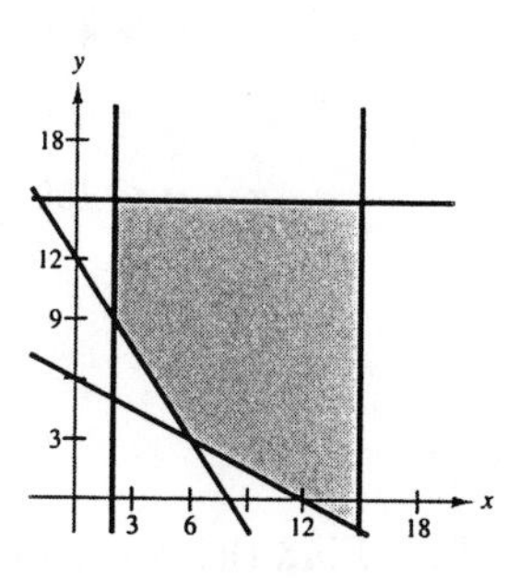

45. Sketch the graph of the solution set of the system of inequalities.

$$2x - 3y \geq 0$$
$$2x - y \leq 8$$
$$y \geq 0$$

Solution:

Vertex A
$2x - 3y = 0$
$2x - y = 8$
$(6,\ 4)$

Vertex B
$2x - 3y = 0$
$y = 0$
$(0,\ 0)$

Vertex C
$2x - y = 8$
$y = 0$
$(4,\ 0)$

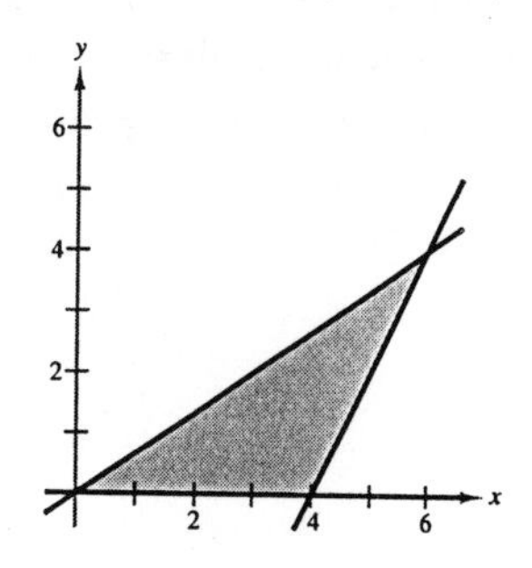

49. A Pennsylvania fruit grower has 1500 bushels of apples that are to be divided between markets in Harrisburg and Philadelphia. These two markets need at least 400 bushels and 600 bushels, respectively. Determine a system of inequalities and sketch a graph of the solution of the system.

Solution:

Let $x =$ the number of bushels for Harrisburg, and
$y =$ the number of bushels for Philadelphia.

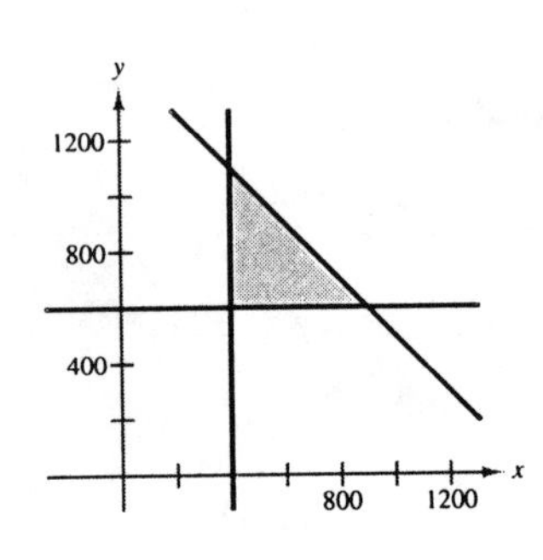

$$x \geq 400$$
$$y \geq 600$$
$$x + y \leq 1500$$

51. Find the consumer surplus and producer surplus.

Demand: $p = 160 - 0.0001x$
Supply: $p = 70 + 0.0002x$

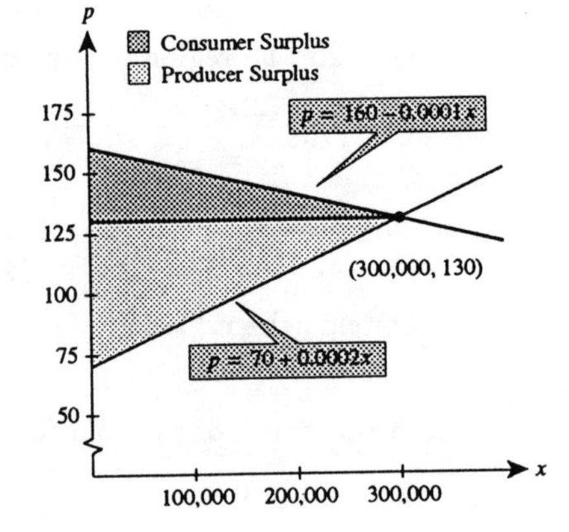

Solution:

$$\text{Demand} = \text{Supply}$$
$$160 - 0.0001x = 70 + 0.0002x$$
$$90 = 0.0003x$$
$$x = 300{,}000 \text{ units}$$
$$p = \$130$$

Point of equilibrium: (300,000, 130)
Consumer surplus: $\frac{1}{2}(300{,}000)(30) = \$4{,}500{,}000$
Producer surplus: $\frac{1}{2}(300{,}000)(60) = \$9{,}000{,}000$

55. Minimize $z = 1.75x + 2.25y$ subject to the following constraints.

$$2x + y \geq 25$$
$$3x + 2y \geq 45$$
$$x \geq 0$$
$$y \geq 0$$

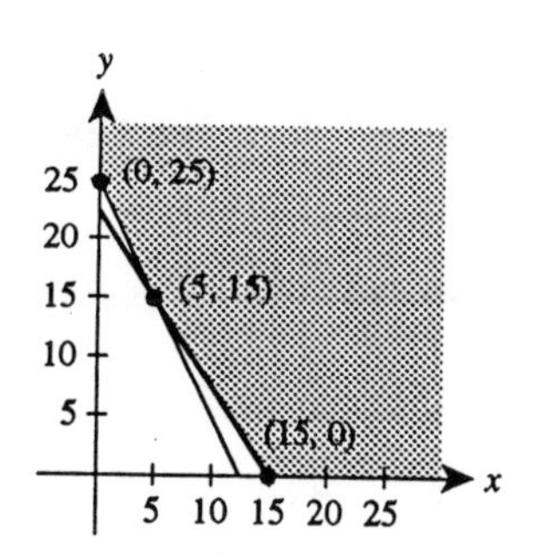

Solution:

Vertex	Value of $z = 1.75x + 2.25y$
(0, 25)	$z = 56.25$
(5, 15)	$z = 42.5$
(15, 0)	$z = 26.25$, minimum value

59. A pet supply company mixes two brands of dry dog food. Brand X costs \$15 per bag and contains 8 units of nutritional element A, 1 unit of nutritional element B, and 2 units of nutritional element C. Brand Y costs \$30 per bag and contains 2 units of nutritional element A, 1 unit of nutritional element B, and 7 units of nutritional element C. Each bag of dog food must contain at least 16 units, 5 units, and 20 units of nutritional elements A, B, and C, respectively. Find the number of bags of brands X and Y that should be mixed to produce a mixture meeting the minimum nutritional requirements and having a minimum cost per bag.

Solution:

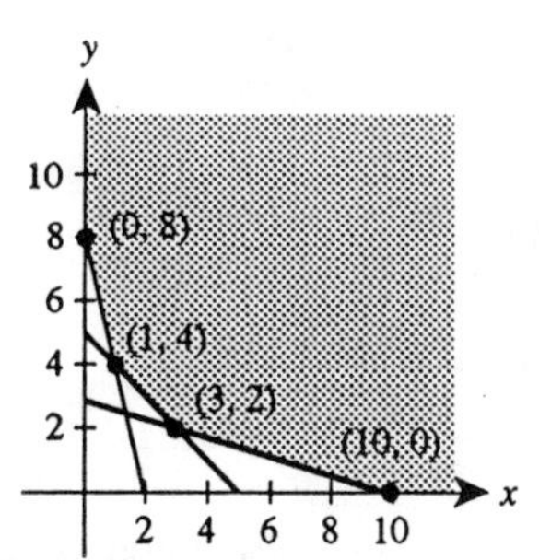

Let $x =$ the number of bags of Brand X, and
$y =$ the number of bags of Brand Y.

Objective function: Minimize $C = 15x + 30y$

Constraints: $8x + 2y \geq 16$

$$x + y \geq 5$$
$$2x + 7y \geq 20$$
$$x \geq 0, \ y \geq 0$$

Vertex	Value of $C = 15x + 30y$
(0, 8)	$C = 15(0) + 30(8) = 240$
(1, 4)	$C = 15(1) + 30(4) = 135$
(3, 2)	$C = 15(3) + 30(2) = 105$, minimum value
(10, 0)	$C = 15(10) + 30(0) = 150$

To minimize cost, use three bags of Brand X and two bags of Brand Y.

Practice Test for Chapter 7

For Exercises 1–3, solve the given system by the method of substitution.

1. $$\begin{aligned} x + y &= 1 \\ 3x - y &= 15 \end{aligned}$$

2. $$\begin{aligned} x - 3y &= -3 \\ x^2 + 6y &= 5 \end{aligned}$$

3. $$\begin{aligned} x + y + z &= 6 \\ 2x - y + 3z &= 0 \\ 5x + 2y - z &= -3 \end{aligned}$$

4. Find two numbers whose sum is 110 and product is 2800.

5. Find the dimensions of a rectangle if its perimeter is 170 feet and its area is 2800 square feet.

For Exercises 6–8, solve the linear system by elimination.

6. $$\begin{aligned} 2x + 15y &= 4 \\ x - 3y &= 23 \end{aligned}$$

7. $$\begin{aligned} x + y &= 2 \\ 38x - 19y &= 7 \end{aligned}$$

8. $$\begin{aligned} 0.4x + 0.5y &= 0.112 \\ 0.3x - 0.7y &= -0.131 \end{aligned}$$

9. Herbert invests \$17,000 in two funds that pay 11% and 13% simple interest, respectively. If he receives \$2080 in yearly interest, how much is invested in each fund?

10. Find the least squares regression line for the points $(4, 3)$, $(1, 1)$, $(-1, -2)$, and $(-2, -1)$.

For Exercises 11–13, solve the system of equations.

11. $$\begin{aligned} x + y &= -2 \\ 2x - y + z &= 11 \\ 4y - 3z &= -20 \end{aligned}$$

12. $4x - \ \ y + 5z = 4$

$2x + \ \ y - \ \ z = 0$

$2x + 4y + 8z = 0$

13. $3x + 2y - \ \ z = 5$

$6x - \ \ y + 5z = 2$

14. Find the equation of the parabola $y = ax^2 + bx + c$ passing through the points $(0, -1)$, $(1, 4)$ and $(2, 13)$.

15. Find the position equation $s = \frac{1}{2}at^2 + v_0t + s_0$ given that $s = 12$ feet after 1 second, $s = 5$ feet after 2 seconds, and $s = 4$ feet after 3 seconds.

16. Graph $x^2 + y^2 \geq 9$.

17. Graph the solution of the system.

$$x + y \leq 6$$
$$x \geq 2$$
$$y \geq 0$$

18. Derive a set of inequalities to describe the triangle with vertices $(0, 0)$, $(0, 7)$, and $(2, 3)$.

19. Find the maximum value of the objective function, $z = 30x + 26y$, subject to the following constraints.

$$x \geq 0$$
$$y \geq 0$$
$$2x + 3y \leq 21$$
$$5x + 3y \leq 30$$

20. Graph the system of inequalities.

$$x^2 + y^2 \leq 4$$
$$(x - 2)^2 + y^2 \geq 4$$

CHAPTER 8
Matrices and Determinants

SECTION 8.1

Matrices and Systems of Linear Equations

■ You should be able to transform an augmented matrix into reduced row-echelon form. This is called Gauss-Jordan elimination. Then use back-substitution to solve the system.

Solutions to Selected Exercises

3. Determine the order of the matrix.

$$\begin{bmatrix} -9 \\ 2 \\ 36 \\ 11 \\ 3 \end{bmatrix}$$

Solution:

Since the matrix has five rows and one column, its order is 5×1.

9. Determine whether the following matrix is in row-echelon form.

$$\begin{bmatrix} 2 & 0 & 4 & 0 \\ 0 & -1 & 3 & 6 \\ 0 & 0 & 1 & 5 \end{bmatrix}$$

Solution:

The first nonzero entries in rows one and two are not 1; the matrix is *not* in row-echelon form.

13. Fill in the blanks using elementary row operations to form a row equivalent matrix.

$$\begin{bmatrix} 1 & 1 & 4 & -1 \\ 3 & 8 & 10 & 3 \\ -2 & 1 & 12 & 6 \end{bmatrix} \rightarrow \begin{bmatrix} 1 & 1 & 4 & -1 \\ 0 & 5 & \_ & \_ \\ 0 & 3 & \_ & \_ \end{bmatrix} \rightarrow \begin{bmatrix} 1 & 1 & 4 & -1 \\ 0 & 1 & \_ & \_ \\ 0 & 3 & 20 & 4 \end{bmatrix}$$

Solution:

$$\begin{bmatrix} 1 & 1 & 4 & -1 \\ 3 & 8 & 10 & 3 \\ -2 & 1 & 12 & 6 \end{bmatrix} \qquad \begin{matrix} \\ -3R_1 + R_2 \rightarrow \\ 2R_1 + R_3 \rightarrow \end{matrix} \begin{bmatrix} 1 & 1 & 4 & -1 \\ 0 & 5 & -2 & 6 \\ 0 & 3 & 20 & 4 \end{bmatrix}$$

$$\frac{1}{5}R_2 \rightarrow \begin{bmatrix} 1 & 1 & 4 & -1 \\ 0 & 1 & -\frac{2}{5} & \frac{6}{5} \\ 0 & 3 & 20 & 4 \end{bmatrix}$$

19. Write the following matrix in row-echelon form. Remember that the row-echelon form of a matrix is not unique.

$$\begin{bmatrix} 1 & -1 & -1 & 1 \\ 5 & -4 & 1 & 8 \\ -6 & 8 & 18 & 0 \end{bmatrix}$$

Solution:

$$\begin{bmatrix} 1 & -1 & -1 & 1 \\ 5 & -4 & 1 & 8 \\ -6 & 8 & 18 & 0 \end{bmatrix} \quad \begin{matrix} \\ -5R_1 + R_2 \rightarrow \\ 6R_1 + R_3 \rightarrow \end{matrix} \begin{bmatrix} 1 & -1 & -1 & 1 \\ 0 & 1 & 6 & 3 \\ 0 & 2 & 12 & 6 \end{bmatrix}$$

$$\begin{matrix} \\ \\ -2R_2 + R_3 \rightarrow \end{matrix} \begin{bmatrix} 1 & -1 & -1 & 1 \\ 0 & 1 & 6 & 3 \\ 0 & 0 & 0 & 0 \end{bmatrix}$$

23. Write the following matrix in *reduced* row-echelon form.

$$\begin{bmatrix} 1 & 2 & 3 & -5 \\ 1 & 2 & 4 & -9 \\ -2 & -4 & -4 & 3 \\ 4 & 8 & 11 & -14 \end{bmatrix}$$

Solution:

$$\begin{bmatrix} 1 & 2 & 3 & -5 \\ 1 & 2 & 4 & -9 \\ -2 & -4 & -4 & 3 \\ 4 & 8 & 11 & -14 \end{bmatrix} \quad \begin{matrix} \\ -R_1 + R_2 \rightarrow \\ 2R_1 + R_3 \rightarrow \\ -4R_1 + R_4 \rightarrow \end{matrix} \begin{bmatrix} 1 & 2 & 3 & -5 \\ 0 & 0 & 1 & -4 \\ 0 & 0 & 2 & -7 \\ 0 & 0 & -1 & 6 \end{bmatrix}$$

$$\begin{matrix} -3R_2 + R_1 \rightarrow \\ \\ -2R_2 + R_3 \rightarrow \\ R_2 + R_4 \rightarrow \end{matrix} \begin{bmatrix} 1 & 2 & 0 & 7 \\ 0 & 0 & 1 & -4 \\ 0 & 0 & 0 & 1 \\ 0 & 0 & 0 & 2 \end{bmatrix}$$

$$\begin{matrix} -7R_3 + R_1 \rightarrow \\ 4R_3 + R_2 \rightarrow \\ \\ -2R_3 + R_4 \rightarrow \end{matrix} \begin{bmatrix} 1 & 2 & 0 & 0 \\ 0 & 0 & 1 & 0 \\ 0 & 0 & 0 & 1 \\ 0 & 0 & 0 & 0 \end{bmatrix}$$

27. Write the system of linear equations represented by the augmented matrix.

$$\left[\begin{array}{ccc|c} 1 & 0 & 2 & -10 \\ 0 & 3 & -1 & 5 \\ 4 & 2 & 0 & 3 \end{array}\right]$$

Solution:

Row 1: $1x + 0y + 2z = -10 \Rightarrow x + 2z = -10$

Row 2: $0x + 3y - 1z = 5 \Rightarrow 3y - z = 5$

Row 3: $4x + 2y + 0z = 3 \Rightarrow 4x + 2y = 3$

31. Write the system of linear equations represented by the augmented matrix. Then use back-substitution to find the solution.

$$\left[\begin{array}{ccc|c} 1 & -1 & 2 & 4 \\ 0 & 1 & -1 & 2 \\ 0 & 0 & 1 & -2 \end{array}\right]$$

Solution:

$$\begin{aligned} \text{Row 1: } x - y + 2z &= 4 \\ \text{Row 2: } \quad y - z &= 2 \\ \text{Row 3: } \quad z &= -2 \\ y &= 2 + z = 2 + (-2) = 0 \\ x &= 4 + y - 2z = 4 + 0 - 2(-2) = 8 \end{aligned}$$

Answer: $(8,\ 0,\ -2)$

35. Write the solution represented by the augmented matrix.

$$\left[\begin{array}{ccc|c} 1 & 0 & 0 & -4 \\ 0 & 1 & 0 & -8 \\ 0 & 0 & 1 & 2 \end{array}\right]$$

Solution:

Row 1: $x = -4$

Row 2: $y = -8$

Row 3: $z = 2$

Answer: $(-4,\ -8,\ 2)$

39. Solve the system of equations. Use Gaussian elimination with back-substitution or Gauss-Jordan elimination.

$$\begin{aligned} -3x + 5y &= -22 \\ 3x + 4y &= 4 \\ 4x - 8y &= 32 \end{aligned}$$

39. –CONTINUED–

Solution:

$$\begin{bmatrix} -3 & 5 & \vdots & -22 \\ 3 & 4 & \vdots & 4 \\ 4 & -8 & \vdots & 32 \end{bmatrix} \begin{matrix} -\frac{1}{3}R_1 \rightarrow \\ -3R_1 + R_2 \rightarrow \\ -4R_1 + R_3 \rightarrow \end{matrix} \begin{bmatrix} 1 & -\frac{5}{3} & \vdots & \frac{22}{3} \\ 0 & 9 & \vdots & -18 \\ 0 & -\frac{4}{3} & \vdots & \frac{8}{3} \end{bmatrix}$$

$$\begin{matrix} \frac{5}{3}R_2 + R_1 \rightarrow \\ \frac{1}{9}R_2 \rightarrow \\ \frac{4}{3}R_2 + R_3 \rightarrow \end{matrix} \begin{bmatrix} 1 & 0 & \vdots & 4 \\ 0 & 1 & \vdots & -2 \\ 0 & 0 & \vdots & 0 \end{bmatrix}$$

Answer: $(4, \ -2)$

43. Solve the system of equations. Use Gaussian elimination with back-substitution or Gauss-Jordan elimination.

$$-x + 2y = 1.5$$

$$2x - 4y = 3$$

Solution:

$$\begin{bmatrix} -1 & 2 & \vdots & 1.5 \\ 2 & -4 & \vdots & 3 \end{bmatrix} \begin{matrix} -R_1 \rightarrow \\ -2R_1 + R_2 \rightarrow \end{matrix} \begin{bmatrix} 1 & -2 & \vdots & -1.5 \\ 0 & 0 & \vdots & 6 \end{bmatrix}$$

The second line says $0 = 6$. This is inconsistent.

47. Solve the system of equations. Use Gaussian elimination with back-substitution or Gauss-Jordan elimination.

$$x + y - 5z = 3$$

$$x \quad\quad - 2z = 1$$

$$2x - y - z = 0$$

Solution:

$$\begin{bmatrix} 1 & 1 & -5 & \vdots & 3 \\ 1 & 0 & -2 & \vdots & 1 \\ 2 & -1 & -1 & \vdots & 0 \end{bmatrix} \begin{matrix} \\ -R_1 + R_2 \rightarrow \\ -2R_1 + R_3 \rightarrow \end{matrix} \begin{bmatrix} 1 & 1 & -5 & \vdots & 3 \\ 0 & -1 & 3 & \vdots & -2 \\ 0 & -3 & 9 & \vdots & -6 \end{bmatrix}$$

$$\begin{matrix} R_2 + R_1 \rightarrow \\ -R_2 \rightarrow \\ 3R_2 + R_3 \rightarrow \end{matrix} \begin{bmatrix} 1 & 0 & -2 & \vdots & 1 \\ 0 & 1 & -3 & \vdots & 2 \\ 0 & 0 & 0 & \vdots & 0 \end{bmatrix}$$

Thus, $x - 2z = 1$ and $y - 3z = 2$. By letting $z = a$, we have $y = 3a + 2$ and $x = 2a + 1$.

Answer: $(2a + 1, \ 3a + 2, \ a)$

53. Solve the system of equations. Use Gaussian elimination with back-substitution or Gauss-Jordan elimination.

$$\begin{aligned} 2x + y - z + 2w &= -6 \\ 3x + 4y \qquad + w &= 1 \\ x + 5y + 2z + 6w &= -3 \\ 5x + 2y - z - w &= 3 \end{aligned}$$

Solution:

$$\begin{bmatrix} 2 & 1 & -1 & 2 & \vdots & -6 \\ 3 & 4 & 0 & 1 & \vdots & 1 \\ 1 & 5 & 2 & 6 & \vdots & -3 \\ 5 & 2 & -1 & -1 & \vdots & 3 \end{bmatrix} \qquad \begin{matrix} R_3 \rightarrow \\ \\ R_1 \rightarrow \\ \\ \end{matrix} \begin{bmatrix} 1 & 5 & 2 & 6 & \vdots & -3 \\ 3 & 4 & 0 & 1 & \vdots & 1 \\ 2 & 1 & -1 & 2 & \vdots & -6 \\ 5 & 2 & -1 & -1 & \vdots & 3 \end{bmatrix}$$

$$\begin{matrix} \\ -3R_1 + R_2 \rightarrow \\ -2R_1 + R_3 \rightarrow \\ -5R_1 + R_4 \rightarrow \end{matrix} \begin{bmatrix} 1 & 5 & 2 & 6 & \vdots & -3 \\ 0 & -11 & -6 & -17 & \vdots & 10 \\ 0 & -9 & -5 & -10 & \vdots & 0 \\ 0 & -23 & -11 & -31 & \vdots & 18 \end{bmatrix}$$

$$\begin{matrix} 5R_4 + R_1 \rightarrow \\ -11R_4 + R_2 \rightarrow \\ -9R_4 + R_3 \rightarrow \\ -2R_2 + R_4 \rightarrow \end{matrix} \begin{bmatrix} 1 & 0 & 7 & 21 & \vdots & -13 \\ 0 & 0 & -17 & -50 & \vdots & 32 \\ 0 & 0 & -14 & -37 & \vdots & 18 \\ 0 & -1 & 1 & 3 & \vdots & -2 \end{bmatrix}$$

$$\begin{matrix} -7R_3 + R_1 \rightarrow \\ 17R_3 + R_2 \rightarrow \\ -\frac{1}{14}R_3 \rightarrow \\ -R_4 \rightarrow \end{matrix} \begin{bmatrix} 1 & 0 & 0 & \frac{5}{2} & \vdots & -4 \\ 0 & 0 & 0 & -\frac{71}{14} & \vdots & \frac{71}{7} \\ 0 & 0 & 1 & \frac{37}{14} & \vdots & -\frac{9}{7} \\ 0 & 1 & -1 & -3 & \vdots & 2 \end{bmatrix}$$

$$\begin{matrix} \\ -\frac{14}{71}R_2 \rightarrow \\ \\ \\ \end{matrix} \begin{bmatrix} 1 & 0 & 0 & \frac{5}{2} & \vdots & -4 \\ 0 & 0 & 0 & 1 & \vdots & -2 \\ 0 & 0 & 1 & \frac{37}{14} & \vdots & -\frac{9}{7} \\ 0 & 1 & -1 & -3 & \vdots & 2 \end{bmatrix}$$

$$\begin{matrix} \\ R_4 \rightarrow \\ \\ R_2 \rightarrow \end{matrix} \begin{bmatrix} 1 & 0 & 0 & \frac{5}{2} & \vdots & -4 \\ 0 & 1 & -1 & -3 & \vdots & 2 \\ 0 & 0 & 1 & \frac{37}{14} & \vdots & -\frac{9}{7} \\ 0 & 0 & 0 & 1 & \vdots & -2 \end{bmatrix}$$

$$\begin{aligned} x \qquad + \tfrac{5}{2}w &= -4 \\ y - z - \quad 3w &= 2 \\ z + \tfrac{37}{14}w &= -\tfrac{9}{7} \\ w &= -2 \end{aligned}$$

Thus,

$w = -2$

$x = -4 - \frac{5}{2}(-2) = 1$

$z = -\frac{9}{7} - \frac{37}{14}(-2) = 4$

$y = 2 + 4 + 3(-2) = 0.$

Answer: $(1,\ 0,\ 4,\ -2)$

57. Solve the system of equations. Use Gaussian elimination with back-substitution or Gauss-Jordan elimination.

$$x + y + z = 0$$
$$2x + 3y + z = 0$$
$$3x + 5y + z = 0$$

Solution:

$$\begin{bmatrix} 1 & 1 & 1 & \vdots & 0 \\ 2 & 3 & 1 & \vdots & 0 \\ 3 & 5 & 1 & \vdots & 0 \end{bmatrix} \begin{matrix} \\ -2R_1 + R_2 \rightarrow \\ -3R_1 + R_3 \rightarrow \end{matrix} \begin{bmatrix} 1 & 1 & 1 & \vdots & 0 \\ 0 & 1 & -1 & \vdots & 0 \\ 0 & 2 & -2 & \vdots & 0 \end{bmatrix}$$

$$\begin{matrix} -R_2 + R_1 \rightarrow \\ \\ -2R_2 + R_3 \rightarrow \end{matrix} \begin{bmatrix} 1 & 0 & 2 & \vdots & 0 \\ 0 & 1 & -1 & \vdots & 0 \\ 0 & 0 & 0 & \vdots & 0 \end{bmatrix}$$

Thus, $x + 2z = 0$ and $y - z = 0$. By letting $z = a$, we have $x = -2a$ and $y = a$.

Answer: $(-2a, a, a)$ where a is any real number.

59. A small corporation borrowed \$1,500,000 to expand its product line. Some of the money was borrowed at 8%, some at 9%, and some at 12%. How much was borrowed at each rate if the annual interest was \$133,000 and the amount borrowed at 8% was 4 times the amount borrowed at 12%?

Solution:

Let $x =$ 8% amount, $y =$ 9% amount, and $z =$ 12% amount.

$$x + y + z = 1{,}500{,}000$$
$$0.08x + 0.09y + 0.12z = 133{,}000$$
$$x - 4z = 0$$

$$\begin{bmatrix} 1 & 1 & 1 & \vdots & 1{,}500{,}000 \\ 8 & 9 & 12 & \vdots & 13{,}300{,}000 \\ 1 & 0 & -4 & \vdots & 0 \end{bmatrix}$$

$$\begin{matrix} \\ -8R_1 + R_2 \rightarrow \\ -R_1 + R_3 \rightarrow \end{matrix} \begin{bmatrix} 1 & 1 & 1 & \vdots & 1{,}500{,}000 \\ 0 & 1 & 4 & \vdots & 1{,}300{,}000 \\ 0 & -1 & -5 & \vdots & -1{,}500{,}000 \end{bmatrix}$$

$$\begin{matrix} -R_2 + R_1 \rightarrow \\ \\ R_2 + R_3 \rightarrow \end{matrix} \begin{bmatrix} 1 & 0 & -3 & \vdots & 200{,}000 \\ 0 & 1 & 4 & \vdots & 1{,}300{,}000 \\ 0 & 0 & -1 & \vdots & -200{,}000 \end{bmatrix}$$

$$\begin{matrix} 3R_3 + R_1 \rightarrow \\ -4R_3 + R_2 \rightarrow \\ -R_3 \rightarrow \end{matrix} \begin{bmatrix} 1 & 0 & 0 & \vdots & 800{,}000 \\ 0 & 1 & 0 & \vdots & 500{,}000 \\ 0 & 0 & 1 & \vdots & 200{,}000 \end{bmatrix}$$

Thus, $x =$ \$800,000, $y =$ \$500,000, and $z =$ \$200,000.

63. Find D, E, and F such that (1, 1), (3, 3), and (4, 2) are solution points of the equation $x^2 + y^2 + Dx + Ey + F = 0$.

Solution:

At (1, 1): $(1)^2 + (1)^2 + D(1) + E(1) + F = 0 \Rightarrow D + E + F = -2$

At (3, 3): $(3)^2 + (3)^2 + D(3) + E(3) + F = 0 \Rightarrow 3D + 3E + F = -18$

At (4, 2): $(4)^2 + (2)^2 + D(4) + E(2) + F = 0 \Rightarrow 4D + 2E + F = -20$

$$\begin{bmatrix} 1 & 1 & 1 & \vdots & -2 \\ 3 & 3 & 1 & \vdots & -18 \\ 4 & 2 & 1 & \vdots & -20 \end{bmatrix} \quad \begin{matrix} \\ -3R_1 + R_2 \rightarrow \\ -4R_1 + R_3 \rightarrow \end{matrix} \begin{bmatrix} 1 & 1 & 1 & \vdots & -2 \\ 0 & 0 & -2 & \vdots & -12 \\ 0 & -2 & -3 & \vdots & -12 \end{bmatrix}$$

$$\begin{matrix} \\ R_3 \rightarrow \\ R_2 \rightarrow \end{matrix} \begin{bmatrix} 1 & 1 & 1 & \vdots & -2 \\ 0 & -2 & -3 & \vdots & -12 \\ 0 & 0 & -2 & \vdots & -12 \end{bmatrix}$$

$$\begin{matrix} \\ -\frac{1}{2}R_2 \rightarrow \\ -\frac{1}{2}R_3 \rightarrow \end{matrix} \begin{bmatrix} 1 & 1 & 1 & \vdots & -2 \\ 0 & 1 & \frac{3}{2} & \vdots & 6 \\ 0 & 0 & 1 & \vdots & 6 \end{bmatrix}$$

$$\begin{aligned} D + E + F &= -2 \\ E + \tfrac{3}{2}F &= 6 \\ F &= 6 \end{aligned}$$

Thus,

$F = 6$

$E = 6 - \frac{3}{2}(6) = -3$

$D = -2 - (-3) - 6 = -5$

The equation of the circle is $x^2 + y^2 - 5x - 3y + 6 = 0$.

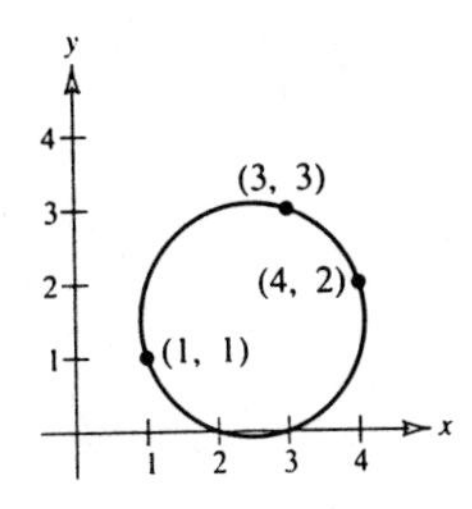

SECTION 8.2

Operations with Matrices

- $A = B$ if and only if they have the same order and $a_{ij} = b_{ij}$.
- You should be able to perform the operations of matrix addition, scalar multiplication, and matrix multiplication.
- Some properties of matrix addition, scalar multiplication, and matrix multiplication are:
  (a) $A + B = B + A$
  (b) $A + (B + C) = (A + B) + C$
  (c) $(cd)A = c(dA)$
  (d) $1A = A$
  (e) $c(A + B) = cA + cB$
  (f) $(c + d)A = cA + dA$
  (g) $A(BC) = (AB)C$
  (h) $A(B + C) = AB + AC$
  (i) $(A + B)C = AC + BC$
  (j) $c(AB) = (cA)B = A(cB)$
- You should remember that $AB \neq BA$ in general.

Solutions to Selected Exercises

3. Find x and y.

$$\begin{bmatrix} 16 & 4 & 5 & 4 \\ -3 & 13 & 15 & 6 \\ 0 & 2 & 4 & 0 \end{bmatrix} = \begin{bmatrix} 16 & 4 & 2x+1 & 4 \\ -3 & 13 & 15 & 3x \\ 0 & 2 & 3y-5 & 0 \end{bmatrix}$$

Solution:

$$\left.\begin{aligned} 5 &= 2x + 1 \\ 6 &= 3x \end{aligned}\right\} \Rightarrow x = 2$$

$$4 = 3y - 5 \quad \Rightarrow y = 3$$

Answer: $x = 2,\ y = 3$

9. Find (a) $A+B$, (b) $A-B$, (c) $3A$, and (d) $3A-2B$.

$$A = \begin{bmatrix} 2 & 2 & -1 & 0 & 1 \\ 1 & 1 & -2 & 0 & -1 \end{bmatrix}, \quad B = \begin{bmatrix} 1 & 1 & -1 & 1 & 0 \\ -3 & 4 & 9 & -6 & -7 \end{bmatrix}$$

Solution:

(a) $$A+B = \begin{bmatrix} 2+1 & 2+1 & -1+(-1) & 0+1 & 1+0 \\ 1+(-3) & 1+4 & -2+9 & 0+(-6) & -1+(-7) \end{bmatrix}$$

$$= \begin{bmatrix} 3 & 3 & -2 & 1 & 1 \\ -2 & 5 & 7 & -6 & -8 \end{bmatrix}$$

(b) $$A-B = \begin{bmatrix} 2-1 & 2-1 & -1-(-1) & 0-1 & 1-0 \\ 1-(-3) & 1-4 & -2-9 & 0-(-6) & -1-(-7) \end{bmatrix}$$

$$= \begin{bmatrix} 1 & 1 & 0 & -1 & 1 \\ 4 & -3 & -11 & 6 & 6 \end{bmatrix}$$

(c) $$3A = \begin{bmatrix} 3(2) & 3(2) & 3(-1) & 3(0) & 3(1) \\ 3(1) & 3(1) & 3(-2) & 3(0) & 3(-1) \end{bmatrix} = \begin{bmatrix} 6 & 6 & -3 & 0 & 3 \\ 3 & 3 & -6 & 0 & -3 \end{bmatrix}$$

(d) $$3A-2B = \begin{bmatrix} 6 & 6 & -3 & 0 & 3 \\ 3 & 3 & -6 & 0 & -3 \end{bmatrix} - \begin{bmatrix} 2 & 2 & -2 & 2 & 0 \\ -6 & 8 & 18 & -12 & -14 \end{bmatrix}$$

$$= \begin{bmatrix} 4 & 4 & -1 & -2 & 3 \\ 9 & -5 & -24 & 12 & 11 \end{bmatrix}$$

13. Find (a) AB, (b) BA, and if possible (c) A^2.

$$A = \begin{bmatrix} 3 & -1 \\ 1 & 3 \end{bmatrix}, \quad B = \begin{bmatrix} 1 & -3 \\ 3 & 1 \end{bmatrix}$$

Solution:

(a) $$AB = \begin{bmatrix} 3 & -1 \\ 1 & 3 \end{bmatrix}\begin{bmatrix} 1 & -3 \\ 3 & 1 \end{bmatrix} = \begin{bmatrix} 3(1)+(-1)(3) & 3(-3)+(-1)(1) \\ 1(1)+3(3) & 1(-3)+3(1) \end{bmatrix} = \begin{bmatrix} 0 & -10 \\ 10 & 0 \end{bmatrix}$$

(b) $$BA = \begin{bmatrix} 1 & -3 \\ 3 & 1 \end{bmatrix}\begin{bmatrix} 3 & -1 \\ 1 & 3 \end{bmatrix} = \begin{bmatrix} 1(3)+(-3)(1) & 1(-1)+(-3)(3) \\ 3(3)+1(1) & 3(-1)+1(3) \end{bmatrix} = \begin{bmatrix} 0 & -10 \\ 10 & 0 \end{bmatrix}$$

(c) $$A^2 = \begin{bmatrix} 3 & -1 \\ 1 & 3 \end{bmatrix}\begin{bmatrix} 3 & -1 \\ 1 & 3 \end{bmatrix} = \begin{bmatrix} 3(3)+(-1)(1) & 3(-1)+(-1)(3) \\ 1(3)+3(1) & 1(-1)+3(3) \end{bmatrix} = \begin{bmatrix} 8 & -6 \\ 6 & 8 \end{bmatrix}$$

17. Find AB, if possible.

$$A = \begin{bmatrix} 2 & 1 \\ -3 & 4 \\ 1 & 6 \end{bmatrix}, \quad B = \begin{bmatrix} 0 & -1 & 0 \\ 4 & 0 & 2 \\ 8 & -1 & 7 \end{bmatrix}$$

Solution:

A is 3×2 and B is 3×3. Since the number of columns of A does not equal the number of rows of B, the multiplication is not possible. **Note:** BA is possible.

21. Find AB, if possible.

$$A = \begin{bmatrix} 5 & 0 & 0 \\ 0 & -8 & 0 \\ 0 & 0 & 7 \end{bmatrix}, \quad B = \begin{bmatrix} \frac{1}{5} & 0 & 0 \\ 0 & -\frac{1}{8} & 0 \\ 0 & 0 & \frac{1}{2} \end{bmatrix}$$

Solution:

$$AB = \begin{bmatrix} 5 & 0 & 0 \\ 0 & -8 & 0 \\ 0 & 0 & 7 \end{bmatrix} \begin{bmatrix} \frac{1}{5} & 0 & 0 \\ 0 & -\frac{1}{8} & 0 \\ 0 & 0 & \frac{1}{2} \end{bmatrix} = \begin{bmatrix} 1 & 0 & 0 \\ 0 & 1 & 0 \\ 0 & 0 & \frac{7}{2} \end{bmatrix}$$

25. Solve for X in $X = 3A - 2B$, given

$$A = \begin{bmatrix} -2 & -1 \\ 1 & 0 \\ 3 & -4 \end{bmatrix} \quad \text{and} \quad B = \begin{bmatrix} 0 & 3 \\ 2 & 0 \\ -4 & -1 \end{bmatrix}.$$

Solution:

$$X = 3A - 2B = \begin{bmatrix} -6 & -3 \\ 3 & 0 \\ 9 & -12 \end{bmatrix} - \begin{bmatrix} 0 & 6 \\ 4 & 0 \\ -8 & -2 \end{bmatrix} = \begin{bmatrix} -6 & -9 \\ -1 & 0 \\ 17 & 10 \end{bmatrix}$$

29. Find matrices A, X, and B such that the given system of linear equations can be written as the matrix equation $AX = B$. Solve the system of equations.

$$-x + y = 4$$
$$-2x + y = 0$$

Solution:

$$A = \begin{bmatrix} -1 & 1 \\ -2 & 1 \end{bmatrix}, \quad X = \begin{bmatrix} x \\ y \end{bmatrix}, \quad B = \begin{bmatrix} 4 \\ 0 \end{bmatrix}$$

By Gauss-Jordan elimination on

$$\left[\begin{array}{cc|c} -1 & 1 & 4 \\ -2 & 1 & 0 \end{array}\right] \quad \begin{array}{r} -R_1 \rightarrow \\ 2R_1 + R_2 \rightarrow \end{array} \left[\begin{array}{cc|c} 1 & -1 & -4 \\ 0 & -1 & -8 \end{array}\right]$$

$$\begin{array}{r} R_2 + R_1 \rightarrow \\ -R_2 \rightarrow \end{array} \left[\begin{array}{cc|c} 1 & 0 & 4 \\ 0 & 1 & 8 \end{array}\right], \quad \text{we have } x = 4 \text{ and } y = 8.$$

33. Find $f(A)$, given

$$f(x) = x^2 - 5x + 2 \quad \text{and} \quad A = \begin{bmatrix} 2 & 0 \\ 4 & 5 \end{bmatrix}.$$

Solution:

$$\begin{aligned} f(A) &= \begin{bmatrix} 2 & 0 \\ 4 & 5 \end{bmatrix}\begin{bmatrix} 2 & 0 \\ 4 & 5 \end{bmatrix} - 5\begin{bmatrix} 2 & 0 \\ 4 & 5 \end{bmatrix} + 2\begin{bmatrix} 1 & 0 \\ 0 & 1 \end{bmatrix} \\ &= \begin{bmatrix} 4 & 0 \\ 28 & 25 \end{bmatrix} - \begin{bmatrix} 10 & 0 \\ 20 & 25 \end{bmatrix} + \begin{bmatrix} 2 & 0 \\ 0 & 2 \end{bmatrix} \\ &= \begin{bmatrix} -4 & 0 \\ 8 & 2 \end{bmatrix} \end{aligned}$$

37. If a, b, and c are real numbers such that $c \neq 0$ and $ac = bc$, then $a = b$. However, if A, B, and C are matrices such that $AC = BC$, then A is *not* necessarily equal to B. Illustrate this, using the following matrices.

$$A = \begin{bmatrix} 1 & 2 & 3 \\ 0 & 5 & 4 \\ 3 & -2 & 1 \end{bmatrix}, \quad B = \begin{bmatrix} 4 & -6 & 3 \\ 5 & 4 & 4 \\ -1 & 0 & 1 \end{bmatrix}, \quad \text{and } C = \begin{bmatrix} 0 & 0 & 0 \\ 0 & 0 & 0 \\ 4 & -2 & 3 \end{bmatrix}$$

Solution:

$$AC = \begin{bmatrix} 1 & 2 & 3 \\ 0 & 5 & 4 \\ 3 & -2 & 1 \end{bmatrix}\begin{bmatrix} 0 & 0 & 0 \\ 0 & 0 & 0 \\ 4 & -2 & 3 \end{bmatrix} = \begin{bmatrix} 12 & -6 & 9 \\ 16 & -8 & 12 \\ 4 & -2 & 3 \end{bmatrix}$$

$$BC = \begin{bmatrix} 4 & -6 & 3 \\ 5 & 4 & 4 \\ -1 & 0 & 1 \end{bmatrix}\begin{bmatrix} 0 & 0 & 0 \\ 0 & 0 & 0 \\ 4 & -2 & 3 \end{bmatrix} = \begin{bmatrix} 12 & -6 & 9 \\ 16 & -8 & 12 \\ 4 & -2 & 3 \end{bmatrix}$$

Thus, $AC = BC$ even though $A \neq B$.

41. A fruit grower raises two crops that are shipped to three outlets. The number of units of product i that are shipped to outlet j is represented by a_{ij} in the matrix

$$A = \begin{bmatrix} 100 & 75 & 75 \\ 125 & 150 & 100 \end{bmatrix}.$$

The profit per unit is represented by the matrix

$$B = [\,\$3.75 \quad \$7.00\,].$$

Find the product BA, and state what each entry of the product represents.

Solution:

$$BA = [\,3.75 \quad 7.00\,]\begin{bmatrix} 100 & 75 & 75 \\ 125 & 150 & 100 \end{bmatrix} = [\,\$1250.00 \quad \$1331.25 \quad \$981.25\,]$$

The entries in the last matrix represent the profit for both crops at each of the three outlets.

45. The matrix

$$P = \overbrace{\begin{matrix} R & D & I \end{matrix}}^{\text{From}}$$

$$P = \begin{bmatrix} 0.6 & 0.1 & 0.1 \\ 0.2 & 0.7 & 0.1 \\ 0.2 & 0.2 & 0.8 \end{bmatrix} \begin{matrix} R \\ D \\ I \end{matrix} \Bigg\} \text{To}$$

is called a stochastic matrix. Each entry p_{ij} $(i \neq j)$ represents the proportion of the voting population that changes from party i to party j, and p_{ii} represents the proportion that remains loyal to the party from one election to the next. Find P^2. (This matrix gives the transition probabilities from the first election to the third.)

Solution:

$$P^2 = \begin{bmatrix} 0.6 & 0.1 & 0.1 \\ 0.2 & 0.7 & 0.1 \\ 0.2 & 0.2 & 0.8 \end{bmatrix} \begin{bmatrix} 0.6 & 0.1 & 0.1 \\ 0.2 & 0.7 & 0.1 \\ 0.2 & 0.2 & 0.8 \end{bmatrix} = \begin{bmatrix} 0.40 & 0.15 & 0.15 \\ 0.28 & 0.53 & 0.17 \\ 0.32 & 0.32 & 0.68 \end{bmatrix}$$

SECTION 8.3

The Inverse of a Square Matrix

- You should be able to find the inverse, if it exists, of a square matrix.
- You should be able to use inverse matrices to solve systems of equations.

Solutions to Selected Exercises

3. Show that B is the inverse of A where

$$A = \begin{bmatrix} 1 & 2 \\ 3 & 4 \end{bmatrix} \quad \text{and} \quad B = \begin{bmatrix} -2 & 1 \\ \frac{3}{2} & -\frac{1}{2} \end{bmatrix}.$$

Solution:

$$AB = \begin{bmatrix} 1 & 2 \\ 3 & 4 \end{bmatrix} \begin{bmatrix} -2 & 1 \\ \frac{3}{2} & -\frac{1}{2} \end{bmatrix} = \begin{bmatrix} -2+3 & 1-1 \\ -6+6 & 3-2 \end{bmatrix} = \begin{bmatrix} 1 & 0 \\ 0 & 1 \end{bmatrix}$$

$$BA = \begin{bmatrix} -2 & 1 \\ \frac{3}{2} & -\frac{1}{2} \end{bmatrix} \begin{bmatrix} 1 & 2 \\ 3 & 4 \end{bmatrix} = \begin{bmatrix} -2+3 & -4+4 \\ \frac{3}{2}-\frac{3}{2} & 3-2 \end{bmatrix} = \begin{bmatrix} 1 & 0 \\ 0 & 1 \end{bmatrix}$$

7. Show that B is the inverse of A where

$$A = \begin{bmatrix} 2 & 0 & 1 & 1 \\ 3 & 0 & 0 & 1 \\ -1 & 1 & -2 & 1 \\ 4 & -1 & 1 & 0 \end{bmatrix} \quad \text{and} \quad B = \begin{bmatrix} -1 & 2 & -1 & -1 \\ -4 & 9 & -5 & -6 \\ 0 & 1 & -1 & -1 \\ 3 & -5 & 3 & 3 \end{bmatrix}.$$

Solution:

$$AB = \begin{bmatrix} 2 & 0 & 1 & 1 \\ 3 & 0 & 0 & 1 \\ -1 & 1 & -2 & 1 \\ 4 & -1 & 1 & 0 \end{bmatrix} \begin{bmatrix} -1 & 2 & -1 & -1 \\ -4 & 9 & -5 & -6 \\ 0 & 1 & -1 & -1 \\ 3 & -5 & 3 & 3 \end{bmatrix}$$

$$= \begin{bmatrix} -2+0+0+3 & 4+0+1+(-5) & -2+0+(-1)+3 & -2+0+(-1)+3 \\ -3+0+0+3 & 6+0+0+(-5) & -3+0+0+3 & -3+0+0+3 \\ 1+(-4)+0+3 & -2+9+(-2)+(-5) & 1+(-5)+2+3 & 1+(-6)+2+3 \\ -4+4+0+0 & 8+(-9)+1+0 & -4+5+(-1)+0 & -4+6+(-1)+0 \end{bmatrix}$$

$$= \begin{bmatrix} 1 & 0 & 0 & 0 \\ 0 & 1 & 0 & 0 \\ 0 & 0 & 1 & 0 \\ 0 & 0 & 0 & 1 \end{bmatrix}$$

–CONTINUED ON NEXT PAGE–

7. –CONTINUED–

$$BA = \begin{bmatrix} -1 & 2 & -1 & -1 \\ -4 & 9 & -5 & -6 \\ 0 & 1 & -1 & -1 \\ 3 & -5 & 3 & 3 \end{bmatrix} \begin{bmatrix} 2 & 0 & 1 & 1 \\ 3 & 0 & 0 & 1 \\ -1 & 1 & -2 & 1 \\ 4 & -1 & 1 & 0 \end{bmatrix}$$

$$= \begin{bmatrix} -2+6+1+(-4) & 0+0+(-1)+1 & -1+0+2+(-1) & -1+2+(-1)+0 \\ -8+27+5+(-24) & 0+0+(-5)+6 & -4+0+10+(-6) & -4+9+(-5)+0 \\ 0+3+1+(-4) & 0+0+(-1)+1 & 0+0+2+(-1) & 0+1+(-1)+0 \\ 6+(-15)+(-3)+12 & 0+0+3+(-3) & 3+0+(-6)+3 & 3+(-5)+3+0 \end{bmatrix}$$

$$= \begin{bmatrix} 1 & 0 & 0 & 0 \\ 0 & 1 & 0 & 0 \\ 0 & 0 & 1 & 0 \\ 0 & 0 & 0 & 1 \end{bmatrix}$$

11. **Find the inverse of the following matrix (if it exists).**

$$\begin{bmatrix} 1 & -2 \\ 2 & -3 \end{bmatrix}$$

Solution:

$$\left[\begin{array}{cc|cc} 1 & -2 & 1 & 0 \\ 2 & -3 & 0 & 1 \end{array}\right] \quad \begin{array}{r} \\ -2R_1 + R_2 \rightarrow \end{array} \left[\begin{array}{cc|cc} 1 & -2 & 1 & 0 \\ 0 & 1 & -2 & 1 \end{array}\right]$$

$$2R_2 + R_1 \rightarrow \left[\begin{array}{cc|cc} 1 & 0 & -3 & 2 \\ 0 & 1 & -2 & 1 \end{array}\right]$$

$$A^{-1} = \begin{bmatrix} -3 & 2 \\ -2 & 1 \end{bmatrix}$$

15. **Find the inverse of the following matrix (if it exists).**

$$\begin{bmatrix} 2 & 4 \\ 4 & 8 \end{bmatrix}$$

Solution:

$$\left[\begin{array}{cc|cc} 2 & 4 & 1 & 0 \\ 4 & 8 & 0 & 1 \end{array}\right] \quad \begin{array}{r} \frac{1}{2}R_1 \rightarrow \\ -4R_1 + R_2 \rightarrow \end{array} \left[\begin{array}{cc|cc} 1 & 2 & \frac{1}{2} & 0 \\ 0 & 0 & -2 & 1 \end{array}\right]$$

Since the left side does not reduce to I_2, the inverse does not exist.

19. **Find the inverse of the following matrix (if it exists).**

$$\begin{bmatrix} 1 & 1 & 1 \\ 3 & 5 & 4 \\ 3 & 6 & 5 \end{bmatrix}$$

19. –CONTINUED–

Solution:

$$\left[\begin{array}{ccc|ccc} 1 & 1 & 1 & 1 & 0 & 0 \\ 3 & 5 & 4 & 0 & 1 & 0 \\ 3 & 6 & 5 & 0 & 0 & 1 \end{array}\right] \quad \begin{array}{r} \\ -3R_1 + R_2 \rightarrow \\ -3R_1 + R_3 \rightarrow \end{array} \left[\begin{array}{ccc|ccc} 1 & 1 & 1 & 1 & 0 & 0 \\ 0 & 2 & 1 & -3 & 1 & 0 \\ 0 & 3 & 2 & -3 & 0 & 1 \end{array}\right]$$

$$\begin{array}{r} -R_2 + R_1 \rightarrow \\ \frac{1}{2}R_2 \rightarrow \\ -3R_2 + R_3 \rightarrow \end{array} \left[\begin{array}{ccc|ccc} 1 & 0 & \frac{1}{2} & \frac{5}{2} & -\frac{1}{2} & 0 \\ 0 & 1 & \frac{1}{2} & -\frac{3}{2} & \frac{1}{2} & 0 \\ 0 & 0 & \frac{1}{2} & \frac{3}{2} & -\frac{3}{2} & 1 \end{array}\right]$$

$$\begin{array}{r} -R_3 + R_1 \rightarrow \\ -R_3 + R_2 \rightarrow \\ 2R_3 \rightarrow \end{array} \left[\begin{array}{ccc|ccc} 1 & 0 & 0 & 1 & 1 & -1 \\ 0 & 1 & 0 & -3 & 2 & -1 \\ 0 & 0 & 1 & 3 & -3 & 2 \end{array}\right]$$

$$A^{-1} = \begin{bmatrix} 1 & 1 & -1 \\ -3 & 2 & -1 \\ 3 & -3 & 2 \end{bmatrix}$$

23. Find the inverse of the following matrix (if it exists).

$$\begin{bmatrix} 0.1 & 0.2 & 0.3 \\ -0.3 & 0.2 & 0.2 \\ 0.5 & 0.4 & 0.4 \end{bmatrix}$$

Solution:

$$\left[\begin{array}{ccc|ccc} 0.1 & 0.2 & 0.3 & 1 & 0 & 0 \\ -0.3 & 0.2 & 0.2 & 0 & 1 & 0 \\ 0.5 & 0.4 & 0.4 & 0 & 0 & 1 \end{array}\right] \quad \begin{array}{r} 10R_1 \rightarrow \\ 10R_2 \rightarrow \\ 10R_3 \rightarrow \end{array} \left[\begin{array}{ccc|ccc} 1 & 2 & 3 & 10 & 0 & 0 \\ -3 & 2 & 2 & 0 & 10 & 0 \\ 5 & 4 & 4 & 0 & 0 & 10 \end{array}\right]$$

$$\begin{array}{r} \\ 3R_1 + R_2 \rightarrow \\ -5R_1 + R_3 \rightarrow \end{array} \left[\begin{array}{ccc|ccc} 1 & 2 & 3 & 10 & 0 & 0 \\ 0 & 8 & 11 & 30 & 10 & 0 \\ 0 & -6 & -11 & -50 & 0 & 10 \end{array}\right]$$

$$\begin{array}{r} \\ R_3 + R_2 \rightarrow \\ 3R_2 + R_3 \rightarrow \end{array} \left[\begin{array}{ccc|ccc} 1 & 2 & 3 & 10 & 0 & 0 \\ 0 & 2 & 0 & -20 & 10 & 10 \\ 0 & 0 & -11 & -110 & 30 & 40 \end{array}\right]$$

$$\begin{array}{r} -R_2 + R_1 \rightarrow \\ \frac{1}{2}R_2 \rightarrow \\ -\frac{1}{11}R_3 \rightarrow \end{array} \left[\begin{array}{ccc|ccc} 1 & 0 & 3 & 30 & -10 & -10 \\ 0 & 1 & 0 & -10 & 5 & 5 \\ 0 & 0 & 1 & 10 & -\frac{30}{11} & -\frac{40}{11} \end{array}\right]$$

$$\begin{array}{r} -3R_3 + R_1 \rightarrow \\ \\ \\ \end{array} \left[\begin{array}{ccc|ccc} 1 & 0 & 0 & 0 & -\frac{20}{11} & \frac{10}{11} \\ 0 & 1 & 0 & -10 & 5 & 5 \\ 0 & 0 & 1 & 10 & -\frac{30}{11} & -\frac{40}{11} \end{array}\right]$$

$$A^{-1} = \begin{bmatrix} 0 & -\frac{20}{11} & \frac{10}{11} \\ -10 & 5 & 5 \\ 10 & -\frac{30}{11} & -\frac{40}{11} \end{bmatrix} = \frac{5}{11}\begin{bmatrix} 0 & -4 & 2 \\ -22 & 11 & 11 \\ 22 & -6 & -8 \end{bmatrix}$$

25. Find the inverse of $\begin{bmatrix} 1 & 0 & 0 \\ 3 & 4 & 0 \\ 2 & 5 & 5 \end{bmatrix}$ (if it exists).

Solution:

$$\left[\begin{array}{ccc|ccc} 1 & 0 & 0 & 1 & 0 & 0 \\ 3 & 4 & 0 & 0 & 1 & 0 \\ 2 & 5 & 5 & 0 & 0 & 1 \end{array}\right] \begin{array}{r} \\ -3R_1 + R_2 \rightarrow \\ -2R_1 + R_3 \rightarrow \end{array} \left[\begin{array}{ccc|ccc} 1 & 0 & 0 & 1 & 0 & 0 \\ 0 & 4 & 0 & -3 & 1 & 0 \\ 0 & 5 & 5 & -2 & 0 & 1 \end{array}\right]$$

$$\begin{array}{r} \\ \frac{1}{4}R_2 \rightarrow \\ -5R_2 + R_3 \rightarrow \end{array} \left[\begin{array}{ccc|ccc} 1 & 0 & 0 & 1 & 0 & 0 \\ 0 & 1 & 0 & -0.75 & 0.25 & 0 \\ 0 & 0 & 5 & 1.75 & -1.25 & 1 \end{array}\right]$$

$$\begin{array}{r} \\ \\ \frac{1}{5}R_3 \rightarrow \end{array} \left[\begin{array}{ccc|ccc} 1 & 0 & 0 & 1 & 0 & 0 \\ 0 & 1 & 0 & -0.75 & 0.25 & 0 \\ 0 & 0 & 1 & 0.35 & -0.25 & 0.2 \end{array}\right]$$

$$A^{-1} = \begin{bmatrix} 1 & 0 & 0 \\ -0.75 & 0.25 & 0 \\ 0.35 & -0.25 & 0.2 \end{bmatrix}$$

29. Find the inverse of the following matrix.

$$\begin{bmatrix} 1 & -2 & -1 & -2 \\ 3 & -5 & -2 & -3 \\ 2 & -5 & -2 & -5 \\ -1 & 4 & 4 & 11 \end{bmatrix}$$

Solution:

$$\left[\begin{array}{cccc|cccc} 1 & -2 & -1 & -2 & 1 & 0 & 0 & 0 \\ 3 & -5 & -2 & -3 & 0 & 1 & 0 & 0 \\ 2 & -5 & -2 & -5 & 0 & 0 & 1 & 0 \\ -1 & 4 & 4 & 11 & 0 & 0 & 0 & 1 \end{array}\right] \begin{array}{r} \\ -3R_1 + R_2 \rightarrow \\ -2R_1 + R_3 \rightarrow \\ R_1 + R_4 \rightarrow \end{array} \left[\begin{array}{cccc|cccc} 1 & -2 & -1 & -2 & 1 & 0 & 0 & 0 \\ 0 & 1 & 1 & 3 & -3 & 1 & 0 & 0 \\ 0 & -1 & 0 & -1 & -2 & 0 & 1 & 0 \\ 0 & 2 & 3 & 9 & 1 & 0 & 0 & 1 \end{array}\right]$$

$$\begin{array}{r} 2R_2 + R_1 \rightarrow \\ \\ R_2 + R_3 \rightarrow \\ -2R_2 + R_4 \rightarrow \end{array} \left[\begin{array}{cccc|cccc} 1 & 0 & 1 & 4 & -5 & 2 & 0 & 0 \\ 0 & 1 & 1 & 3 & -3 & 1 & 0 & 0 \\ 0 & 0 & 1 & 2 & -5 & 1 & 1 & 0 \\ 0 & 0 & 1 & 3 & 7 & -2 & 0 & 1 \end{array}\right]$$

$$\begin{array}{r} -R_3 + R_1 \rightarrow \\ -R_3 + R_2 \rightarrow \\ \\ -R_3 + R_4 \rightarrow \end{array} \left[\begin{array}{cccc|cccc} 1 & 0 & 0 & 2 & 0 & 1 & -1 & 0 \\ 0 & 1 & 0 & 1 & 2 & 0 & -1 & 0 \\ 0 & 0 & 1 & 2 & -5 & 1 & 1 & 0 \\ 0 & 0 & 0 & 1 & 12 & -3 & -1 & 1 \end{array}\right]$$

$$\begin{array}{r} -2R_4 + R_1 \rightarrow \\ -R_4 + R_2 \rightarrow \\ -2R_4 + R_3 \rightarrow \\ \\ \end{array} \left[\begin{array}{cccc|cccc} 1 & 0 & 0 & 0 & -24 & 7 & 1 & -2 \\ 0 & 1 & 0 & 0 & -10 & 3 & 0 & -1 \\ 0 & 0 & 1 & 0 & -29 & 7 & 3 & -2 \\ 0 & 0 & 0 & 1 & 12 & -3 & -1 & 1 \end{array}\right]$$

$$A^{-1} = \begin{bmatrix} -24 & 7 & 1 & -2 \\ -10 & 3 & 0 & -1 \\ -29 & 7 & 3 & -2 \\ 12 & -3 & -1 & 1 \end{bmatrix}$$

33. Use an inverse matrix to solve the given system of linear equations.

$$x - 2y = 4$$
$$2x - 3y = 2$$

Solution:

$$A = \begin{bmatrix} 1 & -2 \\ 2 & -3 \end{bmatrix}$$

From Exercise 11 we have $A^{-1} = \begin{bmatrix} -3 & 2 \\ -2 & 1 \end{bmatrix}$.

$$X = A^{-1}B = \begin{bmatrix} -3 & 2 \\ -2 & 1 \end{bmatrix} \begin{bmatrix} 4 \\ 2 \end{bmatrix} = \begin{bmatrix} -8 \\ -6 \end{bmatrix}$$

Answer: $(-8,\ -6)$

37. Use an inverse matrix to solve the given system of linear equations.

$$-x + y = 20$$
$$-2x + y = 10$$

Solution:

$$A = \begin{bmatrix} -1 & 1 \\ -2 & 1 \end{bmatrix}$$

From Exercise 13 we have $A^{-1} = \begin{bmatrix} 1 & -1 \\ 2 & -1 \end{bmatrix}$.

$$X = A^{-1}B = \begin{bmatrix} 1 & -1 \\ 2 & -1 \end{bmatrix} \begin{bmatrix} 20 \\ 10 \end{bmatrix} = \begin{bmatrix} 10 \\ 30 \end{bmatrix}$$

Answer: $(10,\ 30)$.

41. Use an inverse matrix to solve the following system.

$$\begin{aligned} x_1 - 2x_2 - x_3 - 2x_4 &= 0 \\ 3x_1 - 5x_2 - 2x_3 - 3x_4 &= 1 \\ 2x_1 - 5x_2 - 2x_3 - 5x_4 &= -1 \\ -x_1 + 4x_2 + 4x_3 + 11x_4 &= 2 \end{aligned}$$

41. –CONTINUED–

Solution:

$$A = \begin{bmatrix} 1 & -2 & -1 & -2 \\ 3 & -5 & -2 & -3 \\ 2 & -5 & -2 & -5 \\ -1 & 4 & 4 & 11 \end{bmatrix}$$

From Exercise 29 we have

$$A^{-1} = \begin{bmatrix} -24 & 7 & 1 & -2 \\ -10 & 3 & 0 & -1 \\ -29 & 7 & 3 & -2 \\ 12 & -3 & -1 & 1 \end{bmatrix}.$$

$$X = A^{-1}B = \begin{bmatrix} -24 & 7 & 1 & -2 \\ -10 & 3 & 0 & -1 \\ -29 & 7 & 3 & -2 \\ 12 & -3 & -1 & 1 \end{bmatrix} \begin{bmatrix} 0 \\ 1 \\ -1 \\ 2 \end{bmatrix} = \begin{bmatrix} 2 \\ 1 \\ 0 \\ 0 \end{bmatrix}$$

Answer: (2, 1, 0, 0)

45. Consider a person who invests in AAA-rated bonds, A-rated bonds, and B-rated bonds. The average yield is 6.5% on AAA-bonds, 7% on A-bonds, and 9% on B-bonds. Suppose the person invests twice as much in B-bonds as in A-bonds. A system of linear equations (where x, y, and z represent the amounts invested in AAA-, A-, and B-bonds, respectively) is as follows.

$$\begin{aligned} x + \quad y + \quad z &= \text{(total investment)} \\ 0.065x + 0.07y + 0.09z &= \text{(annual return)} \\ 2y - \quad z &= 0 \end{aligned}$$

Use the inverse of the coefficient matrix of this system to find the amount invested in each type of bond if the total investment is \$12,000, and the annual return is \$835.

Solution:

$$A = \begin{bmatrix} 1 & 1 & 1 \\ 0.065 & 0.07 & 0.09 \\ 0 & 2 & -1 \end{bmatrix}$$

Using the methods of this section, we have

$$A^{-1} = \frac{1}{11} \begin{bmatrix} 50 & -600 & -4 \\ -13 & 200 & 5 \\ -26 & 400 & -1 \end{bmatrix}.$$

$$X = A^{-1}B = \frac{1}{11} \begin{bmatrix} 50 & -600 & -4 \\ -13 & 200 & 5 \\ -26 & 400 & -1 \end{bmatrix} \begin{bmatrix} 12{,}000 \\ 835 \\ 0 \end{bmatrix} = \begin{bmatrix} 9000 \\ 1000 \\ 2000 \end{bmatrix}$$

Answer: \$9000 in AAA bonds, \$1000 in A bonds, \$2000 in B bonds

SECTION 8.4

The Determinant of a Square Matrix

- You should be able to determine the determinant of a matrix of order 2 or of order 3 by using the products of the diagonals.
- You should be able to use expansion by cofactors to find the determinant of a matrix of order 3 or greater.
- The determinant of a triangular matrix equals the product of the entries on the main diagonal.

Solutions to Selected Exercises

7. Find the determinant of $\begin{bmatrix} -7 & 6 \\ \frac{1}{2} & 3 \end{bmatrix}$.

Solution:

$$\begin{vmatrix} -7 & 6 \\ \frac{1}{2} & 3 \end{vmatrix} = -7(3) - 6(\tfrac{1}{2}) = -24$$

11. Find the determinant of $\begin{bmatrix} 2 & -1 & 0 \\ 4 & 2 & 1 \\ 4 & 2 & 1 \end{bmatrix}$.

Solution:

$$\left|\begin{matrix} 2 & -1 & 0 \\ 4 & 2 & 1 \\ 4 & 2 & 1 \end{matrix}\right|\begin{matrix} 2 & -1 \\ 4 & 2 \\ 4 & 2 \end{matrix} = 4 + (-4) + 0 - 0 - 4 - (-4) = 0$$

15. Find the determinant of $\begin{bmatrix} 1 & 4 & -2 \\ 3 & 6 & -6 \\ -2 & 1 & 4 \end{bmatrix}$.

Solution:

$$\left|\begin{matrix} 1 & 4 & -2 \\ 3 & 6 & -6 \\ -2 & 1 & 4 \end{matrix}\right|\begin{matrix} 1 & 4 \\ 3 & 6 \\ -2 & 1 \end{matrix} = 24 + 48 + (-6) - 24 - (-6) - 48 = 0$$

19. Find the determinant of

$$\begin{bmatrix} -1 & 2 & -5 \\ 0 & 3 & 4 \\ 0 & 0 & 3 \end{bmatrix}$$

Solution:

$$\begin{vmatrix} -1 & 2 & -5 \\ 0 & 3 & 4 \\ 0 & 0 & 3 \end{vmatrix} = (-1)(3)(3) = -9 \quad \text{(Upper Triangular)}$$

23. Find the determinant of

$$\begin{bmatrix} x & y & 1 \\ -2 & -2 & 1 \\ 1 & 5 & 1 \end{bmatrix}.$$

Solution:

$$\begin{vmatrix} x & y & 1 \\ -2 & -2 & 1 \\ 1 & 5 & 1 \end{vmatrix} \begin{matrix} x & y \\ -2 & -2 \\ 1 & 5 \end{matrix} = -2x + y + (-10) - (-2) - 5x - (-2y) = -7x + 3y - 8$$

25. Find (a) all minors, and (b) cofactors for

$$\begin{bmatrix} 3 & 4 \\ 2 & -5 \end{bmatrix}.$$

Solution:

$$\begin{bmatrix} 3 & 4 \\ 2 & -5 \end{bmatrix}$$

(a) $M_{11} = -5$

$M_{12} = 2$

$M_{21} = 4$

$M_{22} = 3$

(b) $C_{11} = M_{11} = -5$

$C_{12} = -M_{12} = -2$

$C_{21} = -M_{21} = -4$

$C_{22} = M_{22} = 3$

29. Find the determinant of the following matrix by the method of expansion by cofactors. Expand using (a) Row 1 and (b) Column 2.

$$\begin{bmatrix} -3 & 2 & 1 \\ 4 & 5 & 6 \\ 2 & -3 & 1 \end{bmatrix}$$

Solution:

(a) Expansion along the first row:

$$\begin{vmatrix} -3 & 2 & 1 \\ 4 & 5 & 6 \\ 2 & -3 & 1 \end{vmatrix} = -3C_{11} + 2C_{12} + 1C_{13}$$
$$= -3(23) + 2(8) + 1(-22) = -69 + 16 - 22 = -75$$

(b) Expansion along the second column:

$$\begin{vmatrix} -3 & 2 & 1 \\ 4 & 5 & 6 \\ 2 & -3 & 1 \end{vmatrix} = 2C_{12} + 5C_{22} - 3C_{32}$$
$$= 2(8) + 5(-5) - 3(22) = 16 - 25 - 66 = -75$$

35. Find the determinant of

$$\begin{bmatrix} 1 & 4 & -2 \\ 3 & 2 & 0 \\ -1 & 4 & 3 \end{bmatrix}.$$

Solution:

Expansion along the third column:

$$\begin{vmatrix} 1 & 4 & -2 \\ 3 & 2 & 0 \\ -1 & 4 & 3 \end{vmatrix} = -2\begin{vmatrix} 3 & 2 \\ -1 & 4 \end{vmatrix} + 0 + 3\begin{vmatrix} 1 & 4 \\ 3 & 2 \end{vmatrix}$$
$$= -2(14) + 3(-10) = -28 - 30 = -58$$

39. Find the determinant of

$$\begin{bmatrix} 3 & 6 & -5 & 4 \\ -2 & 0 & 6 & 0 \\ 1 & 1 & 2 & 2 \\ 0 & 3 & -1 & -1 \end{bmatrix}.$$

Solution:

Expansion along the second row:

$$\begin{vmatrix} 3 & 6 & -5 & 4 \\ -2 & 0 & 6 & 0 \\ 1 & 1 & 2 & 2 \\ 0 & 3 & -1 & -1 \end{vmatrix} = -(-2)\begin{vmatrix} 6 & -5 & 4 \\ 1 & 2 & 2 \\ 3 & -1 & -1 \end{vmatrix} - 6\begin{vmatrix} 3 & 6 & 4 \\ 1 & 1 & 2 \\ 0 & 3 & -1 \end{vmatrix}$$
$$= 2[6(0) - 1(9) + 3(-18)] - 6[3(-7) - (-18)]$$
$$= 2[-9 - 54] - 6[-21 + 18] = -126 + 18 = -108$$

43. Find the determinant of

$$\begin{bmatrix} 3 & 2 & 4 & -1 & 5 \\ -2 & 0 & 1 & 3 & 2 \\ 1 & 0 & 0 & 4 & 0 \\ 6 & 0 & 2 & -1 & 0 \\ 3 & 0 & 5 & 1 & 0 \end{bmatrix}.$$

Solution:

Expansion along the second column:

$$\begin{vmatrix} 3 & 2 & 4 & -1 & 5 \\ -2 & 0 & 1 & 3 & 2 \\ 1 & 0 & 0 & 4 & 0 \\ 6 & 0 & 2 & -1 & 0 \\ 3 & 0 & 5 & 1 & 0 \end{vmatrix} = -2\begin{vmatrix} -2 & 1 & 3 & 2 \\ 1 & 0 & 4 & 0 \\ 6 & 2 & -1 & 0 \\ 3 & 5 & 1 & 0 \end{vmatrix}$$

$$= -2(-2)\begin{vmatrix} 1 & 0 & 4 \\ 6 & 2 & -1 \\ 3 & 5 & 1 \end{vmatrix} \quad \text{Expansion along the fourth column}$$

$$= 4[1(7) - 0 + 4(24)] = 4[7 + 96] = 412$$

45. Solve for x.

$$\begin{vmatrix} x-1 & 2 \\ 3 & x-2 \end{vmatrix} = 0$$

Solution:

$$\begin{vmatrix} x-1 & 2 \\ 3 & x-2 \end{vmatrix} = 0$$

$$(x-1)(x-2) - 6 = 0$$

$$x^2 - 3x - 4 = 0$$

$$(x+1)(x-4) = 0$$

$$x = -1 \text{ or } x = 4$$

47. Evaluate the determinant of $\begin{bmatrix} 4u & -1 \\ -1 & 2v \end{bmatrix}$. Determinants of this type occur in calculus.

Solution:

$$\begin{vmatrix} 4u & -1 \\ -1 & 2v \end{vmatrix} = 8uv - 1$$

49. Evaluate the determinant of $\begin{bmatrix} e^{2x} & e^{3x} \\ 2e^{2x} & 3e^{3x} \end{bmatrix}$. Determinants of this type occur in calculus.

Solution:

$$\begin{vmatrix} e^{2x} & e^{3x} \\ 2e^{2x} & 3e^{3x} \end{vmatrix} = 3e^{5x} - 2e^{5x} = e^{5x}$$

SECTION 8.5

Properties of Determinants

- You should know what effect each elementary row (column) operation has on the determinant of a matrix.
- You should know what conditions yield a determinant of zero.
- You should be able to use determinants to determine if a matrix has an inverse.

Solutions to Selected Exercises

5. State the property of determinants that verifies the equation.

$$\begin{vmatrix} 1 & 3 & 4 \\ -7 & 2 & -5 \\ 6 & 1 & 2 \end{vmatrix} = -\begin{vmatrix} 1 & 4 & 3 \\ -7 & -5 & 2 \\ 6 & 2 & 1 \end{vmatrix}$$

Solution:

Interchanging Columns 2 and 3 results in a change of sign of the determinant.

9. State the property of determinants that verifies the equation.

$$\begin{vmatrix} 5 & 0 & 10 \\ 25 & -30 & 40 \\ -15 & 5 & 20 \end{vmatrix} = 5^3 \begin{vmatrix} 1 & 0 & 2 \\ 5 & -6 & 8 \\ -3 & 1 & 4 \end{vmatrix}$$

Solution:

Multiplying the entries of all three rows by 5 produces a determinant that is 5^3 the determinant of the second matrix.

13. State the property of determinants that verifies the equation.

$$\begin{vmatrix} 3 & 2 & 4 \\ -2 & 1 & 5 \\ 5 & -7 & -20 \end{vmatrix} = \begin{vmatrix} 7 & 2 & -6 \\ 0 & 1 & 0 \\ -9 & -7 & 15 \end{vmatrix}$$

Solution:

Adding multiples of Column 2 to Columns 1 and 3 leaves the determinant unchanged.

19. Use elementary row (or column) operations as aids for evaluating

$$\begin{vmatrix} 3 & 8 & -7 \\ 0 & -5 & 4 \\ 8 & 1 & 6 \end{vmatrix}.$$

Solution:

$$\begin{vmatrix} 3 & 8 & -7 \\ 0 & -5 & 4 \\ 8 & 1 & 6 \end{vmatrix} = 3(-34) + 8(-3) = -126$$

Expansion along Column 1

23. Use elementary row (or column) operations as aids for evaluating

$$\begin{vmatrix} 7 & 0 & -14 \\ -2 & 5 & 4 \\ -6 & 2 & 12 \end{vmatrix}.$$

Solution:

$$\begin{vmatrix} 7 & 0 & -14 \\ -2 & 5 & 4 \\ -6 & 2 & 12 \end{vmatrix} = \begin{vmatrix} 7 & 0 & 0 \\ -2 & 5 & 0 \\ 6 & 2 & 0 \end{vmatrix} = 0$$

29. Use elementary row (or column) operations as aids for evaluating

$$\begin{vmatrix} 3 & -2 & 4 & 3 & 1 \\ -1 & 0 & 2 & 1 & 0 \\ 5 & -1 & 0 & 3 & 2 \\ 4 & 7 & -8 & 0 & 0 \\ 1 & 2 & 3 & 0 & 2 \end{vmatrix}.$$

29. –CONTINUED–

Solution:

$$\begin{vmatrix} 3 & -2 & 4 & 3 & 1 \\ -1 & 0 & 2 & 1 & 0 \\ 5 & -1 & 0 & 3 & 2 \\ 4 & 7 & -8 & 0 & 0 \\ 1 & 2 & 3 & 0 & 2 \end{vmatrix} = \begin{vmatrix} 3 & -2 & 4 & 3 & 1 \\ -1 & 0 & 2 & 1 & 0 \\ -1 & 3 & -8 & -3 & 0 \\ 4 & 7 & -8 & 0 & 0 \\ -4 & 3 & 3 & -3 & 0 \end{vmatrix} \begin{matrix} \\ \\ \leftarrow -2R_1 + R_3 \\ \\ \leftarrow -R_3 + R_5 \end{matrix}$$

$$= \begin{vmatrix} -1 & 0 & 2 & 1 \\ -1 & 3 & -8 & -3 \\ 4 & 7 & -8 & 0 \\ -4 & 3 & 3 & -3 \end{vmatrix} \quad \text{(Expansion along Column 5)}$$

$$= \begin{vmatrix} -1 & 0 & 2 & 1 \\ -4 & 3 & -2 & 0 \\ 4 & 7 & -8 & 0 \\ -3 & 0 & 11 & 0 \end{vmatrix} \begin{matrix} \\ \leftarrow 3R_1 + R_2 \\ \\ \leftarrow -R_2 + R_4 \end{matrix}$$

$$= -\begin{vmatrix} -4 & 3 & -2 \\ 4 & 7 & -8 \\ -3 & 0 & 11 \end{vmatrix} \quad \text{(Expansion along Column 4)}$$

$$= -\begin{vmatrix} -4 & 3 & -2 \\ 0 & 10 & -10 \\ -3 & 0 & 11 \end{vmatrix} \leftarrow R_1 + R_2$$

$$= -[-4(110) - 3(-10)] \quad \text{(Expansion along Column 1)}$$

$$= -[-440 + 30]$$

$$= 410$$

33. Use a determinant to determine whether the following matrix is invertible.

$$\begin{bmatrix} 14 & 7 & 0 \\ 2 & 3 & 0 \\ 1 & -5 & 2 \end{bmatrix}$$

Solution:

$$\begin{vmatrix} 14 & 7 & 0 \\ 2 & 3 & 0 \\ 1 & -5 & 2 \end{vmatrix} = 2\begin{vmatrix} 14 & 7 \\ 2 & 3 \end{vmatrix} = 2(42 - 14) = 56 \neq 0$$

The matrix *is* invertible.

37. Find the value(s) of k so that $\begin{bmatrix} k-1 & 3 \\ 2 & k-2 \end{bmatrix}$ is singular.

Solution:

$$\begin{vmatrix} k-1 & 3 \\ 2 & k-2 \end{vmatrix} = 0$$
$$(k-1)(k-2) - 6 = 0$$
$$k^2 - 3k - 4 = 0$$
$$(k+1)(k-4) = 0$$
$$k = -1 \text{ or } k = 4$$

39. Verify $\begin{vmatrix} w & x \\ y & z \end{vmatrix} = -\begin{vmatrix} y & z \\ w & x \end{vmatrix}$.

Solution:

$$\begin{vmatrix} w & x \\ y & z \end{vmatrix} = wz - xy$$
$$-\begin{vmatrix} y & z \\ w & x \end{vmatrix} = -[xy - wz] = wz - xy$$

Therefore, $\begin{vmatrix} w & x \\ y & z \end{vmatrix} = -\begin{vmatrix} y & z \\ w & x \end{vmatrix}$.

43. Show that $\begin{vmatrix} 1 & x & x^2 \\ 1 & y & y^2 \\ 1 & z & z^2 \end{vmatrix} = (y-x)(z-x)(z-y)$.

Solution:

$$\begin{vmatrix} 1 & x & x^2 \\ 1 & y & y^2 \\ 1 & z & z^2 \end{vmatrix} = \begin{vmatrix} y & y^2 \\ z & z^2 \end{vmatrix} - \begin{vmatrix} x & x^2 \\ z & z^2 \end{vmatrix} + \begin{vmatrix} x & x^2 \\ y & y^2 \end{vmatrix}$$
$$= (yz^2 - y^2z) - (xz^2 - x^2z) + (xy^2 - x^2y)$$
$$= yz^2 - xz^2 - y^2z + x^2z + xy(y-x)$$
$$= z^2(y-x) - z(y^2 - x^2) + xy(y-x)$$
$$= z^2(y-x) - z(y-x)(y+x) + xy(y-x)$$
$$= (y-x)[z^2 - z(y+x) + xy]$$
$$= (y-x)[z^2 - zy - zx + xy]$$
$$= (y-x)[z^2 - zx - zy + xy]$$
$$= (y-x)[z(z-x) - y(z-x)]$$
$$= (y-x)(z-x)(z-y)$$

47. Find (a) $|A|$, (b) $|B|$, (c) AB, and (d) $|AB|$.

$$A = \begin{bmatrix} -1 & 2 & 1 \\ 1 & 0 & 1 \\ 0 & 1 & 0 \end{bmatrix}, \qquad B = \begin{bmatrix} -1 & 0 & 0 \\ 0 & 2 & 0 \\ 0 & 0 & 3 \end{bmatrix}$$

Solution:

(a) $$|A| = \begin{vmatrix} -1 & 2 & 1 \\ 1 & 0 & 1 \\ 0 & 1 & 0 \end{vmatrix} = \begin{vmatrix} -1 & 2 & 1 \\ 0 & 2 & 2 \\ 0 & 1 & 0 \end{vmatrix} \leftarrow R_1 + R_2$$

$$= -1\begin{vmatrix} 2 & 2 \\ 1 & 0 \end{vmatrix} = 2$$

(b) $$|B| = \begin{vmatrix} -1 & 0 & 0 \\ 0 & 2 & 0 \\ 0 & 0 & 3 \end{vmatrix} = (-1)(2)(3) = -6 \quad \text{(Triangular)}$$

(c) $$AB = \begin{bmatrix} -1 & 2 & 1 \\ 1 & 0 & 1 \\ 0 & 1 & 0 \end{bmatrix}\begin{bmatrix} -1 & 0 & 0 \\ 0 & 2 & 0 \\ 0 & 0 & 3 \end{bmatrix} = \begin{bmatrix} 1 & 4 & 3 \\ -1 & 0 & 3 \\ 0 & 2 & 0 \end{bmatrix}$$

(d) $$|AB| = \begin{vmatrix} 1 & 4 & 3 \\ -1 & 0 & 3 \\ 0 & 2 & 0 \end{vmatrix} = -2\begin{vmatrix} 1 & 3 \\ -1 & 3 \end{vmatrix} = -12 \quad \text{(Expansion along Row 3)}$$

49. Find square matrices A and B to demonstrate that $|A+B| \neq |A| + |B|$.

Solution:

Let $A = \begin{bmatrix} 1 & 0 \\ 0 & 1 \end{bmatrix}$ and $B = \begin{bmatrix} -1 & 0 \\ 0 & -1 \end{bmatrix}$.

Then $A + B = \begin{bmatrix} 0 & 0 \\ 0 & 0 \end{bmatrix}$.

$$|A+B| = 0 \quad \text{and} \quad |A| + |B| = 1 + 1 = 2$$

Thus, $|A+B| \neq |A| + |B|$.

SECTION 8.6

Applications of Determinants and Matrices

- You should be able to use Cramer's Rule to solve a system of linear equations.
- Now you should be able to solve a system of linear equations by substitution, elimination, elementary row operations on an augmented matrix, using the inverse matrix, or Cramer's Rule.
- You should be able to find the area of a triangle in the xy-plane given the vertices.
- You should be able to use determinants to determine if three points are collinear.
- You should be able to use determinants to find the equation of a line through two distinct points.

Solutions to Selected Exercises

5. Use Cramer's Rule to solve the system of equations.

$$20x + 8y = 11$$
$$12x - 24y = 21$$

Solution:

$$x = \frac{\begin{vmatrix} 11 & 8 \\ 21 & -24 \end{vmatrix}}{\begin{vmatrix} 20 & 8 \\ 12 & -24 \end{vmatrix}} = \frac{-432}{-576} = \frac{3}{4}$$

$$y = \frac{\begin{vmatrix} 20 & 11 \\ 12 & 21 \end{vmatrix}}{\begin{vmatrix} 20 & 8 \\ 12 & -24 \end{vmatrix}} = \frac{288}{-576} = -\frac{1}{2}$$

Answer: $\left(\frac{3}{4}, -\frac{1}{2}\right)$

9. Use Cramer's Rule to solve the system of equations.

$$3x + 6y = 5$$
$$6x + 14y = 11$$

Solution:

$$x = \frac{\begin{vmatrix} 5 & 6 \\ 11 & 14 \end{vmatrix}}{\begin{vmatrix} 3 & 6 \\ 6 & 14 \end{vmatrix}} = \frac{4}{6} = \frac{2}{3}, \quad y = \frac{\begin{vmatrix} 3 & 5 \\ 6 & 11 \end{vmatrix}}{\begin{vmatrix} 3 & 6 \\ 6 & 14 \end{vmatrix}} = \frac{3}{6} = \frac{1}{2}$$

Answer: $\left(\frac{2}{3}, \frac{1}{2}\right)$

13. Use Cramer's Rule to solve the system of equations for x.

$$3x + 4y + 4z = 11$$
$$4x - 4y + 6z = 11$$
$$6x - 6y \quad\quad = 3$$

Solution:

$$x = \frac{\begin{vmatrix} 11 & 4 & 4 \\ 11 & -4 & 6 \\ 3 & -6 & 0 \end{vmatrix}}{\begin{vmatrix} 3 & 4 & 4 \\ 4 & -4 & 6 \\ 6 & -6 & 0 \end{vmatrix}} = \frac{252}{252} = 1$$

19. Use Cramer's Rule to solve the system of equations for x.

$$7x - 3y \quad + 2w = 41$$
$$-2x + y \quad - w = -13$$
$$4x \quad + z - 2w = 12$$
$$-x + y \quad - w = -8$$

Solution:

$$x = \frac{\begin{vmatrix} 41 & -3 & 0 & 2 \\ -13 & 1 & 0 & -1 \\ 12 & 0 & 1 & -2 \\ -8 & 1 & 0 & -1 \end{vmatrix}}{\begin{vmatrix} 7 & -3 & 0 & 2 \\ -2 & 1 & 0 & -1 \\ 4 & 0 & 1 & -2 \\ -1 & 1 & 0 & -1 \end{vmatrix}} = \frac{\begin{vmatrix} 41 & -3 & 2 \\ -13 & 1 & -1 \\ -8 & 1 & -1 \end{vmatrix}}{\begin{vmatrix} 7 & -3 & 2 \\ -2 & 1 & -1 \\ -1 & 1 & -1 \end{vmatrix}} = \frac{5}{1} = 5 \quad \text{(Expansion along Column 3)}$$

23. The maximum Social Security contributions for an employee between 1981 and 1989 are shown in the accompanying figure. (The figure shows the amount contributed by the *employee*. This amount is matched by the employer.) The least squares regression line $y = a + bt$ for this data is found by solving the system

$$9a + 45b = 24.983$$
$$45a + 285b = 137.012$$

where y is the contribution in 1000s of dollars and t is the calendar year with $t = 1$ corresponding to 1981. Use Cramer's Rule to solve this system, and use the result to approximate the maximum Social Security contribution in 1992. (*Source:* U.S. Social Security Administration)

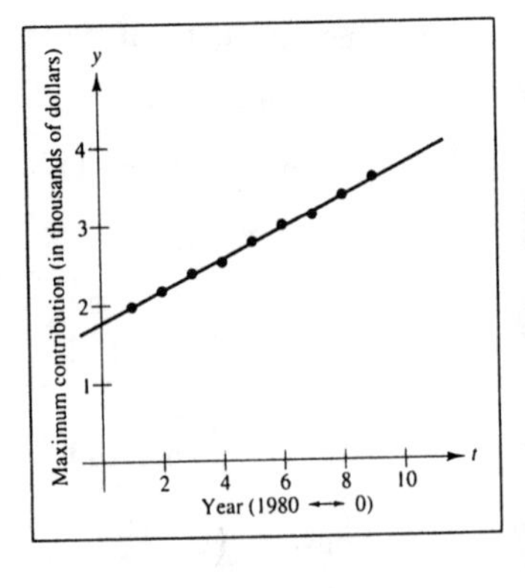

Solution:

$$a = \frac{\begin{vmatrix} 24.983 & 45 \\ 137.012 & 285 \end{vmatrix}}{\begin{vmatrix} 9 & 45 \\ 45 & 285 \end{vmatrix}} = \frac{954.615}{540} \approx 1.768$$

Using back-substitution in the first equation we have $b \approx 0.202$. Thus, $y \approx 1.768 + 0.202t$. When $t = 12$, $y \approx 4.2$ thousand dollars.

27. Use a determinant to find the area of the triangle with vertices $(-2, -3)$, $(2, -3)$, and $(0, 4)$.

Solution:

$$A = \frac{1}{2}\begin{vmatrix} -2 & -3 & 1 \\ 2 & -3 & 1 \\ 0 & 4 & 1 \end{vmatrix} = 14 \text{ square units}$$

31. Use a determinant to find the area of the triangle with vertices $(-2, 4)$, $(2, 3)$, $(-1, 5)$.

Solution:

$$A = \frac{1}{2}\begin{vmatrix} -2 & 4 & 1 \\ 2 & 3 & 1 \\ -1 & 5 & 1 \end{vmatrix} = \frac{5}{2} \text{ square units}$$

35. A large region of forest has been infested with gypsy moths. The region is roughly triangular, as shown in the figure. From the northernmost vertex A of the region, the distance to Vertex B is 25 miles south and 10 miles east, and the distance to vertex C is 20 miles south and 28 miles east. Approximate the number of square miles in this region.

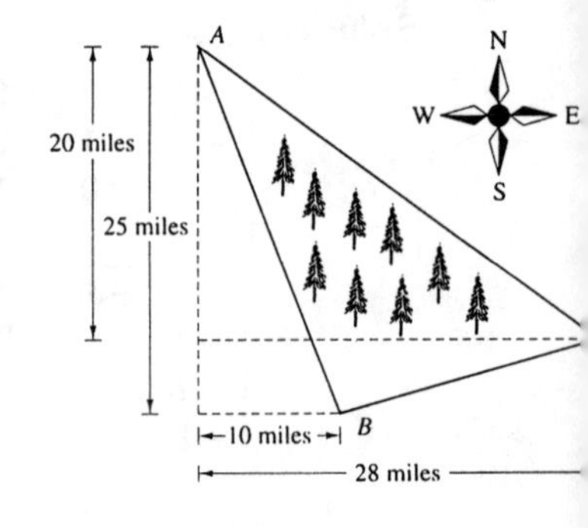

Solution: Vertices: $(0, 25)$, $(10, 0)$, $(28, 5)$

$$A = \frac{1}{2}\begin{vmatrix} 0 & 25 & 1 \\ 10 & 0 & 1 \\ 28 & 5 & 1 \end{vmatrix} = \frac{500}{2} = 250 \text{ square miles}$$

39. Use a determinant to determine if the points $(2, -\frac{1}{2})$, $(-4, 4)$, and $(6, -3)$ are collinear.

Solution:

The points are *not* collinear since $\begin{vmatrix} 2 & -\frac{1}{2} & 1 \\ -4 & 4 & 1 \\ 6 & -3 & 1 \end{vmatrix} = -3 \neq 0.$

45. Use a determinant to find an equation of the line through the points $(-4, 3)$ and $(2, 1)$.

Solution:

$$\begin{vmatrix} x & y & 1 \\ -4 & 3 & 1 \\ 2 & 1 & 1 \end{vmatrix} = 0$$

$$2x + 6y - 10 = 0$$

$$x + 3y - 5 = 0$$

49. Write a cryptogram for LANDING SUCCESSFUL using the following matrix.

$$A = \begin{bmatrix} 1 & 2 & 2 \\ 3 & 7 & 9 \\ -1 & -4 & -7 \end{bmatrix}$$

Solution:

L A N D I N G _ S U C C E S S F U L

[12 1 14] [4 9 14] [7 0 19] [21 3 3] [5 19 19] [6 21 12]

$$[12 \quad 1 \quad 14]A = [1 \quad -25 \quad -65]$$

$$[4 \quad 9 \quad 14]A = [17 \quad 15 \quad -9]$$

$$[7 \quad 0 \quad 19]A = [-12 \quad -62 \quad -119]$$

$$[21 \quad 3 \quad 3]A = [27 \quad 51 \quad 48]$$

$$[5 \quad 19 \quad 19]A = [43 \quad 67 \quad 48]$$

$$[6 \quad 21 \quad 12]A = [57 \quad 111 \quad 117]$$

Cryptogram: 1 −25 −65 17 15 −9 −12 −62 −119 27 51 48 43 67 48 57 111 117

53. Decode the cryptogram

20 17 −15 −12 −56 −104 1 −25 −65 62 143 181

using the inverse of the following matrix.

$$A = \begin{bmatrix} 1 & 2 & 2 \\ 3 & 7 & 9 \\ -1 & -4 & -7 \end{bmatrix}$$

Solution:

To find A^{-1}, use Gauss-Jordan elimination.

$$\left[\begin{array}{ccc|ccc} 1 & 2 & 2 & 1 & 0 & 0 \\ 3 & 7 & 9 & 0 & 1 & 0 \\ -1 & -4 & -7 & 0 & 0 & 1 \end{array}\right] \rightarrow \left[\begin{array}{ccc|ccc} 1 & 0 & 0 & -13 & 6 & 4 \\ 0 & 1 & 0 & 12 & -5 & -3 \\ 0 & 0 & 1 & -5 & 2 & 1 \end{array}\right]$$

$$A^{-1} = \begin{bmatrix} -13 & 6 & 4 \\ 12 & -5 & -3 \\ -5 & 2 & 1 \end{bmatrix}$$

$$[20 \quad 17 \quad -15]A^{-1} = [19 \quad 5 \quad 14]$$

$$[-12 \quad -56 \quad -104]A^{-1} = [4 \quad 0 \quad 16]$$

$$[1 \quad -25 \quad -65]A^{-1} = [12 \quad 1 \quad 14]$$

$$[62 \quad 143 \quad 181]A^{-1} = [5 \quad 19 \quad 0]$$

19	5	14	4	0	16	12	1	14	5	19	0
S	E	N	D	_	P	L	A	N	E	S	_

REVIEW EXERCISES FOR CHAPTER 8

Solutions to Selected Exercises

3. Use matrices and elementary row operations to solve the system of equations.

$$\begin{aligned} 0.2x - 0.1y &= 0.07 \\ 0.4x - 0.5y &= -0.01 \end{aligned}$$

Solution:

$$\begin{bmatrix} 0.2 & -0.1 & \vdots & 0.07 \\ 0.4 & -0.5 & \vdots & -0.01 \end{bmatrix} \quad \begin{matrix} 5R_1 \rightarrow \\ -2R_1 + R_2 \rightarrow \end{matrix} \begin{bmatrix} 1 & -0.5 & \vdots & 0.35 \\ 0 & -0.3 & \vdots & -0.15 \end{bmatrix}$$

$$\begin{matrix} 0.5R_2 + R_1 \rightarrow \\ -\frac{1}{0.3}R_2 \rightarrow \end{matrix} \begin{bmatrix} 1 & 0 & \vdots & 0.6 \\ 0 & 1 & \vdots & 0.5 \end{bmatrix}$$

$x = 0.6$

$y = 0.5$

Answer: $(0.6, 0.5)$

7. Use matrices and elementary row operations to solve the system of equations.

$$\begin{aligned} 2x + 3y + 3z &= 3 \\ 6x + 6y + 12z &= 13 \\ 12x + 9y - z &= 2 \end{aligned}$$

Solution:

$$\begin{bmatrix} 2 & 3 & 3 & \vdots & 3 \\ 6 & 6 & 12 & \vdots & 13 \\ 12 & 9 & -1 & \vdots & 2 \end{bmatrix} \quad \begin{matrix} \\ -3R_1 + R_2 \rightarrow \\ -2R_2 + R_3 \rightarrow \end{matrix} \begin{bmatrix} 2 & 3 & 3 & \vdots & 3 \\ 0 & -3 & 3 & \vdots & 4 \\ 0 & -3 & -25 & \vdots & -24 \end{bmatrix}$$

$$\begin{matrix} R_2 + R_1 \rightarrow \\ \\ -R_2 + R_3 \rightarrow \end{matrix} \begin{bmatrix} 2 & 0 & 6 & \vdots & 7 \\ 0 & -3 & 3 & \vdots & 4 \\ 0 & 0 & -28 & \vdots & -28 \end{bmatrix}$$

$$\begin{matrix} \frac{1}{2}R_1 \rightarrow \\ -\frac{1}{3}R_2 \rightarrow \\ -\frac{1}{28}R_3 \rightarrow \end{matrix} \begin{bmatrix} 1 & 0 & 3 & \vdots & \frac{7}{2} \\ 0 & 1 & -1 & \vdots & -\frac{4}{3} \\ 0 & 0 & 1 & \vdots & 1 \end{bmatrix}$$

$$\begin{aligned} z &= 1 \\ x - 3z &= \frac{7}{2} \Rightarrow x = \frac{1}{2} \\ y - z &= -\frac{4}{3} \Rightarrow y = -\frac{1}{3} \end{aligned}$$

Answer: $\left(\frac{1}{2}, -\frac{1}{3}, 1\right)$

11. Use matrices and elementary row operations to solve the system of equations.

$$\begin{aligned} x + 2y + 6z &= 1 \\ 2x + 5y + 15z &= 4 \\ 3x + y + 3z &= -6 \end{aligned}$$

Solution:

$$\begin{bmatrix} 1 & 2 & 6 & \vdots & 1 \\ 2 & 5 & 15 & \vdots & 4 \\ 3 & 1 & 3 & \vdots & -6 \end{bmatrix} \begin{matrix} \\ -2R_1 + R_2 \to \\ -3R_1 + R_3 \to \end{matrix} \begin{bmatrix} 1 & 2 & 6 & \vdots & 1 \\ 0 & 1 & 3 & \vdots & 2 \\ 0 & -5 & -15 & \vdots & -9 \end{bmatrix}$$

$$\begin{matrix} -2R_2 + R_1 \to \\ \\ 5R_2 + R_3 \to \end{matrix} \begin{bmatrix} 1 & 0 & 0 & \vdots & -3 \\ 0 & 1 & 3 & \vdots & 2 \\ 0 & 0 & 0 & \vdots & 1 \end{bmatrix}$$

$$x = -3$$

$$y + 3z = 2$$

$0 = 1$, Inconsistent, no solution

15. Perform the indicated matrix operation.

$$\begin{bmatrix} 1 & 2 \\ 5 & -4 \\ 6 & 0 \end{bmatrix} \begin{bmatrix} 6 & -2 & 8 \\ 4 & 0 & 0 \end{bmatrix}$$

Solution:

$$\begin{bmatrix} 1 & 2 \\ 5 & -4 \\ 6 & 0 \end{bmatrix} \begin{bmatrix} 6 & -2 & 8 \\ 4 & 0 & 0 \end{bmatrix} = \begin{bmatrix} 1(6) + 2(4) & 1(-2) + 2(0) & 1(8) + 2(0) \\ 5(6) + (-4)(4) & 5(-2) + (-4)(0) & 5(8) + (-4)(0) \\ 6(6) + (0)(4) & 6(-2) + (0)(0) & 6(8) + (0)(0) \end{bmatrix}$$

$$= \begin{bmatrix} 14 & -2 & 8 \\ 14 & -10 & 40 \\ 36 & -12 & 48 \end{bmatrix}$$

19. Perform the indicated matrix operation.

$$\begin{bmatrix} 1 & 3 & 2 \\ 0 & 2 & -4 \\ 0 & 0 & 3 \end{bmatrix} \begin{bmatrix} 4 & -3 & 2 \\ 0 & 3 & -1 \\ 0 & 0 & 2 \end{bmatrix}$$

Solution:

$$\begin{bmatrix} 1 & 3 & 2 \\ 0 & 2 & -4 \\ 0 & 0 & 3 \end{bmatrix} \begin{bmatrix} 4 & -3 & 2 \\ 0 & 3 & -1 \\ 0 & 0 & 2 \end{bmatrix} = \begin{bmatrix} 1(4) & 1(-3) + 3(3) & 1(2) + 3(-1) + 2(2) \\ 0 & 2(3) & 2(-1) + (-4)(2) \\ 0 & 0 & 3(2) \end{bmatrix}$$

$$= \begin{bmatrix} 4 & 6 & 3 \\ 0 & 6 & -10 \\ 0 & 0 & 6 \end{bmatrix}$$

23. Solve for X in $3X + 2A = B$, given

$$A = \begin{bmatrix} -4 & 0 \\ 1 & -5 \\ -3 & 2 \end{bmatrix} \quad \text{and} \quad B = \begin{bmatrix} 1 & 2 \\ -2 & 1 \\ 4 & 4 \end{bmatrix}.$$

Solution:

$$X = \frac{1}{3}[B - 2A] = \frac{1}{3}\left(\begin{bmatrix} 1 & 2 \\ -2 & 1 \\ 4 & 4 \end{bmatrix} - 2\begin{bmatrix} -4 & 0 \\ 1 & -5 \\ -3 & 2 \end{bmatrix}\right) = \frac{1}{3}\begin{bmatrix} 9 & 2 \\ -4 & 11 \\ 10 & 0 \end{bmatrix}$$

25. Write the system of linear equations represented by the matrix equation.

$$\begin{bmatrix} 5 & 4 \\ -1 & 1 \end{bmatrix}\begin{bmatrix} x \\ y \end{bmatrix} = \begin{bmatrix} 2 \\ -22 \end{bmatrix}$$

Solution:

$$\begin{bmatrix} 5 & 4 \\ -1 & 1 \end{bmatrix}\begin{bmatrix} x \\ y \end{bmatrix} = \begin{bmatrix} 2 \\ -22 \end{bmatrix}$$

$$\begin{bmatrix} 5x + 4y \\ -x + y \end{bmatrix} = \begin{bmatrix} 2 \\ -22 \end{bmatrix}$$

$$5x + 4y = 2$$

$$-x + y = -22$$

29. Find the inverse of the following matrix.

$$\begin{bmatrix} 2 & 0 & 3 \\ -1 & 1 & 1 \\ 2 & -2 & 1 \end{bmatrix}$$

Solution:

$$\left[\begin{array}{ccc|ccc} 2 & 0 & 3 & 1 & 0 & 0 \\ -1 & 1 & 1 & 0 & 1 & 0 \\ 2 & -2 & 1 & 0 & 0 & 1 \end{array}\right] \begin{array}{r} R_2 + R_1 \rightarrow \\ R_1 + R_2 \rightarrow \\ -2R_1 + R_3 \rightarrow \end{array} \left[\begin{array}{ccc|ccc} 1 & 1 & 4 & 1 & 1 & 0 \\ 0 & 2 & 5 & 1 & 2 & 0 \\ 0 & -4 & -7 & -2 & -2 & 1 \end{array}\right]$$

$$\begin{array}{r} -R_2 + R_1 \rightarrow \\ \frac{1}{2}R_2 \rightarrow \\ 4R_2 + R_3 \rightarrow \end{array} \left[\begin{array}{ccc|ccc} 1 & 0 & \frac{3}{2} & \frac{1}{2} & 0 & 0 \\ 0 & 1 & \frac{5}{2} & \frac{1}{2} & 1 & 0 \\ 0 & 0 & 3 & 0 & 2 & 1 \end{array}\right]$$

$$\begin{array}{r} -\frac{3}{2}R_3 + R_1 \rightarrow \\ -\frac{5}{2}R_3 + R_2 \rightarrow \\ \frac{1}{3}R_3 \rightarrow \end{array} \left[\begin{array}{ccc|ccc} 1 & 0 & 0 & \frac{1}{2} & -1 & -\frac{1}{2} \\ 0 & 1 & 0 & \frac{1}{2} & -\frac{2}{3} & -\frac{5}{6} \\ 0 & 0 & 1 & 0 & \frac{2}{3} & \frac{1}{3} \end{array}\right]$$

Inverse: $\begin{bmatrix} \frac{1}{2} & -1 & -\frac{1}{2} \\ \frac{1}{2} & -\frac{2}{3} & -\frac{5}{6} \\ 0 & \frac{2}{3} & \frac{1}{3} \end{bmatrix}$

33. Evaluate the determinant.

$$\begin{vmatrix} 3 & 0 & -4 & 0 \\ 0 & 8 & 1 & 2 \\ 6 & 1 & 8 & 2 \\ 0 & 3 & -4 & 1 \end{vmatrix}$$

Solution:

$$\begin{vmatrix} 3 & 0 & -4 & 0 \\ 0 & 8 & 1 & 2 \\ 6 & 1 & 8 & 2 \\ 0 & 3 & -4 & 1 \end{vmatrix} = 3\begin{vmatrix} 8 & 1 & 2 \\ 1 & 8 & 2 \\ 3 & -4 & 1 \end{vmatrix} + (-4)\begin{vmatrix} 0 & 8 & 2 \\ 6 & 1 & 2 \\ 0 & 3 & 1 \end{vmatrix} \quad \text{(Expansion along Row 1)}$$

$$= 3[8(8-(-8)) - 1(1-6) + 2(-4-24)] - 4[0 - 6(8-6) + 0]$$

$$= 3[128 + 5 - 56] - 4[-12]$$

$$= 279$$

37. Solve the system of linear equations using (a) the inverse of the coefficient matrix and (b) Cramer's Rule.

$$\begin{aligned} -3x - 3y - 4z &= 2 \\ y + z &= -1 \\ 4x + 3y + 4z &= -1 \end{aligned}$$

Solution:

(a) $$\left[\begin{array}{ccc|ccc} -3 & -3 & -4 & 1 & 0 & 0 \\ 0 & 1 & 1 & 0 & 1 & 0 \\ 4 & 3 & 4 & 0 & 0 & 1 \end{array}\right] \begin{array}{l} R_3 + R_1 \rightarrow \\ \\ -4R_1 + R_3 \rightarrow \end{array} \left[\begin{array}{ccc|ccc} 1 & 0 & 0 & 1 & 0 & 1 \\ 0 & 1 & 1 & 0 & 1 & 0 \\ 0 & 3 & 4 & -4 & 0 & -3 \end{array}\right]$$

$$\begin{array}{l} \\ -R_3 + R_2 \rightarrow \\ -3R_2 + R_3 \rightarrow \end{array} \left[\begin{array}{ccc|ccc} 1 & 0 & 0 & 1 & 0 & 1 \\ 0 & 1 & 0 & 4 & 4 & 3 \\ 0 & 0 & 1 & -4 & -3 & -3 \end{array}\right]$$

$$\begin{bmatrix} x \\ y \\ z \end{bmatrix} = \begin{bmatrix} 1 & 0 & 1 \\ 4 & 4 & 3 \\ -4 & -3 & -3 \end{bmatrix} \begin{bmatrix} 2 \\ -1 \\ -1 \end{bmatrix} = \begin{bmatrix} 1 \\ 1 \\ -2 \end{bmatrix}$$

Answer: $(1, 1, -2)$

37. –CONTINUED–

$$\text{(b) } x = \frac{\begin{vmatrix} 2 & -3 & -4 \\ -1 & 1 & 1 \\ -1 & 3 & 4 \end{vmatrix}}{\begin{vmatrix} -3 & -3 & -4 \\ 0 & 1 & 1 \\ 4 & 3 & 4 \end{vmatrix}} = \frac{1}{1} = 1$$

$$y = \frac{\begin{vmatrix} -3 & 2 & -4 \\ 0 & -1 & 1 \\ 4 & -1 & 4 \end{vmatrix}}{\begin{vmatrix} -3 & -3 & -4 \\ 0 & 1 & 1 \\ 4 & 3 & 4 \end{vmatrix}} = \frac{1}{1} = 1$$

$$z = \frac{\begin{vmatrix} -3 & -3 & 2 \\ 0 & 1 & -1 \\ 4 & 3 & -1 \end{vmatrix}}{\begin{vmatrix} -3 & -3 & -4 \\ 0 & 1 & 1 \\ 4 & 3 & 4 \end{vmatrix}} = \frac{-2}{1} = -2$$

Answer: $(1, 1, -2)$

43. Use a determinant to find the area of the triangle with vertices $(1, 0)$, $(5, 0)$, and $(5, 8)$.

Solution:

$$\text{Area} = \frac{1}{2}\begin{vmatrix} 1 & 0 & 1 \\ 5 & 0 & 1 \\ 5 & 8 & 1 \end{vmatrix} = \frac{1}{2}(32) = 16 \text{ square units}$$

47. Use a determinant to find the equation of the line through the points $(-4, 0)$ and $(4, 4)$.

Solution:

$$\begin{vmatrix} x & y & 1 \\ -4 & 0 & 1 \\ 4 & 4 & 1 \end{vmatrix} = 0$$

$$-4x + 8y - 16 = 0$$

$$x - 2y + 4 = 0$$

51. A florist wants to arrange a dozen flowers consisting of two varieties—carnations and roses. Carnations cost \$0.75 each and roses cost \$1.50 each. How many of each should the florist use in order for the arrangement to cost \$12.00?

Solution:

Let $x =$ the number of carnations, and $y =$ the number of roses. Then,

$$x + y = 12$$
$$0.75x + 1.50y = \$12.00.$$

By Cramer's Rule we have

$$x = \frac{\begin{vmatrix} 12 & 1 \\ 12 & 1.50 \end{vmatrix}}{\begin{vmatrix} 1 & 1 \\ 0.75 & 1.50 \end{vmatrix}} = \frac{6}{0.75} = 8$$

Using back-substitution in the first equation yields $y = 4$. The florist should use 8 carnations and 4 roses.

55. If A is a 3×3 matrix and $|A| = 2$, what is the value of $|4A|$? Give a reason for your answer.

Solution:

$|4A| = 4^3(2) = 128$, since $4A$ means that each of the three rows of A was multiplied by 4.

Practice Test for Chapter 8

1. Put the matrix in reduced echelon form.

$$\begin{bmatrix} 1 & -2 & 4 \\ 3 & -5 & 9 \end{bmatrix}$$

For Exercises 2–4, use matrices to solve the system of equations.

2. $\begin{aligned} 3x + 5y &= 3 \\ 2x - y &= -11 \end{aligned}$

3. $\begin{aligned} 2x + 3y &= -3 \\ 3x + 2y &= 8 \\ x + y &= 1 \end{aligned}$

4. $\begin{aligned} x \quad\quad + 3z &= -5 \\ 2x + y \quad\quad &= 0 \\ 3x + y - z &= 3 \end{aligned}$

5. Multiply $\begin{bmatrix} 1 & 4 & 5 \\ 2 & 0 & -3 \end{bmatrix}\begin{bmatrix} 1 & 6 \\ 0 & -7 \\ -1 & 2 \end{bmatrix}$.

6. Given $A = \begin{bmatrix} 9 & 1 \\ -4 & 8 \end{bmatrix}$ and $B = \begin{bmatrix} 6 & -2 \\ 3 & 5 \end{bmatrix}$, find $3A - 5B$.

7. Find $f(A)$:

$$f(x) = x^2 - 7x + 8, \quad A = \begin{bmatrix} 3 & 0 \\ 7 & 1 \end{bmatrix}.$$

8. True or false:

$(A + B)(A + 3B) = A^2 + 4AB + 3B^2$ where A and B are matrices.

(Assume that A^2, AB, and B^2 exist.)

For Exercises 9–10, find the inverse of the matrix, if it exists.

9. $\begin{bmatrix} 1 & 2 \\ 3 & 5 \end{bmatrix}$

10. $\begin{bmatrix} 1 & 1 & 1 \\ 3 & 6 & 5 \\ 6 & 10 & 8 \end{bmatrix}$

11. Use an inverse matrix to solve the systems.

(a) $\begin{aligned} x + 2y &= 4 \\ 3x + 5y &= 1 \end{aligned}$

(b) $\begin{aligned} x + 2y &= 3 \\ 3x + 5y &= -2 \end{aligned}$

For Exercises 12–14, find the determinant of the matrix.

12. $\begin{bmatrix} 6 & -1 \\ 3 & 4 \end{bmatrix}$

13. $\begin{bmatrix} 1 & 3 & -1 \\ 5 & 9 & 0 \\ 6 & 2 & -5 \end{bmatrix}$

14. $\begin{bmatrix} 1 & 4 & 2 & 3 \\ 0 & 1 & -2 & 0 \\ 3 & 5 & -1 & 1 \\ 2 & 0 & 6 & 1 \end{bmatrix}$

15. True or false:

$$\begin{vmatrix} 3 & 0 & 0 \\ 0 & 3 & 0 \\ 0 & 0 & 3 \end{vmatrix} = -3^3 \begin{vmatrix} 1 & 0 & 0 \\ 0 & 0 & 1 \\ 0 & 1 & 0 \end{vmatrix}$$

16. Evaluate $\begin{bmatrix} 6 & 4 & 3 & 0 & 6 \\ 0 & 5 & 1 & 4 & 8 \\ 0 & 0 & 2 & 7 & 3 \\ 0 & 0 & 0 & 9 & 2 \\ 0 & 0 & 0 & 0 & 1 \end{bmatrix}$.

17. Use a determinant to find the area of the triangle with vertices (0, 7), (5, 0), and (3, 9).

For Exercises 18–20, use Cramer's Rule to find the indicated value.

18. Find x.

$$\begin{aligned} 6x - 7y &= 4 \\ 2x + 5y &= 11 \end{aligned}$$

19. Find z.

$$\begin{aligned} 3x \qquad + z &= 1 \\ y + 4z &= 3 \\ x - y \qquad &= 2 \end{aligned}$$

20. Find y.

$$\begin{aligned} 721.4x - 29.1y &= 33.77 \\ 45.9x + 105.6y &= 19.85 \end{aligned}$$

CHAPTER 9

Sequences, Counting Principles, and Probability

SECTION 9.1

Sequences and Summation Notation

- Given the general nth term in a sequence, you should be able to find, or list, some of the terms.
- You should be able to find an expression for the nth term of a sequence.
- You should be able to use and evaluate factorials.
- You should be able to use sigma notation for a sum.

Solutions to Selected Exercises

3. Write the first five terms of the following sequence. (Assume n begins with 1.)

$$a_n = 2^n$$

Solution:

$$a_n = 2^n$$
$$a_1 = 2^1 = 2$$
$$a_2 = 2^2 = 4$$
$$a_3 = 2^3 = 8$$
$$a_4 = 2^4 = 16$$
$$a_5 = 2^5 = 32$$

Terms: 2, 4, 8, 16, 32

7. Write the first five terms of the following sequence. (Assume n begins with 1.)

$$a_n = \frac{1+(-1)^n}{n}$$

Solution:

$$a_n = \frac{1+(-1)^n}{n}$$
$$a_1 = \frac{1+(-1)}{1} = \frac{0}{1} = 0$$
$$a_2 = \frac{1+(-1)^2}{2} = \frac{2}{2} = 1$$
$$a_3 = \frac{1+(-1)^3}{3} = \frac{0}{3} = 0$$
$$a_4 = \frac{1+(-1)^4}{4} = \frac{2}{4} = \frac{1}{2}$$
$$a_5 = \frac{1+(-1)^5}{5} = \frac{0}{5} = 0$$

Terms: 0, 1, 0, $\frac{1}{2}$, 0

13. Write the first five terms of the following sequence. (Assume n begins with 1.)

$$a_n = \frac{3^n}{n!}$$

Solution:

$$a_n = \frac{3^n}{n!}$$

$$a_1 = \frac{3^1}{1!} = 3$$

$$a_2 = \frac{3^2}{2!} = \frac{9}{2}$$

$$a_3 = \frac{3^3}{3!} = \frac{27}{6} = \frac{9}{2}$$

$$a_4 = \frac{3^4}{4!} = \frac{81}{24} = \frac{27}{8}$$

$$a_5 = \frac{3^5}{5!} = \frac{243}{120} = \frac{81}{40}$$

Terms: $3, \frac{9}{2}, \frac{9}{2}, \frac{27}{8}, \frac{81}{40}$

17. Write the first five terms of the sequence $a_1 = 3$ and $a_{k+1} = 2(a_k - 1)$. (Assume n begins with 1.)

Solution:

$$a_1 = 3 \text{ and } a_{k+1} = 2(a_k - 1)$$

$$a_1 = 3$$

$$a_2 = 2(3-1) = 4$$

$$a_3 = 2(4-1) = 6$$

$$a_4 = 2(6-1) = 10$$

$$a_5 = 2(10-1) = 18$$

Terms: 3, 4, 6, 10, 18

19. Simplify the ratio $\frac{4!}{6!}$.

Solution:

$$\frac{4!}{6!} = \frac{4!}{6 \cdot 5 \cdot 4!} = \frac{1}{6 \cdot 5} = \frac{1}{30}$$

23. Simplify the ratio

$$\frac{(2n-1)!}{(2n+1)!}.$$

Solution:

$$\frac{(2n-1)!}{(2n+1)!} = \frac{(2n-1)!}{(2n+1)(2n)(2n-1)!} = \frac{1}{(2n+1)(2n)} = \frac{1}{2n(2n+1)}$$

27. Write an expression for the nth term of the sequence 0, 3, 8, 15, 24, ... (Assume n begins with 1.)

Solution:

$$a_1 = 0 = 1^2 - 1$$
$$a_2 = 3 = 2^2 - 1$$
$$a_3 = 8 = 3^2 - 1$$
$$a_4 = 15 = 4^2 - 1$$
$$a_5 = 24 = 5^2 - 1$$

Therefore, $a_n = n^2 - 1$.

31. Write an expression for the nth term of the sequence $1+\frac{1}{1}, 1+\frac{1}{2}, 1+\frac{1}{3}, 1+\frac{1}{4}, 1+\frac{1}{5}, \ldots$. (Assume n begins with 1.)

Solution:

$$a_1 = 1 + \tfrac{1}{1}$$
$$a_2 = 1 + \tfrac{1}{2}$$
$$a_3 = 1 + \tfrac{1}{3}$$
$$a_4 = 1 + \tfrac{1}{4}$$
$$a_5 = 1 + \tfrac{1}{5}$$

Therefore, $a_n = 1 + \dfrac{1}{n}$.

35. Write an expression for the nth term of the sequence 1, −1, 1, −1, 1, ... (Assume n begins with 1.)

Solution:

$$a_1 = 1 = (-1)^{1-1} \quad \text{or} \quad (-1)^{1+1}$$
$$a_2 = -1 = (-1)^{2-1} \quad \text{or} \quad (-1)^{2+1}$$
$$a_3 = 1 = (-1)^{3-1} \quad \text{or} \quad (-1)^{3+1}$$
$$a_4 = -1 = (-1)^{4-1} \quad \text{or} \quad (-1)^{4+1}$$
$$a_5 = 1 = (-1)^{5-1} \quad \text{or} \quad (-1)^{5+1}$$

Therefore, $a_n = (-1)^{n-1}$ or $a_n = (-1)^{n+1}$.

39. Find the sum.

$$\sum_{k=1}^{4} 10$$

Solution:

$$\sum_{k=1}^{4} 10 = 10 + 10 + 10 + 10$$
$$= 40$$

43. Find the sum.

$$\sum_{k=0}^{3} \frac{1}{k^2+1}$$

Solution:

$$\sum_{k=0}^{3} \frac{1}{k^2+1} = \frac{1}{(0)^2+1} + \frac{1}{(1)^2+1} + \frac{1}{(2)^2+1} + \frac{1}{(3)^2+1}$$
$$= 1 + \frac{1}{2} + \frac{1}{5} + \frac{1}{10}$$
$$= \frac{10+5+2+1}{10}$$
$$= \frac{18}{10} = \frac{9}{5}$$

47. Find the sum.

$$\sum_{i=1}^{4}(9+2i)$$

Solution:

$$\sum_{i=1}^{4}(9+2i) = (9+2)+(9+4)+(9+6)+(9+8) = 56$$

51. Use sigma notation to write the sum

$$\frac{1}{3(1)}+\frac{1}{3(2)}+\frac{1}{3(3)}+\cdots+\frac{1}{3(9)}.$$

Solution:

$$\frac{1}{3(1)}+\frac{1}{3(2)}+\frac{1}{3(3)}+\cdots+\frac{1}{3(9)} = \sum_{i=1}^{9}\frac{1}{3i}$$

55. Use sigma notation to write the sum $3-9+27-81+243-729$.

Solution:

$$3-9+27-81+243-729 = 3^1-3^2+3^3-3^4+3^5-3^6 = \sum_{i=1}^{6}(-1)^{i+1}3^i$$

59. Use sigma notation to write the sum

$$\frac{1}{4}+\frac{3}{8}+\frac{7}{16}+\frac{15}{32}+\frac{31}{64}.$$

Solution:

$$\frac{1}{4}+\frac{3}{8}+\frac{7}{16}+\frac{15}{32}+\frac{31}{64} = \frac{2^1-1}{2^2}+\frac{2^2-1}{2^3}+\frac{2^3-1}{2^4}+\frac{2^4-1}{2^5}+\frac{2^5-1}{2^6}$$

$$= \sum_{i=1}^{5}\frac{2^i-1}{2^{i+1}}$$

63. The average cost of a day in a hospital from 1980 to 1987 is given by the model

$$a_n = 242.67 + 42.67n, \quad n = 0, 1, 2, \ldots, 7$$

where a_n is the average cost in dollars and n is the year with $n = 0$ corresponding to 1980. (*Source:* American Hospital Association) Find the terms of this finite sequence and construct a bar graph that represents the sequence.

Solution:

$$a_n = 242.67 + 42.67n, \; n = 0, 1, 2, \ldots, 7$$

$$a_0 = 242.67$$

$$a_1 = 285.34$$

$$a_2 = 328.01$$

$$a_3 = 370.68$$

$$a_4 = 413.35$$

$$a_5 = 456.02$$

$$a_6 = 498.69$$

$$a_7 = 541.36$$

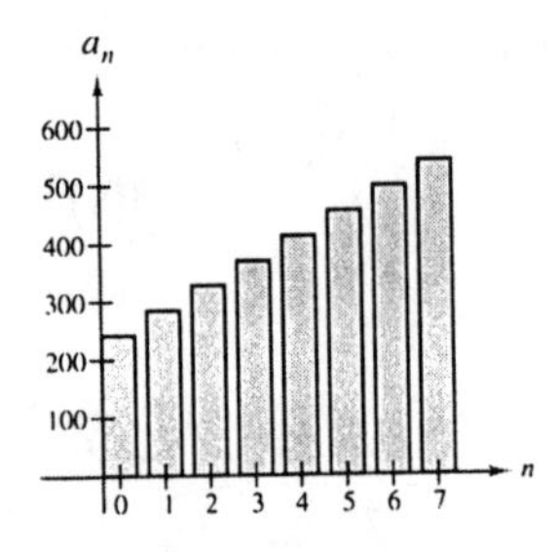

67. Prove that

$$\sum_{i=1}^{n}(x_i - \overline{x}) = 0, \quad \text{where } \overline{x} = \frac{1}{n}\sum_{i=1}^{n} x_i.$$

Solution:

$$\begin{aligned}\sum_{i=1}^{n}(x_i - \overline{x}) &= \sum_{i=1}^{n} x_i - \sum_{i=1}^{n} \overline{x} \\ &= \sum_{i=1}^{n} x_i - n\overline{x} \\ &= \sum_{i=1}^{n} x_i - n\left(\frac{1}{n}\sum_{i=1}^{n} x_i\right) \\ &= 0\end{aligned}$$

SECTION 9.2

Arithmetic Sequences

- You should be able to recognize an arithmetic sequence, find its common difference, and find its nth term.
- You should be able to find the nth partial sum of an arithmetic sequence with common difference d using the formula

$$S_n = \frac{n}{2}(a_1 + a_n).$$

- You should know that the arithmetic mean of a and b is $\frac{a+b}{2}$.

Solutions to Selected Exercises

5. Determine whether the sequence $\frac{9}{4}$, 2, $\frac{7}{4}$, $\frac{3}{2}$, $\frac{5}{4}$, ... is arithmetic. If it is, find the common difference.

Solution:

$$\frac{9}{4}, 2, \frac{7}{4}, \frac{3}{2}, \frac{5}{4}, \ldots = \frac{9}{4}, \frac{8}{4}, \frac{7}{4}, \frac{6}{4}, \frac{5}{4}, \ldots$$

$$a_n = \frac{10}{4} - \frac{1}{4}n$$

Therefore, the sequence *is* arithmetic with $d = -\frac{1}{4}$.

9. Determine whether the sequence 5.3, 5.7, 6.1, 6.5, 6.9, ... is arithmetic. If it is, find the common difference.

Solution:

$$5.3,\ 5.7,\ 6.1,\ 6.5,\ 6.9,\ \ldots = 4.9 + 0.4,\ 4.9 + 2(0.4),\ 4.9 + 3(0.4),\ 4.9 + 4(0.4),\ 4.9 + 5(0.4),\ \ldots$$

$$a_n = 4.9 + 0.4n$$

Therefore, the sequence *is* arithmetic with $d = 0.4$.

13. Write the first five terms of the following sequence. Determine whether the sequence is arithmetic, and if it is, find the common difference.

$$a_n = \frac{1}{n+1}$$

Solution:

$$a_n = \frac{1}{n+1}$$

$$a_1 = \frac{1}{1+1} = \frac{1}{2}$$

$$a_2 = \frac{1}{2+1} = \frac{1}{3}$$

$$a_3 = \frac{1}{3+1} = \frac{1}{4}$$

$$a_4 = \frac{1}{4+1} = \frac{1}{5}$$

$$a_5 = \frac{1}{5+1} = \frac{1}{6}$$

The sequence is *not* arithmetic.

17. Write the first five terms of the sequence $a_1 = 1,\ a_2 = 1,\ a_n = a_{n-1} + a_{n-2},\ n \geq 3$. Determine whether the sequence is arithmetic, and if it is, find the common difference.

Solution:

$$a_1 = 1,\ a_2 = 1,\ a_n = a_{n-1} + a_{n-2},\ n \geq 3$$

$$a_1 = 1$$

$$a_2 = 1$$

$$a_3 = a_2 + a_1 = 1 + 1 = 2$$

$$a_4 = a_3 + a_2 = 2 + 1 = 3$$

$$a_5 = a_4 + a_3 = 3 + 2 = 5$$

The sequence 1, 1, 2, 3, 5, ... is *not* arithmetic.

21. Find a formula for a_n for the arithmetic sequence

$$a_1 = 100, \; d = -8.$$

Solution:

$$a_1 = 100, \; d = -8$$
$$a_n = dn + c = -8n + c$$
$$a_1 = 100 = -8(1) + c \Rightarrow c = 108$$

Thus, $a_n = -8n + 108$.

25. Find a formula for a_n for the given arithmetic sequence

$$4, \; \frac{3}{2}, \; -1, \; -\frac{7}{2}, \; \ldots.$$

Solution:

$$4, \; \frac{3}{2}, \; -1, \; -\frac{7}{2}, \; \ldots \; d = -\frac{5}{2}$$
$$a_n = dn + c = -\frac{5}{2}n + c$$
$$a_1 = 4 = -\frac{5}{2}(1) + c \quad \Rightarrow \quad c = \frac{13}{2}$$

Thus, $a_n = -\frac{5}{2}n + \frac{13}{2}$.

29. Find a formula for a_n for the given arithmetic sequence

$$a_3 = 94, \; a_6 = 85.$$

Solution:

$$a_n = dn + c$$
$$a_3 = 94 = d(3) + c \qquad a_6 = 85 = d(6) + c$$
$$3d + c = 94$$
$$6d + c = 85$$

Solving this system of equations yields $d = -3$ and $c = 103$.

Thus, $a_n = -3n + 103$.

33. Write the first five terms of the arithmetic sequence given $a_1 = -2.6$ and $d = -0.4$.

Solution:

$$a_1 = -2.6, \; d = -0.4$$

$$a_2 = -2.6 - 0.4, \; = -3$$

$$a_3 = -3 - 0.4 = -3.4$$

$$a_4 = -3.4 - 0.4 = -3.8$$

$$a_5 = -3.8 - 0.4 = -4.2$$

Terms: $-2.6, \; -3, \; -3.4, \; -3.8, \; -4.2$

39. Write the first five terms of the arithmetic sequence given $a_8 = 26$ and $a_{12} = 42$.

Solution:

$$a_8 = 26, \; a_{12} = 42$$

$$a_8 = 26 = d(8) + c, \qquad a_{12} = 42 = d(12) + c$$

$$8d + c = 26$$

$$12d + c = 42$$

Solving this system yields $d = 4$ and $c = -6$.

Thus, $a_n = 4n - 6$.

$$a_1 = -2$$

$$a_2 = 2$$

$$a_3 = 6$$

$$a_4 = 10$$

$$a_5 = 14$$

Terms: $-2, \; 2, \; 6, \; 10, \; 14$

43. Find the nth partial sum of the arithmetic sequence $-6, \; -2, \; 2, \; 6, \; \ldots, \; n = 50$.

Solution:

$$-6, \; -2, \; 2, \; 6, \; \ldots, \; n = 50$$

$$a_n = 4n - 10$$

$$a_1 = -6 \text{ and } a_{50} = 190$$

$$S_{50} = \tfrac{50}{2}(-6 + 190) = 4600$$

47. Find the nth partial sum of the arithmetic sequence $a_1 = 100,\ a_{25} = 220,\ n = 25$.

Solution:

$$a_1 = 100,\ a_{25} = 220,\ n = 25$$

$$S_{25} = \tfrac{25}{2}(100 + 220) = \tfrac{25}{2}(320) = 4000$$

51. Find the sum.

$$\sum_{n=1}^{100} 5n$$

Solution:

$$a_n = 5n$$

$$a_1 = 5,\ a_{100} = 500$$

$$S_{100} = \tfrac{100}{2}(5 + 500) = 25{,}250$$

55. Find the sum.

$$\sum_{n=1}^{500} (n + 3)$$

Solution:

$$a_n = n + 3$$

$$a_1 = 4,\ a_{500} = 503$$

$$S_{500} = \tfrac{500}{2}(4 + 503)$$

$$= 126{,}750$$

59. Find the sum.

$$\sum_{n=0}^{50} (1000 - 5n)$$

Solution:

$$\sum_{n=0}^{50}(1000 - 5n) = 1000 + \sum_{n=1}^{50}(1000 - 5n)$$

$$= 1000 + \frac{50}{2}(995 + 750) = 1000 + 43{,}625 = 44{,}625$$

63. Insert three arithmetic means between the pair of numbers 3 and 6.

Solution:

$$3,\ 6;\ k = 3$$

$$3,\ m_1,\ m_2,\ m_3,\ 6$$

$$a_5 = 6 = 3 + 4d$$

$$d = \frac{3}{4}$$

$$m_1 = 3 + \frac{3}{4} = \frac{15}{4}$$

$$m_2 = \frac{15}{4} + \frac{3}{4} = \frac{18}{4} = \frac{9}{2}$$

$$m_3 = \frac{18}{4} + \frac{3}{4} = \frac{21}{4}$$

67. A person accepts a position with a company and will receive a salary of \$27,500 for the first year. The person is guaranteed a raise of \$1,500 per year for the first five years.

(a) Determine the person's salary during the sixth year of employment.

(b) Determine the person's total compensation from the company through six full years of employment.

Solution:

(a) $$\begin{aligned} a_n &= a_1 + (n-1)d \\ &= 27{,}500 + (n-1)(1500) \\ &= 1500n + 26{,}000 \\ a_6 &= 1500(6) + 26{,}000 \\ &= \$35{,}000 \end{aligned}$$

(b) $$\begin{aligned} S_6 &= \tfrac{6}{2}(27{,}500 + 35{,}000) \\ &= \$187{,}500 \end{aligned}$$

69. Determine the seating capacity of an auditorium with 30 rows of seats if there are 20 seats in the first row, 24 seats in the second row, 28 seats in the third row, and so on.

Solution:

$$\begin{aligned} a_n &= 16 + 4n \\ a_1 &= 20, \quad a_{30} = 136 \\ S_{30} &= \tfrac{30}{2}(20 + 136) = 2340 \text{ seats} \end{aligned}$$

SECTION 9.3

Geometric Sequences

- You should be able to identify a geometric sequence, find its common ratio, and find the nth term.
- You should be able to find the nth partial sum of a geometric sequence with common ratio r using the formula

$$S_n = \frac{a_1(1-r^n)}{1-r}, \quad r \neq 1.$$

- You should know that if $|r| < 1$, then

$$\sum_{n=0}^{\infty} a_1 r^n = \sum_{n=1}^{\infty} a_1 r^{n-1} = \frac{a_1}{1-r}.$$

Solutions to Selected Exercises

3. Determine whether the sequence 3, 12, 21, 30, ... is geometric. If it is, find its common ratio.

Solution:

$$3,\ 12,\ 21,\ 30,\ \ldots$$

$$a_n = -6 + 9n$$

This is an arithmetic sequence, *not* a geometric sequence.

7. Determine whether the sequence $\frac{1}{2}, \frac{2}{3}, \frac{3}{4}, \frac{4}{5}, \ldots$ is geometric. If it is, find its common ratio.

Solution:

$$\frac{1}{2}, \frac{2}{3}, \frac{3}{4}, \frac{4}{5}, \ldots$$

$$a_n = \frac{n}{n+1}$$

This is *not* a geometric sequence.

11. Write the first five terms of the geometric sequence $a_1 = 2,\ r = 3$.

Solution:

$$a_1 = 2,\ r = 3$$
$$a_2 = 2(3) = 6$$
$$a_3 = 2(3)^2 = 18$$
$$a_4 = 2(3)^3 = 54$$
$$a_5 = 2(3)^4 = 162$$

Terms: 2, 6, 18, 54, 162

15. Write the first five terms of the geometric sequence $a_1 = 5,\ r = -\frac{1}{10}$.

Solution:

$$a_1 = 5,\ r = -\frac{1}{10}$$
$$a_2 = 5\left(-\frac{1}{10}\right) = -\frac{1}{2}$$
$$a_3 = 5\left(-\frac{1}{10}\right)^2 = \frac{1}{20}$$
$$a_4 = 5\left(-\frac{1}{10}\right)^3 = -\frac{1}{200}$$
$$a_5 = 5\left(-\frac{1}{10}\right)^4 = \frac{1}{2000}$$

Terms: $5,\ -\frac{1}{2},\ \frac{1}{20},\ -\frac{1}{200},\ \frac{1}{2000}$

21. Find the nth term of the geometric sequence $a_1 = 4,\ r = \frac{1}{2},\ n = 10$.

Solution:

$$a_1 = 4,\ r = \frac{1}{2},\ n = 10$$
$$a_{10} = 4\left(\frac{1}{2}\right)^9 = \frac{1}{128} = \left(\frac{1}{2}\right)^7$$

25. Find the nth term of the geometric sequence $a_1 = 100,\ r = e^x,\ n = 9$.

Solution:

$$a_1 = 100,\ r = e^x,\ n = 9$$
$$a_9 = 100(e^x)^8 = 100e^{8x}$$

29. Find the nth term of the geometric sequence $a_1 = 16,\ a_4 = \frac{27}{4},\ n = 3$.

Solution:

$$a_1 = 16,\ a_4 = \frac{27}{4},\ n = 3$$
$$a_4 = 16r^3 = \frac{27}{4}$$
$$r^3 = \frac{27}{64}$$
$$r = \frac{3}{4}$$
$$a_3 = 16\left(\frac{3}{4}\right)^2 = 16\left(\frac{9}{16}\right) = 9$$

33. A sum of \$1000 is invested at 10% interest. Find the amount after 10 years if the interest is compounded (a) annually, (b) semiannually, (c) quarterly, (d) monthly, and (e) daily.

Solution:

$$A = P\left(1+\frac{r}{n}\right)^{nt} = 1000\left(1+\frac{0.10}{n}\right)^{n(10)}$$

(a) $n = 1,\quad A = 1000(1+0.10)^{10} \approx \$2593.74$

(b) $n = 2,\quad A = 1000\left(1+\frac{0.10}{2}\right)^{2(10)} \approx \$2653.30$

(c) $n = 4,\quad A = 1000\left(1+\frac{0.10}{4}\right)^{4(10)} \approx \$2685.06$

(d) $n = 12,\quad A = 1000\left(1+\frac{0.10}{12}\right)^{12(10)} \approx \$2707.04$

(e) $n = 365,\quad A = 1000\left(1+\frac{0.10}{365}\right)^{365(10)} \approx \$2717.91$

37. Find the sum.

$$\sum_{n=1}^{9} 2^{n-1}$$

Solution:

$$\sum_{n=1}^{9} 2^{n-1} = \frac{1(1-2^9)}{1-2} = 511$$

43. Find the sum.

$$\sum_{n=0}^{20} 3\left(\frac{3}{2}\right)^n$$

Solution:

$$\sum_{n=0}^{20} 3\left(\frac{3}{2}\right)^n = \frac{3(1-(3/2)^{21})}{1-(3/2)} \approx 29{,}921.31$$

47. A deposit of \$100 is made at the beginning of each month for five years in an account that pays 10%, compounded monthly. What is the balance A in the account at the end of five years?

$$A = 100\left(1+\frac{0.01}{12}\right)^1 + \cdots + 100\left(1+\frac{0.10}{12}\right)^{60}$$

Solution:

$$A = \sum_{n=1}^{60} 100\left(1+\frac{0.10}{12}\right)^n = 100\left(1+\frac{0.10}{12}\right) \cdot \frac{\left[1-\left(1+\frac{0.10}{12}\right)^{60}\right]}{\left[1-\left(1+\frac{0.10}{12}\right)\right]} \approx \$7808.24$$

49. A deposit of P dollars is made at the beginning of each month in an account at an annual interest rate r compounded monthly. The balance A after t years is

$$A = P\left(1+\frac{r}{12}\right) + P\left(1+\frac{r}{12}\right)^2 + \cdots + P\left(1+\frac{r}{12}\right)^{12t}.$$

Show that the balance is given by

$$A = P\left[\left(1+\frac{r}{12}\right)^{12t} - 1\right]\left(1+\frac{12}{r}\right).$$

Solution:

Let $N = 12t$ be the total number of deposits.

$$\begin{aligned}
A &= P\left(1+\frac{r}{12}\right) + P\left(1+\frac{r}{12}\right)^2 + \cdots + P\left(1+\frac{r}{12}\right)^N \\
&= \left(1+\frac{r}{12}\right)\left[P + P\left(1+\frac{r}{12}\right) + \cdots + P\left(1+\frac{r}{12}\right)^{N-1}\right] \\
&= P\left(1+\frac{r}{12}\right)\sum_{n=1}^{N}\left(1+\frac{r}{12}\right)^{n-1} \\
&= P\left(1+\frac{r}{12}\right)\frac{1-\left(1+\frac{r}{12}\right)^N}{1-\left(1+\frac{r}{12}\right)} \\
&= P\left(1+\frac{r}{12}\right)\left(-\frac{12}{r}\right)\left[1-\left(1+\frac{r}{12}\right)^N\right] \\
&= P\left(\frac{12}{r}+1\right)\left[-1+\left(1+\frac{r}{12}\right)^N\right] \\
&= P\left[\left(1+\frac{r}{12}\right)^N - 1\right]\left(1+\frac{12}{r}\right) \\
&= P\left[\left(1+\frac{r}{12}\right)^{12t} - 1\right]\left(1+\frac{12}{r}\right)
\end{aligned}$$

57. You accept a job with a salary of \$30,000 for the first year. Suppose that during the next 39 years you receive a 5% raise each year. What would your total compensation be over the 40-year period?

Solution:

$$T = \sum_{n=0}^{39} 30{,}000(1.05)^n = \frac{30{,}000(1 - 1.05^{40})}{1 - 1.05} \approx \$3{,}623{,}993.23$$

59. Find the sum of the infinite geometric series.

$$\sum_{n=0}^{\infty} \left(\frac{1}{2}\right)^n = 1 + \frac{1}{2} + \frac{1}{4} + \frac{1}{8} + \cdots$$

Solution:

$$\sum_{n=0}^{\infty} \left(\frac{1}{2}\right)^n = 1 + \frac{1}{2} + \frac{1}{4} + \frac{1}{8} + \cdots = \frac{1}{1 - (1/2)} = 2$$

63. Find the sum of the infinite geometric series.

$$\sum_{n=0}^{\infty} 4\left(\frac{1}{4}\right)^n = 4 + 1 + \frac{1}{4} + \frac{1}{16} + \cdots$$

Solution:

$$\sum_{n=0}^{\infty} 4\left(\frac{1}{4}\right)^n = 4 + 1 + \frac{1}{4} + \frac{1}{16} + \cdots = \frac{4}{1 - (1/4)} = \frac{16}{3}$$

67. Find the sum of the infinite geometric series $4 - 2 + 1 - \frac{1}{2} + \cdots$.

Solution:

$$4 - 2 + 1 - \frac{1}{2} + \cdots = \sum_{n=0}^{\infty} 4\left(-\frac{1}{2}\right)^n = \frac{4}{1 - (-1/2)} = \frac{8}{3}$$

69. A ball is dropped from a height of 16 feet. Each time it drops h feet, it rebounds $0.81h$ feet. Find the total distance traveled by the ball.

Solution:

$$\text{Total distance} = \left[\sum_{n=0}^{\infty} 32(0.81)^n\right] - 16 = \frac{32}{1 - 0.81} - 16 \approx 152.42 \text{ feet}$$

SECTION 9.4

Mathematical Induction

- You should be sure that you understand the principle of mathematical induction. If P_n is a statement involving the positive integer n, where P_1 is true and the truth of P_k implies the truth of P_{k+1}, then P_n is true for all positive integers n.
- You should be able to verify (by induction) the formulas for the sums of powers of integers and be able to use these formulas.

Solutions to Selected Exercises

1. Find the following sum using the formulas for the sums of powers of integers.

$$\sum_{n=1}^{20} n$$

Solution:

$$\sum_{n=1}^{N} n = \frac{N(N+1)}{2}$$

$$\sum_{n=1}^{20} n = \frac{20(21)}{2} = 210$$

7. Find the following sum using the formulas for the sums of powers of integers.

$$\sum_{n=1}^{6} n^4$$

Solution:

$$\sum_{n=1}^{N} n^4 = \frac{N(N+1)(2N+1)(3N^2+3N-1)}{30}$$

$$\sum_{n=1}^{6} n^4 = \frac{6(7)(13)(125)}{30} = 2275$$

11. Find S_{k+1} for $S_k = \dfrac{5}{k(k+1)}$.

Solution:

$$S_k = \frac{5}{k(k+1)}$$

$$S_{k+1} = \frac{5}{(k+1)((k+1)+1)}$$

$$= \frac{5}{(k+1)(k+2)}$$

13. Find S_{k+1} for $S_k = \dfrac{k^2(k+1)^2}{4}$.

Solution:

$$S_k = \frac{k^2(k+1)^2}{4}$$

$$S_{k+1} = \frac{(k+1)^2((k+1)+1)^2}{4}$$

$$= \frac{(k+1)^2(k+2)^2}{4}$$

17. Use mathematical induction to prove the formula for every positive integer n.

$$2 + 7 + 12 + 17 + \cdots + (5n-3) = \frac{n}{2}(5n-1)$$

Solution:

1. When $n = 1$,

$$S_1 = 2 = \frac{1}{2}(5(1) - 1).$$

2. Assume that

$$S_k = 2 + 7 + 12 + 17 + \cdots + (5k-3) = \frac{k}{2}(5k-1).$$

Then,

$$S_{k+1} = 2 + 7 + 12 + 17 + \cdots + (5k-3) + [5(k+1) - 3]$$

$$= S_k + (5k + 5 - 3)$$

$$= \frac{k}{2}(5k-1) + 5k + 2$$

$$= \frac{5k^2 - k + 10k + 4}{2}$$

$$= \frac{5k^2 + 9k + 4}{2}$$

$$= \frac{(k+1)(5k+4)}{2}$$

$$= \frac{(k+1)}{2}[5(k+1) - 1].$$

We conclude by mathematical induction that the formula is valid for all positive integer values of n.

21. Use mathematical induction to prove the formula for every positive integer n.

$$1 + 2 + 3 + 4 + \cdots + n = \frac{n(n+1)}{2}$$

Solution:

1. When $n = 1$, $S_1 = 1 = \dfrac{1(1+1)}{2}$.
2. Assume that

$$S_k = 1 + 2 + 3 + 4 + \cdots + k = \frac{k(k+1)}{2}.$$

Then,

$$S_{k+1} = 1 + 2 + 3 + 4 + \cdots + k + (k+1)$$

$$= S_k + (k+1) = \frac{k(k+1)}{2} + \frac{2(k+1)}{2} = \frac{(k+1)(k+2)}{2}.$$

Therefore, we conclude that this formula holds for all positive integer values of n.

23. Use mathematical induction to prove the formula for every positive integer n.

$$1^3 + 2^3 + 3^3 + 4^3 + \cdots + n^3 = \frac{n^2(n+1)^2}{4}$$

Solution:

1. When $n = 1$, $S_1 = 1^3 = 1 = \dfrac{1(1+1)^2}{4}$.
2. Assume that

$$S_k = 1^3 + 2^3 + 3^3 + 4^3 + \cdots + k^3 = \frac{k^2(k+1)^2}{4}.$$

Then,

$$S_{k+1} = 1^3 + 2^3 + 3^3 + 4^3 + \cdots + k^3 + (k+1)^3$$

$$= S_k + (k+1)^3$$

$$= \frac{k^2(k+1)^2}{4} + (k+1)^3$$

$$= \frac{k^2(k+1)^2 + 4(k+1)^3}{4}$$

$$= \frac{(k+1)^2[k^2 + 4(k+1)]}{4}$$

$$= \frac{(k+1)^2(k^2 + 4k + 4)}{4} = \frac{(k+1)^2(k+2)^2}{4}.$$

Therefore, we conclude that this formula holds for all positive integer values of n.

27. Use mathematical induction to prove the formula for every positive integer n.

$$\sum_{i=1}^{n} i(i+1) = \frac{n(n+1)(n+2)}{3}$$

Solution:

1. When $n = 1$, $S_1 = 2 = \frac{1(2)(3)}{3}$.

2. Assume that

$$S_k = 1(2) + 2(3) + 3(4) + \cdots + k(k+1) = \frac{k(k+1)(k+2)}{3}.$$

Then,

$$\begin{aligned} S_{k+1} &= 1(2) + 2(3) + 3(4) + \cdots + k(k+1) + (k+1)(k+2) \\ &= S_k + (k+1)(k+2) \\ &= \frac{k(k+1)(k+2)}{3} + \frac{3(k+1)(k+2)}{3} \\ &= \frac{(k+1)(k+2)(k+3)}{3}. \end{aligned}$$

Thus, this formula is valid for all positive integer values of n.

29. Use mathematical induction to prove the inequality, $\left(\frac{4}{3}\right)^n > n$, $n \geq 7$.

Solution:

1. When $n = 7$, $\left(\frac{4}{3}\right)^7 \approx 7.4915 > 7$.

2. Assume that

$$\left(\frac{4}{3}\right)^k > k, \; k > 7.$$

Then,

$$\left(\frac{4}{3}\right)^{k+1} = \left(\frac{4}{3}\right)^k \left(\frac{4}{3}\right) > k\left(\frac{4}{3}\right) = k + \frac{k}{3} > k + 1 \text{ for } k > 7.$$

Thus,

$$\left(\frac{4}{3}\right)^{k+1} > k + 1.$$

Therefore,

$$\left(\frac{4}{3}\right)^n > n.$$

33. Use mathematical induction to prove the property $(ab)^n = a^n b^n$ for all positive integers n.

Solution:

1. When $n = 1$, $(ab)^1 = a^1 b^1 = ab$.
2. Assume that $(ab)^k = a^k b^k$.

Then,
$$\begin{aligned}(ab)^{k+1} &= (ab)^k(ab)\\ &= a^k b^k ab\\ &= a^{k+1} b^{k+1}.\end{aligned}$$

Thus, $(ab)^n = a^n b^n$.

37. Use mathematical induction to prove the Generalized Distributive Law:

$$x(y_1 + y_2 + \cdots + y_n) = xy_1 + xy_2 + \cdots + xy_n.$$

Solution:

1. When $n = 1$, $x(y_1) = xy_1$.
2. Assume that

$$x(y_1 + y_2 + \cdots + y_k) = xy_1 + xy_2 + \cdots + xy_k.$$

Then,

$$\begin{aligned}xy_1 + xy_2 + \cdots + xy_k + xy_{k+1} &= x(y_1 + y_2 + \cdots + y_k) + xy_{k+1}\\ &= x[(y_1 + y_2 + \cdots + y_k) + y_{k+1}]\\ &= x(y_1 + y_2 + \cdots + y_k + y_{k+1}).\end{aligned}$$

Hence, the formula holds.

41. Use mathematical induction to prove that a factor of $\left(2^{2n-1}+3^{2n-1}\right)$ is 5.

Solution:

1. When $n=1$, $\left(2^{2(1)-1}+3^{2(1)-1}\right)=2+3=5$ and 5 is a factor.
2. Assume that

   5 is a factor of $\left(2^{2k-1}+3^{2k-1}\right)$.

Then,

$$\begin{aligned}\left(2^{2(k+1)-1}+3^{2(k+1)-1}\right) &= \left(2^{2k+2-1}+3^{2k+2-1}\right)\\ &= \left(2^{2k-1}2^2+3^{2k-1}3^2\right)\\ &= \left(4\cdot 2^{2k-1}+9\cdot 3^{2k-1}\right)\\ &= \left(2^{2k-1}+3^{2k-1}\right)+\left(2^{2k-1}+3^{2k-1}\right)\\ &\quad+\left(2^{2k-1}+3^{2k-1}\right)+\left(2^{2k-1}+3^{2k-1}\right)\\ &\quad+5\cdot 3^{2k-1}.\end{aligned}$$

Since 5 is a factor of each set of parenthesis and 5 is a factor of $5\cdot 3^{2k-1}$, then 5 is a factor of the whole sum.

Thus, 5 is a factor of $\left(2^{2n-1}+3^{2n-1}\right)$ for every positive integer n.

SECTION 9.5

The Binomial Theorem

- You should be able to use the formula

$$(x+y)^n = x^n + nx^{n-1}y + \frac{n(n-1)}{2!}x^{n-2}y^2 + \cdots + {}_nC_r{}^n x^{n-r}y^r + \cdots + y^n$$

where ${}_nC_r = \frac{n!}{(n-r)!r!}$, to expand $(x+y)^n$.

- You should be able to use Pascal's Triangle in binomial expansion.

Solutions to Selected Exercises

3. Evaluate ${}_{12}C_0$.

Solution:

$$ {}_{12}C_0 = \frac{12!}{(12-0)!0!} = \frac{12!}{(12!)(1)} = 1$$

7. Evaluate ${}_{100}C_{98}$.

Solution:

$$ {}_{100}C_{98} = \frac{100!}{(100-98)!98!} = \frac{100 \cdot 99 \cdot 98!}{2! \cdot 98!} = \frac{100 \cdot 99}{2} = 4950$$

11. Use the Binomial Theorem to expand $(x+1)^4$. Simplify your answer.

Solution:

$$\begin{aligned}(x+1)^4 &= {}_4C_0x^4 + {}_4C_1x^3(1) + {}_4C_2x^2(1)^2 + {}_4C_3x(1)^3 + {}_4C_4(1)^4 \\ &= x^4 + 4x^3 + 6x^2 + 4x + 1\end{aligned}$$

17. Use the Binomial Theorem to expand $(x+y)^5$. Simplify your answer.

Solution:

$$\begin{aligned}(x+y)^5 &= {}_5C_0x^5 + {}_5C_1x^4y + {}_5C_2x^3y^2 + {}_5C_3x^2y^3 + {}_5C_4xy^4 + {}_5C_5y^5 \\ &= x^5 + 5x^4y + 10x^3y^2 + 10x^2y^3 + 5xy^4 + y^5\end{aligned}$$

19. Use the Binomial Theorem to expand $(r+3s)^6$. Simplify your answer.

Solution:

$$\begin{aligned}(r+3s)^6 &= {}_6C_0r^6 + {}_6C_1r^5(3s) + {}_6C_2r^4(3s)^2 + {}_6C_3r^3(3s)^3 + {}_6C_4r^2(3s)^4 \\ &\quad + {}_6C_5r(3s)^5 + {}_6C_6(3s)^6 \\ &= r^6 + 18r^5s + 135r^4s^2 + 540r^3s^3 + 1215r^2s^4 + 1458rs^5 + 729s^6\end{aligned}$$

23. Use the Binomial Theorem to expand $(1-2x)^3$. Simplify your answer.

Solution:

$$\begin{aligned}(1-2x)^3 &= [1+(-2x)]^3 \\ &= {}_3C_01^3 + {}_3C_1(1)^2(-2x) + {}_3C_2(1)(-2x)^2 + {}_3C_3(-2x)^3 \\ &= 1 - 6x + 12x^2 - 8x^3\end{aligned}$$

27. Use the Binomial Theorem to expand the following. Simplify your answer.

$$\left(\frac{1}{x}+y\right)^5$$

Solution:

$$\begin{aligned}\left(\frac{1}{x}+y\right)^5 &= {}_5C_0\left(\frac{1}{x}\right)^5 + {}_5C_1\left(\frac{1}{x}\right)^4y + {}_5C_2\left(\frac{1}{x}\right)^3y^2 + {}_5C_3\left(\frac{1}{x}\right)^2y^3 \\ &\quad + {}_5C_4\left(\frac{1}{x}\right)y^4 + {}_5C_5y^5 \\ &= \frac{1}{x^5} + \frac{5y}{x^4} + \frac{10y^2}{x^3} + \frac{10y^3}{x^2} + \frac{5y^4}{x} + y^5\end{aligned}$$

33. Use the Binomial Theorem to expand $(2-3i)^6$. Simplify your answer by recalling that $i^2 = -1$.

Solution:

$$\begin{aligned}(2-3i)^6 &= {}_6C_02^6 - {}_6C_1(2)^5(3i) + {}_6C_2(2)^4(3i)^2 - {}_6C_3(2)^3(3i)^3 + {}_6C_4(2)^2(3i)^4 \\ &\quad - {}_6C_5(2)(3i)^5 + {}_6C_6(3i)^6 \\ &= 64 - 576i - 2160 + 4320i + 4860 - 2916i - 729 \\ &= 2035 + 828i\end{aligned}$$

37. Expand $(2t - s)^5$, using Pascal's Triangle to determine the coefficients.

Solution:

$$\begin{array}{c} 1 \\ 1 \quad 1 \\ 1 \quad 2 \quad 1 \\ 1 \quad 3 \quad 3 \quad 1 \\ 1 \quad 4 \quad 6 \quad 4 \quad 1 \\ 1 \quad 5 \quad 10 \quad 10 \quad 5 \quad 1 \end{array}$$

$$\begin{aligned}(2t - s)^5 &= (2t)^5 - 5(2t)^4 s + 10(2t)^3 s^2 - 10(2t)^2 s^3 + 5(2t)s^4 - s^5 \\ &= 32t^5 - 80t^4 s + 80t^3 s^2 - 40t^2 s^3 + 10ts^4 - s^5\end{aligned}$$

41. Find the coefficient a of the term ax^5 in the expansion of $(x+3)^{12}$.

Solution:

$${}_{12}C_7 x^5 (3)^7 = \frac{12!3^7x^5}{(12-7)!7!} = 1{,}732{,}104x^5$$

$a = 1{,}732{,}104$

45. Find the coefficient a of the term ax^4y^5 in the expansion of $(3x - 2y)^9$.

Solution:

$${}_9C_5(3x)^4(-2y)^5 = \frac{9!}{5!4!}(81x^4)(-32y^5) = -326{,}592x^4y^5$$

$a = -326{,}592$

51. Use the Binomial Theorem to expand $\left(\frac{1}{3} + \frac{2}{3}\right)^8$. In the study of probability, it is sometimes necessary to use the expansion of $(p+q)^n$, where $p + q = 1$.

Solution:

$$\begin{aligned}\left(\frac{1}{3} + \frac{2}{3}\right)^8 &= \left(\frac{1}{3}\right)^8 + 8\left(\frac{1}{3}\right)^7\left(\frac{2}{3}\right) + 28\left(\frac{1}{3}\right)^6\left(\frac{2}{3}\right)^2 + 56\left(\frac{1}{3}\right)^5\left(\frac{2}{3}\right)^3 + 70\left(\frac{1}{3}\right)^4\left(\frac{2}{3}\right)^4 \\ &\quad + 56\left(\frac{1}{3}\right)^3\left(\frac{2}{3}\right)^5 + 28\left(\frac{1}{3}\right)^2\left(\frac{2}{3}\right)^6 + 8\left(\frac{1}{3}\right)\left(\frac{2}{3}\right)^7 + \left(\frac{2}{3}\right)^8 \\ &= \frac{1}{6561} + \frac{16}{6561} + \frac{112}{6561} + \frac{448}{6561} + \frac{1120}{6561} + \frac{1792}{6561} + \frac{1792}{6561} + \frac{1024}{6561} + \frac{256}{6561}\end{aligned}$$

55. Use the Binomial Theorem to approximate $(1.02)^8$ accurate to three decimal places. For example, $(1.02)^8 = (1+0.02)^8 = 1 + 8(0.02) + 28(0.02)^2 + \cdots$

Solution:

$$\begin{aligned}(1.02)^8 = (1+0.02)^8 &= 1 + 8(0.02) + 28(0.02)^2 + 56(0.02)^3 + 70(0.02)^4 + 56(0.02)^5 \\ &\quad + 28(0.02)^6 + 8(0.02)^7 + (0.02)^8 \\ &= 1 + 0.16 + 0.0112 + 0.000448 + \cdots \quad \approx 1.172\end{aligned}$$

59. Shift the graph of $f(x) = -x^2 + 3x + 2$ four units to the left to form the graph of g. Then write the polynomial function g in standard form.

Solution:

$$f(x) = -x^2 + 3x + 2$$

$$\begin{aligned}g(x) &= f(x+4) \\ &= -(x+4)^2 + 3(x+4) + 2 \\ &= -(x^2 + 8x + 16) + 3x + 12 + 2 \\ &= -x^2 - 5x - 2\end{aligned}$$

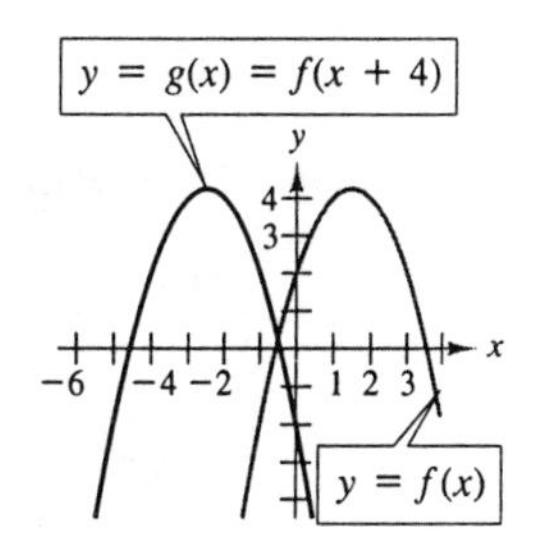

63. The average amount of life insurance per household (in households that carry life insurance) from 1970 through 1988 can be approximated by the model

$$f(t) = 0.2187t^2 + 0.6715t + 26.67, \quad 0 \le t \le 18.$$

In this model, $f(t)$ represents the amount of life insurance (in 1000s of dollars) and t represents the calendar year with $t = 0$ corresponding to 1970 (see figure). You want to adjust this model so that $t = 0$ corresponds to 1980 rather than 1970. To do this, you shift the graph of f ten units to the *left* and obtain

$$g(t) = f(t+10) = 0.2187(t+10)^2 + 0.6715(t+10) + 26.67.$$

Write this new polynomial function in standard form. (*Source:* American Council of Life Insurance)

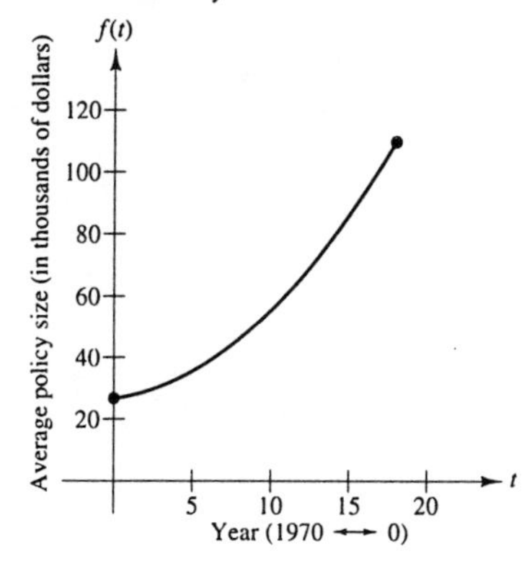

63. –CONTINUED–

Solution:

$f(t) = 0.2187t^2 + 0.6715t + 26.67$

$$\begin{aligned} g(t) &= f(t+10) \\ &= 0.2187(t+10)^2 + 0.6715(t+10) + 26.67 \\ &= 0.2187(t^2 + 20t + 100) + 0.6715t + 6.715 + 26.67 \\ &= 0.2187t^2 + 5.0455t + 55.255 \end{aligned}$$

67. Prove ${}_{n+1}C_m = {}_nC_m + {}_nC_{m-1}$ for all integers m and n, $0 \le m \le n$.

Solution:

$$\begin{aligned} {}_nC_m + {}_nC_{m-1} &= \frac{n!}{(n-m)!m!} + \frac{n!}{(n-m+1)!(m-1)!} \\ &= \frac{n!(n-m+1)!(m-1)! + n!(n-m)!m!}{(n-m)!m!(n-m+1)!(m-1)!} \\ &= \frac{n![(n-m+1)!(m-1)! + m!(n-m)!]}{(n-m)!m!(n-m+1)!(m-1)!} \\ &= \frac{n!\cancel{(m-1)!}[(n-m+1)! + m(n-m)!]}{(n-m)!m!(n-m+1)!\cancel{(m-1)!}} \\ &= \frac{n!\cancel{(n-m)!}[(n-m+1) + m]}{\cancel{(n-m)!}m!(n-m+1)!} \\ &= \frac{n![n+1]}{m!(n-m+1)!} \\ &= \frac{(n+1)!}{[(n+1)-m]!m!} \\ &= {}_{n+1}C_m \end{aligned}$$

SECTION 9.6

Counting Principles, Permutations, Combinations

- You should know The Fundamental Principle of Counting.
- ${}_nP_r = \frac{n!}{(n-r)!}$ is the number of permutations of n elements taken r at a time.
- Given a set of n objects that has n_1 of one kind, n_2 of a second kind, and so on, the number of distinguishable permutations is

$$\frac{n!}{n_1!n_2!\ldots n_k!}.$$

- ${}_nC_r = \frac{n!}{(n-r)!r!}$ is the number of combinations of n elements taken r at a time.

Solutions to Selected Exercises

3. A small college needs two additional faculty members, a chemist and a statistician. In how many ways can these positions be filled if there are three applicants for the chemistry position and four applicants for the position in statistics?

Solution:

$3 \cdot 4 = 12$ ways to fill the positions

7. In a certain state the automobile license plates consist of two letters followed by a four-digit number. How many distinct license plate numbers can be formed?

Solution:

$26 \cdot 26 \cdot 10 \cdot 10 \cdot 10 \cdot 10 = 6,760,000$ distinct license plate numbers

11. How many three-digit numbers can be formed under the following conditions?

(a) The leading digit cannot be zero.

(b) The leading digit cannot be zero and no repetition of digits is allowed.

(c) The leading digit cannot be zero and the number must be a multiple of 5.

(d) The number is at least 400.

Solution:

(a) $9 \cdot 10 \cdot 10 = 900$ ways

(b) $9 \cdot 9 \cdot 8 = 648$ ways

(c) $9 \cdot 10 \cdot 2 = 180$ ways (The last digit must be 0 or 5.)

(d) $6 \cdot 10 \cdot 10 = 600$ ways (The first digit must be greater than or equal to 4.)

15. Three couples have reserved seats in a given row for a concert. In how many different ways can they be seated, given the following conditions?

(a) There are no seating restrictions.

(b) The two members of each couple wish to sit together.

Solution:

(a) $6! = 720$ different ways

(b) $6 \cdot 4 \cdot 2 = 48$ different ways

19. Evaluate ${}_8P_3$.

Solution:

$${}_8P_3 = \frac{8!}{(8-3)!} = \frac{8!}{5!} = 8 \cdot 7 \cdot 6 = 336$$

23. Evaluate ${}_{100}P_2$.

Solution:

$${}_{100}P_2 = \frac{100!}{(100-2)!} = \frac{100!}{98!} = 100 \cdot 99 = 9900$$

29. In how many ways can five children line up in one row to have their picture taken?

Solution:

$5! = 120$ ways

31. From a pool of 12 candidates, the offices of president, vice-president, secretary, and treasurer will be filled. In how many different ways can the offices be filled, if each of the 12 candidates can hold any office?

Solution:

$${}_{12}P_4 = 12 \cdot 11 \cdot 10 \cdot 9 = 11{,}880 \text{ ways}$$

35. Find the number of distinguishable permutations of the letters A, A, Y, Y, Y, Y, X, X, X.

Solution:

$$\frac{9!}{2!4!3!} = \frac{9 \cdot 8 \cdot 7 \cdot 6 \cdot 5}{2 \cdot 3 \cdot 2} = 1260 \text{ distinguishable permutations}$$

41. In order to conduct a certain experiment, four students are randomly selected from a class of 20. How many different groups of four students are possible?

Solution:

$${}_{20}C_4 = \frac{20!}{(20-4)!4!} = \frac{20!}{16!4!} = \frac{20 \cdot 19 \cdot 18 \cdot 17}{4 \cdot 3 \cdot 2} = 4845 \text{ different groups}$$

45. How many subsets of four elements can be formed from a set of 100 elements?

Solution:

$${}_{100}C_4 = \frac{100!}{(100-4)!4!} = \frac{100!}{96!4!} = \frac{100 \cdot 99 \cdot 98 \cdot 97}{4 \cdot 3 \cdot 2} = 3{,}921{,}225 \text{ subsets}$$

49. An employer interviews eight people for four openings in the company. Three of the eight people are women. If all eight are qualified, in how many ways could the employer fill the four positions if (a) the selection is random and (b) exactly two are women?

Solution:

(a) $${}_8C_4 = \frac{8!}{(8-4)!4!} = \frac{8!}{4!4!} = \frac{8 \cdot 7 \cdot 6 \cdot 5}{4 \cdot 3 \cdot 2} = 70 \text{ ways}$$

(b) $${}_3C_2 \cdot {}_5C_2 = \frac{3!}{(3-2)!2!} \cdot \frac{5!}{(5-2)!2!} = 3 \cdot 10 = 30 \text{ ways}$$

51. Four people are to be selected at random from a group of four couples. In how many ways can this be done, given the following conditions?

(a) There are no restrictions.

(b) There is to be at least one couple in the group of four.

(c) The selection must include one member from each couple.

Solution:

(a) ${}_8C_4 = \dfrac{8!}{4!4!} = 70$ ways

(b) There are 16 ways that a group of four can be formed without any couples in the group. Therefore, if at least one couple is to be in the group, there are $70 - 16 = 54$ ways that could occur.

(c) $2 \cdot 2 \cdot 2 \cdot 2 = 16$ ways

55. Find the number of diagonals of an octagon.

Solution:

$${}_8C_2 - 8 = \frac{8!}{6!2!} - 8 = 28 - 8 = 20 \text{ diagonals}$$

61. Prove ${}_nC_{n-1} = {}_nC_1$.

Solution:

$${}_nC_{n-1} = \frac{n!}{(n-(n-1))!(n-1)!} = \frac{n!}{(1)!(n-1)!} = \frac{n!}{(n-1)!1!} = {}_nC_1$$

SECTION 9.7

Probability

You should know the following basic principles of probability.

- If an event A has $n(A)$ equally likely outcomes and its sample space has $n(S)$ equally likely outcomes, then the probability of event A is

$$P(A) = \frac{n(A)}{n(S)}, \text{ where } 0 \leq P(A) \leq 1.$$

- If A and B are mutually exclusive events, then $P(A \cup B) = P(A) + P(B)$.

  If A and B are not mutually exclusive events, then $P(A \cup B) = P(A) + P(B) - P(A \cap B)$.

- If A and B are independent events, then the probability that both A and B will occur is $P(A)P(B)$.

- The complement of an event A is $P(A') = 1 - P(A)$.

Solutions to Selected Exercises

5. Two county supervisors are selected from five supervisors, A, B, C, D, and E, to study a recycling plan. Determine the sample space.

Solution:

S = {AB, AC, AD, AE, BC, BD, BE, CD, CE, DE}

9. A coin is tossed three times. Find the probability of getting at least one head.

Solution:

$$S = \{HHH,\ HHT,\ HTH,\ HTT,\ THH,\ THT,\ TTH,\ TTT\}$$

$$P(TTT) = \frac{1}{8}$$

$$P(\text{at least one head}) = 1 - \frac{1}{8} = \frac{7}{8}$$

13. One card is selected from a standard deck of 52 playing cards. Find the probability of getting a black card that is not a face card.

Solution:

Twenty–six of the cards are black. Six of these are face cards (J, Q, K of clubs and spades). Therefore, there are 20 black cards that are not face cards.

$$P(E) = \frac{20}{52} = \frac{5}{13}$$

15. A six-sided die is tossed twice. Find the probability that the sum is 4.

Solution:

$$n(S) = 6(6) = 36$$

$$E = \{(1,\ 3),\ (2,\ 2),\ (3,\ 1)\}$$

$$P(E) = \frac{3}{36} = \frac{1}{12}$$

19. A six-sided die is tossed twice. Find the probability that the sum is odd and no more than 7.

Solution:

$$n(S) = 6 \cdot 6 = 36$$

$$E = \{(1,\ 2),\ (1,\ 4),\ (1,\ 6),\ (2,\ 1),\ (2,\ 3),\ (2,\ 5),\ (3,\ 2),\ (3,\ 4),$$
$$(4,\ 1),\ (4,\ 3),\ (5,\ 2),\ (6,\ 1)\}$$

$$P(E) = \frac{12}{36} = \frac{1}{3}$$

23. Two marbles are drawn (the first is *not* replaced before the second is drawn) from a bag containing one green, two yellow, and three red marbles. Find the probability of drawing neither yellow marble.

Solution:

$$P(E) = \frac{{}_4C_2}{{}_6C_2} = \frac{\frac{4!}{2!2!}}{\frac{6!}{4!2!}} = \frac{6}{15} = \frac{2}{5}$$

27. The probability that an event *will not* happen is $p = 0.15$. Find the probability that the event *will* happen.

Solution:

$$1 - p = 1 - 0.15 = 0.85$$

31. Taylor, Moore, and Jenkins are candidates for public office. It is estimated that Moore and Jenkins have about the same probability of winning, and Taylor is believed to be twice as likely to win as either of the others. Find the probability of each candidate winning the election.

Solution:

$$p + p + 2p = 1$$
$$p = 0.25$$

Taylor: 0.50
Moore: 0.25
Jenkins: 0.25

35. Four letters and envelopes are addressed to four different people. If the letters are randomly inserted into the envelopes, what is the probability that (a) exactly one will be inserted in the correct envelope and (b) at least one will be inserted in the correct envelope?

Solution:

Total ways to insert letters: $4! = 24$ ways

4 correct: 1 way
3 correct: not possible
2 correct: 6 ways
1 correct: 8 ways
0 correct: 9 ways

(a) $\dfrac{8}{24} = \dfrac{1}{3}$

(b) $\dfrac{8+6+1}{24} = \dfrac{15}{24} = \dfrac{5}{8}$

39. Two cards are selected at random from an ordinary deck of 52 playing cards. Find the probability that two aces are selected, given the following conditions.

(a) The cards are drawn in sequence, with the first card being replaced and the deck reshuffled prior to the second drawing.

(b) The two cards are drawn consecutively, without replacement.

Solution:

(a) $\left(\dfrac{4}{52}\right)\left(\dfrac{4}{52}\right) = \dfrac{1}{169}$

(b) $\left(\dfrac{4}{52}\right)\left(\dfrac{3}{51}\right) = \dfrac{1}{221}$

43. Two integers (between 1 and 30 inclusive) are chosen by a random number generator on a computer. What is the probability that (a) the numbers are both even, (b) one number is even and one is odd, (c) both numbers are less than 10, and (d) the same number is chosen twice?

Solution:

(a) $P(EE) = \dfrac{15}{30} \bullet \dfrac{15}{30} = \dfrac{1}{4}$

(b) $P(EO \text{ or } OE) = 2\left(\dfrac{15}{30}\right)\left(\dfrac{15}{30}\right) = \dfrac{1}{2}$

(c) $P(N_1 < 10,\ N_2 < 10) = \dfrac{9}{30} \bullet \dfrac{9}{30} = \dfrac{9}{100}$

(d) $P(N_1 N_1) = \dfrac{30}{30} \bullet \dfrac{1}{30} = \dfrac{1}{30}$

45. A space vehicle has an independent back-up system for one of its communication networks. The probability that either system will function satisfactorily for the duration of a flight is 0.985. What is the probability that during a given flight (a) both systems function satisfactorily, (b) at least one system functions satisfactorily, and (c) both systems fail?

Solution:

(a) $P(SS) = (0.985)^2 \approx 0.9702$

(b) $P(S) = 1 - P(FF) = 1 - (0.015)^2 \approx 0.9998$

(c) $P(FF) = (0.015)^2 \approx 0.0002$

49. Assume that the probability of the birth of a child of a particular sex is 50%. In a family with four children, what is the probability that (a) all the children are boys, (b) all the children are the same sex, and (c) there is at least one boy?

Solution:

(a) $P(BBBB) = (\frac{1}{2})(\frac{1}{2})(\frac{1}{2})(\frac{1}{2}) = \frac{1}{16}$

(b) $P(BBBB) + P(GGGG) = \frac{1}{16} + \frac{1}{16} = \frac{1}{8}$

(c) $1 - P(GGGG) = 1 - \frac{1}{16} = \frac{15}{16}$

REVIEW EXERCISES FOR CHAPTER 9

Solutions to Selected Exercises

3. Use sigma notation to write the sum $\frac{1}{2} + \frac{2}{3} + \frac{3}{4} + \cdots + \frac{9}{10}$.

Solution:

$$\frac{1}{2} + \frac{2}{3} + \frac{3}{4} + \cdots + \frac{9}{10} = \sum_{k=1}^{9} \frac{k}{k+1}$$

7. Find the sum.

$$\sum_{i=0}^{6} 2^i$$

Solution:

$$\sum_{i=0}^{6} 2^i = 2^0 + 2^1 + 2^2 + 2^3 + 2^4 + 2^5 + 2^6 = 1 + 2 + 4 + 8 + 16 + 32 + 64 = 127$$

13. Find the sum.

$$\sum_{n=0}^{10} (n^2 + 3)$$

Solution:

$$\sum_{n=0}^{10} (n^2 + 3) = \sum_{n=0}^{10} n^2 + \sum_{n=0}^{10} 3 = \frac{10(11)(21)}{6} + 3(11) = 418$$

17. Write out the first five terms of the arithmetic sequence given $a_4 = 10$ and $a_{10} = 28$.

Solution:

$$a_4 = 10 = d(4) + c, \qquad a_{10} = 28 = d(10) + c$$

$$4d + c = 10$$

$$10d + c = 28$$

Solving this system yields $d = 3$ and $c = -2$.

Thus, $a_n = 3n - 2$

$a_1 = 1$

$a_2 = 4$

$a_3 = 7$

$a_4 = 10$

$a_5 = 13$

21. Find the sum for the first 100 multiples of 5.

Solution:

$$\sum_{i=1}^{100} 5i = 5\sum_{i=1}^{100} i = 5\left(\frac{100(101)}{2}\right) = 25{,}250$$

23. Write the first five terms of the geometric sequence, $a_1 = 4$, $r = -\frac{1}{4}$.

Solution:

$$a_1 = 4$$

$$a_2 = 4\left(-\frac{1}{4}\right) = -1$$

$$a_3 = 4\left(-\frac{1}{4}\right)^2 = \frac{1}{4}$$

$$a_4 = 4\left(-\frac{1}{4}\right)^3 = -\frac{1}{16}$$

$$a_5 = 4\left(-\frac{1}{4}\right)^4 = \frac{1}{64}$$

Terms: $4, -1, \frac{1}{4}, -\frac{1}{16}, \frac{1}{64}$

27. Write an expression for the nth term of the geometric sequence with $a_1 = 16$ and $a_2 = -8$. Then find the sum of the first 20 terms of the sequence.

Solution:

$$a_1 = 16,\ a_2 = -8 \Rightarrow r = -\frac{1}{2}$$

$$a_n = 16\left(-\frac{1}{2}\right)^{n-1}$$

$$S_{20} = 16\left[\frac{1-(-1/2)^{20}}{1-(-1/2)}\right] \approx 10.667$$

31. A deposit of \$200 is made at the beginning of each month for two years into an account that pays 6%, compounded monthly. What is the balance in the account at the end of two years?

Solution:

$$A = \sum_{i=1}^{24} 200\left(1 + \frac{0.06}{12}\right)^i = \sum_{i=1}^{24} 200(1.005)^i$$

$$= 200(1.005)\left[\frac{1-(1.005)^{24}}{1-1.005}\right]$$

$$\approx \$5111.82$$

33. Use mathematical induction to prove the formula $1 + 4 + \cdots + (3n-2) = (n/2)(3n-1)$ for every positive integer n.

Solution:

1. When $n = 1$,

$$1 = \frac{1}{2}[3(1) - 1] = 1.$$

2. Assume that

$$1 + 4 + 7 + \cdots + (3k-2) = \frac{k}{2}(3k-1).$$

Then,

$$\begin{aligned} 1 + 4 + 7 + \cdots + (3k-2) + [3(k+1) - 2] &= [1 + 4 + 7 + \cdots + (3k-2)] + (3k+1) \\ &= \frac{k}{2}(3k-1) + (3k+1) \\ &= \frac{k(3k-1)}{2} + \frac{2(3k+1)}{2} \\ &= \frac{3k^2 + 5k + 2}{2} \\ &= \frac{(k+1)(3k+2)}{2} \\ &= \frac{(k+1)}{2}[3(k+1) - 1]. \end{aligned}$$

Thus, the formula holds for all positive integers n.

37. Evaluate ${}_6C_4$.

Solution:

$${}_6C_4 = \frac{6!}{(6-4)!4!} = \frac{6 \cdot 5}{2} = 15$$

43. Use the Binomial Theorem to expand the following binomial. Simplify your answer.

$$\left(\frac{2}{x} - 3x\right)^6$$

Solution:

$$\left(\frac{2}{x} - 3x\right)^6 = \left(\frac{2}{x}\right)^6 + 6\left(\frac{2}{x}\right)^5(-3x) + 15\left(\frac{2}{x}\right)^4(-3x)^2 + 20\left(\frac{2}{x}\right)^3(-3x)^3$$

$$+ 15\left(\frac{2}{x}\right)^2(-3x)^4 + 6\left(\frac{2}{x}\right)(-3x)^5 + (-3x)^6$$

$$= \frac{64}{x^6} - \frac{576}{x^4} + \frac{2160}{x^2} - 4320 + 4860x^2 - 2916x^4 + 729x^6$$

47. The complexity of interpersonal relationships increases dramatically as the size of a group increases. Determine the number of different two-person relationships in a family of (a) 2, (b) 4, and (c) 6.

Solution:

(a) ${}_2C_2 = 1$

(b) ${}_4C_2 = 6$

(c) ${}_6C_2 = 15$

51. A man has five pairs of socks (no two pairs are the same color). If he randomly selects two socks from the drawer, what is the probability that he gets a matched pair?

Solution:

$$P(\text{pair}) = \frac{10}{10} \cdot \frac{1}{9} = \frac{1}{9}$$

55. Find the probability of obtaining at least one tail when a coin is tossed five times.

Solution:

$$1 - P(HHHHH) = 1 - \left(\frac{1}{2}\right)^5 = \frac{31}{32}$$

57. Five cards are drawn from an ordinary deck of 52 playing cards. Find the probability of getting two pairs. (For example, the hand could be A-A-5-5-Q or 4-4-7-7-K.)

Solution:

$$P(2 \text{ pairs}) = \frac{({}_{13}C_2)({}_4C_2)({}_4C_2)({}_{44}C_1)}{({}_{52}C_5)} = 0.0475$$

Practice Test for Chapter 9

1. Write out the first five terms of the sequence $a_n = \dfrac{2n}{(n+2)!}$.

2. Write an expression for the nth term of the sequence $\{\frac{4}{3}, \frac{5}{9}, \frac{6}{27}, \frac{7}{81}, \frac{8}{243}, \ldots\}$.

3. Find the sum $\sum_{i=1}^{6}(2i-1)$.

4. Write out the first five terms of the arithmetic sequence where $a_1 = 23$ and $d = -2$.

5. Find a_n for the arithmetic sequence with $a_1 = 12$, $d = 3$, and $n = 50$.

6. Find the sum of the first 200 positive integers.

7. Write out the first five terms of the geometric sequence with $a_1 = 7$ and $r = 2$.

8. Evaluate $\sum_{n=0}^{9} 6\left(\frac{2}{3}\right)^n$.

9. Evaluate $\sum_{n=0}^{\infty}(0.03)^n$.

10. Use mathematical induction to prove that $1 + 2 + 3 + 4 + \cdots + n = \dfrac{n(n+1)}{2}$.

11. Use mathematical induction to prove that $n! > 2^n$, $n \geq 4$.

12. Evaluate ${}_{13}C_4$.

13. Expand $(x+3)^5$.

14. Find the term involving x^7 in $(x-2)^{12}$.

15. Evaluate ${}_{30}P_4$.

16. How many ways can six people sit at a table with six chairs?

17. Twelve cars run in a race. How many different ways can they come in first, second, and third place? (Assume that there are no ties.)

18. Two six-sided dice are tossed. Find the probability that the total of the two dice is less than 5.

19. Two cards are selected at random from a deck of 52 playing cards without replacement. Find the probability that the first card is a King and the second card is a black ten.

20. A manufacturer has determined that for every 1000 units it produces, 3 will be faulty. What is the probability that an order of 50 units will have one or more faulty units?

CHAPTER 1

Practice Test Solutions

1. $\dfrac{|-42|-20}{15-|-4|}=\dfrac{42-20}{15-4}=\dfrac{22}{11}=2$

2. $\dfrac{x}{z}-\dfrac{z}{y}=\dfrac{x}{z}\cdot\dfrac{y}{y}-\dfrac{z}{y}\cdot\dfrac{z}{z}=\dfrac{xy-z^2}{yz}$

3. $|x-7|\le 4$

4. $10(-5)^3=10(-125)=-1250$

5. $(-4x^3)(-2x^{-5})\left(\dfrac{1}{16}x\right)=(-4)(-2)\left(\dfrac{1}{16}\right)x^{3+(-5)+1}=\dfrac{8}{16}x^{-1}=\dfrac{1}{2x}$

6. $0.0000412=4.12\times 10^{-5}$

7. $125^{2/3}=(\sqrt[3]{125})^2=(5)^2=25$

8. $\sqrt[4]{64x^7y^9}=\sqrt[4]{16\cdot 4x^4x^3y^8y}$
$=2xy^2\sqrt[4]{4x^3y}$

9. $\dfrac{6}{\sqrt{12}}=\dfrac{6}{2\sqrt{3}}\cdot\dfrac{\sqrt{3}}{\sqrt{3}}=\dfrac{6\sqrt{3}}{6}=\sqrt{3}$

10. $3\sqrt{80}-7\sqrt{500}=3(4\sqrt{5})-7(10\sqrt{5})$
$=12\sqrt{5}-70\sqrt{5}=-58\sqrt{5}$

11. $(8x^4-9x^2+2x-1)-(3x^3+5x+4)=8x^4-3x^3-9x^2-3x-5$

12. $(x-3)(x^2+x-7)=x^3+x^2-7x-3x^2-3x+21=x^3-2x^2-10x+21$

13. $[(x-2)-y]^2=(x-2)^2-2y(x-2)+y^2$
$=x^2-4x+4-2xy+4y+y^2=x^2+y^2-2xy-4x+4y+4$

14. $16x^4-1=(4x^2+1)(4x^2-1)=(4x^2+1)(2x+1)(2x-1)$

15. $6x^2+5x-4=(2x-1)(3x+4)$

16. $x^3-64=x^3-4^3=(x-4)(x^2+4x+16)$

17. $-\dfrac{3}{x}+\dfrac{x}{x^2+2}=\dfrac{-3(x^2+2)+x^2}{x(x^2+2)}=\dfrac{-2x^2-6}{x(x^2+2)}=-\dfrac{2(x^2+3)}{x(x^2+2)}$

18. $\dfrac{x-3}{4x}\div\dfrac{x^2-9}{x^2}=\dfrac{\cancel{x-3}}{\cancel{4x}}\cdot\dfrac{\cancel{x^2}}{(x+3)\cancel{(x-3)}}=-\dfrac{x}{4(x+3)}$

19. $$\frac{1-\frac{1}{x}}{1-\frac{1}{1-(1/x)}} = \frac{\frac{x-1}{x}}{1-\frac{1}{(x-1)/x}} = \frac{\frac{x-1}{x}}{1-\frac{x}{x-1}} = \frac{\frac{x-1}{x}}{\frac{-1}{x-1}} = \frac{x-1}{x} \cdot \frac{x-1}{-1} = \frac{-(x-1)^2}{x}$$

20. $$\frac{1}{3}(x-1)^{5/2} - \frac{1}{6}(x-1)^{1/2} = \frac{1}{6}(x-1)^{1/2}[2(x-1)^2 - 1]$$
$$= \frac{1}{6}(x-1)^{1/2}(2x^2 - 4x + 2 - 1) = \frac{1}{6}(x-1)^{1/2}(2x^2 - 4x + 1)$$

CHAPTER 2

Practice Test Solutions

1.
$$5x+4=7x-8$$
$$4+8=7x-5x$$
$$12=2x$$
$$x=6$$

2.
$$\frac{x}{3}-5=\frac{x}{5}+1$$
$$15\left(\frac{x}{3}-5\right)=15\left(\frac{x}{5}+1\right)$$
$$5x-75=3x+15$$
$$2x=90$$
$$x=45$$

3.
$$\frac{3x+1}{6x-7}=\frac{2}{5}$$
$$5(3x+1)=2(6x-7)$$
$$15x+5=12x-14$$
$$3x=-19$$
$$x=-\frac{19}{3}$$

4.
$$(x-3)^2+4=(x+1)^2$$
$$x^2-6x+9+4=x^2+2x+1$$
$$-8x=-12$$
$$x=\frac{-12}{-8}$$
$$x=\frac{3}{2}$$

5.
$$A=\frac{1}{2}(a+b)h$$
$$2A=ah+bh$$
$$2A-bh=ah$$
$$\frac{2A-bh}{h}=a$$

6.
$$x+(x+1)+(x+2)=132$$
$$3x+3=132$$
$$3x=129$$
$$x=43$$
$$x+1=44$$
$$x+2=45$$

7. Percent $=\frac{301}{4300}=0.07=7\%$

8. Let $x=$ number of quarters.
Then $53-x=$ number of nickels.
$$25x+5(53-x)=605$$
$$20x+265=605$$
$$20x=340$$
$$x=17 \text{ quarters}$$
$$53-x=36 \text{ nickels}$$

9. Let $x =$ amount in $9\frac{1}{2}\%$ fund.
Then $15,000 - x =$ amount in 11% fund.

$$0.095x + 0.11(15,000 - x) = 1582.50$$
$$-0.015x + 1650 = 1582.50$$
$$-0.015x = -67.5$$
$$x = \$4500 \text{ at } 9\tfrac{1}{2}\%$$
$$15,000 - x = \$10,500 \text{ at } 11\%$$

10.
$$28 + 5x - 3x^2 = 0$$
$$(4 - x)(7 + 3x) = 0$$
$$4 - x = 0 \rightarrow x = 4$$
$$7 + 3x = 0 \rightarrow x = -\tfrac{7}{3}$$

11.
$$(x - 2)^2 = 24$$
$$x - 2 = \pm\sqrt{24}$$
$$x - 2 = \pm 2\sqrt{6}$$
$$x = 2 \pm 2\sqrt{6}$$

12.
$$x^2 - 4x - 9 = 0$$
$$x^2 - 4x + 2^2 = 9 + 2^2$$
$$(x - 2)^2 = 13$$
$$x - 2 = \pm\sqrt{13}$$
$$x = 2 \pm \sqrt{13}$$

13.
$$\frac{1}{x^2 - 6x + 1} = \frac{1}{x^2 - 6x + 3^2 - 3^2 + 1}$$
$$= \frac{1}{(x - 3)^2 - 8}$$

14. $x^2 + 5x - 1 = 0$
$a = 1,\ b = 5,\ c = -1$
$$x = \frac{-5 \pm \sqrt{(5)^2 - 4(1)(-1)}}{2(1)}$$
$$= \frac{-5 \pm \sqrt{25 + 4}}{2} = \frac{-5 \pm \sqrt{29}}{2}$$

15. $3x^2 - 2x + 4 = 0$
$a = 3,\ b = -2,\ c = 4$
$$x = \frac{-(-2) \pm \sqrt{(-2)^2 - 4(3)(4)}}{2(3)}$$
$$= \frac{2 \pm \sqrt{4 - 48}}{6}$$
$$= \frac{2 \pm \sqrt{-44}}{6}$$
$$= \frac{2 \pm 2i\sqrt{11}}{6}$$
$$= \frac{1 \pm i\sqrt{11}}{3} = \frac{1}{3} \pm \frac{\sqrt{11}}{3}i$$

16. $60,000 = xy$

$$y = \frac{60,000}{x}$$

$$2x + 2y = 1100$$

$$2x + 2\left(\frac{60,000}{x}\right) = 1100$$

$$x + \frac{60,000}{x} = 550$$

$$x^2 + 60,000 = 550x$$

$$x^2 - 550x + 60,000 = 0$$

$$(x - 150)(x - 400) = 0$$

$x = 150$ or $x = 400$

$y = 400$ $\quad y = 150$

Length: 400 feet

Width: 150 feet

y

x x

y

17. $x(x + 2) = 624$

$$x^2 + 2x - 624 = 0$$

$$(x - 24)(x + 26) = 0$$

$x = 24$ or $x = -26$, (extraneous solution)

$x + 2 = 26$

18. $x^3 - 10x^2 + 24x = 0$

$$x(x^2 - 10x + 24) = 0$$

$$x(x - 4)(x - 6) = 0$$

$x = 0,\ x = 4,\ x = 6$

19. $\sqrt[3]{6 - x} = 4$

$$6 - x = 64$$

$$-x = 58$$

$$x = -58$$

20. $(x^2 - 8)^{2/5} = 4$

$$x^2 - 8 = \pm 4^{5/2}$$

$x^2 - 8 = 32$ or $x^2 - 8 = -32$

$x^2 = 40$ $\quad x^2 = -24$

$x = \pm\sqrt{40}$ $\quad x = \pm\sqrt{-24}$

$x = \pm 2\sqrt{10}$ $\quad x = \pm 2\sqrt{6}i$

21. $x^4 - x^2 - 12 = 0$

$(x^2 - 4)(x^2 + 3) = 0$

$x^2 = 4$ or $x^2 = -3$

$x^2 = \pm 2$ $\quad x = \pm\sqrt{3}\,i$

22. $4 - 3x > 16$

$-3x > 12$

$x < -4$

23. $\left|\dfrac{x-3}{2}\right| < 5$

$-5 < \dfrac{x-3}{2} < 5$

$-10 < x - 3 < 10$

$-7 < x < 13$

24. $\dfrac{x+1}{x-3} < 2$

$\dfrac{x+1}{x-3} - 2 < 0$

$\dfrac{x+1-2(x-3)}{x-3} < 0$

$\dfrac{7-x}{x-3} < 0$

Critical numbers: $x = 7$ and $x = 3$

Test intervals: $(-\infty, 3)$, $(3, 7)$, $(7, \infty)$

Solution intervals: $(-\infty, 3) \cup (7, \infty)$

25. $|3x - 4| \geq 9$

$3x - 4 \leq -9$ or $3x - 4 \geq 9$

$3x \leq -5$ $\quad 3x \geq 13$

$x \leq -\frac{5}{3}$ $\quad x \geq \frac{13}{3}$

CHAPTER 3

Practice Test Solutions

1. $d = \sqrt{(4-0)^2 + (-1-3)^2}$

$= \sqrt{16+16}$

$= \sqrt{32}$

$= 4\sqrt{2}$

2. Midpoint: $\left(\frac{4+0}{2}, \frac{-1+3}{2}\right) = (2, 1)$

3. $6 = \sqrt{(x-0)^2 + (-2-0)^2}$

$6 = \sqrt{x^2+4}$

$36 = x^2 + 4$

$x^2 = 32$

$x = \pm\sqrt{32}$

$x = \pm 4\sqrt{2}$

4. x-intercept: Let $y = 0$; $0 = \dfrac{x-2}{x+3}$

$0 = x - 2$

$x = 2 \quad (2, 0)$

y-intercept: Let $x = 0$; $y = \dfrac{0-2}{0+3}$

$y = -\frac{2}{3} \quad \left(0, -\frac{2}{3}\right)$

5.

$$xy^2 = 6$$

$x(-y)^2 = 6 \Rightarrow xy^2 = 6$ x-axis symmetry

$(-x)y^2 = 6 \Rightarrow xy^2 = -6$ No y-axis symmetry

$(-x)(-y)^2 = 6 \Rightarrow xy^2 = -6$ No origin symmetry

6. x-intercepts: $(0, 0)$, $(2, 0)$, $(-2, 0)$

Origin symmetry:

x	0	1	−1	2	−2	3
y	0	−3	3	0	0	15

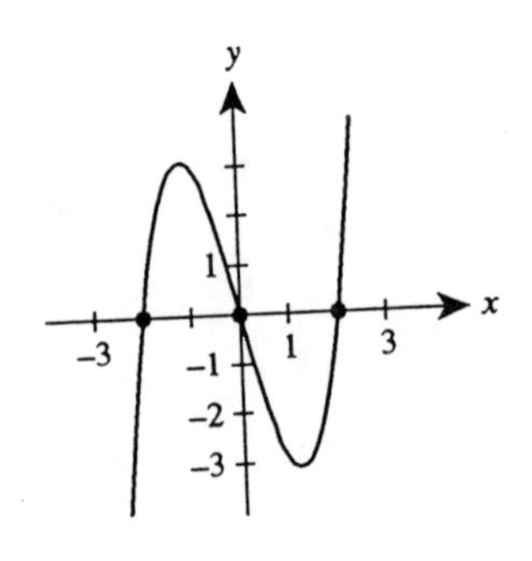

7.
$$x^2 + y^2 - 6x + 2y + 6 = 0$$
$$x^2 - 6x + \underline{9} + y^2 + 2y + \underline{1} = -6 + 9 + 1$$
$$(x-3)^2 + (y+1)^2 = 4$$

Center: $(3, \ -1)$

Radius: 2

8.
$$f(x-3) = (x-3)^2 - 2(x-3) + 1$$
$$= x^2 - 6x + 9 - 2x + 6 + 1$$
$$= x^2 - 8x + 16$$

9.
$$f(3) = 12 - 11 = 1$$
$$\frac{f(x) - f(3)}{x - 3} = \frac{(4x - 11) - 1}{x - 3}$$
$$= \frac{4x - 12}{x - 3}$$
$$= \frac{4(x-3)}{x-3} = 4$$

10. $f(x) = \sqrt{36 - x^2} = \sqrt{(6+x)(6-x)}$

Domain: $[-6, \ 6]$

Range: $[0, \ 6]$

11. (a) $6x - 5y + 4 = 0$

$$y = \frac{6x + 4}{5} \quad \text{function of } x$$

(b) $x^2 + y^2 = 9$

$$y = \pm\sqrt{9 - x^2} \quad \text{not a function of } x$$

(c) $y^3 = x^2 + 6$

$$y = \sqrt[3]{x^2 + 6} \quad \text{function of } x$$

12. Parabola: Vertex $(0, \ -5)$

Intercepts: $(0, \ -5)$, $(\pm\sqrt{5}, \ 0)$

y-axis symmetry

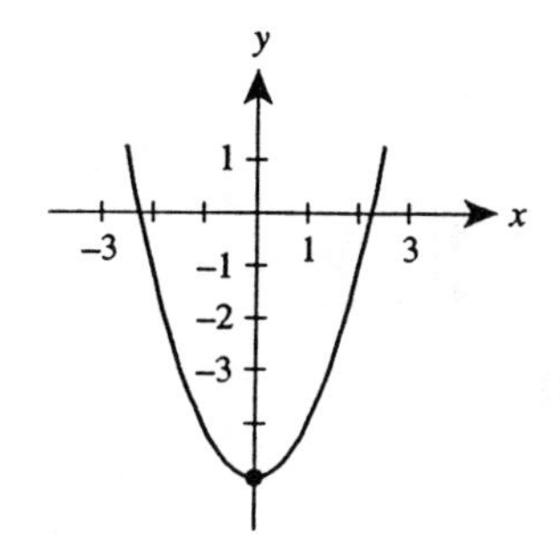

13. Intercepts: (0, 3), (−3, 0)

x	0	1	−1	2	−2	−3	−4
y	3	4	2	5	1	0	1

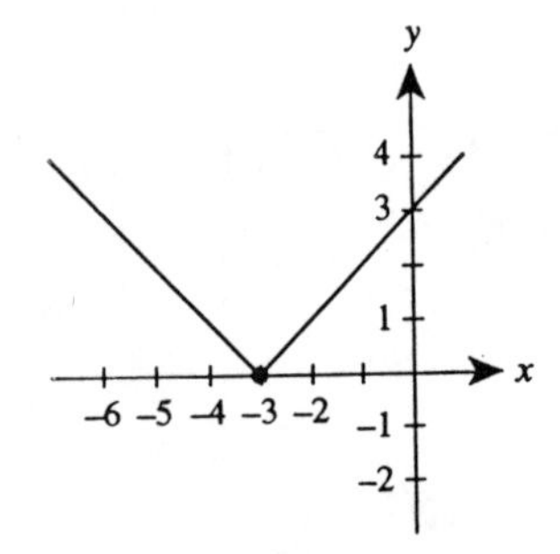

14.

x	0	1	2	3
y	1	3	5	7

x	−1	−2	−3
y	2	6	12

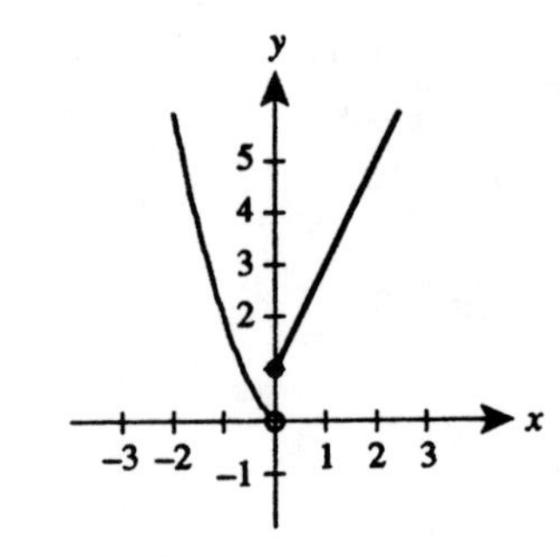

15. $m = \dfrac{-1-4}{3-2} = -5$

$$y - 4 = -5(x - 2)$$
$$y - 4 = -5x + 10$$
$$y = -5x + 14$$

16. $y = \dfrac{4}{3}x - 3$

17. $2x + 3y = 0$

$$y = -\tfrac{2}{3}x$$
$$m_1 = -\tfrac{2}{3}$$
$$\perp m_2 = \tfrac{3}{2} \text{ through } (4,\ 1)$$
$$y - 1 = \tfrac{3}{2}(x - 4)$$
$$y - 1 = \tfrac{3}{2}x - 6$$
$$y = \tfrac{3}{2}x - 5$$

18. (5, 32) and (9, 44)

$$m = \frac{44-32}{9-5} = \frac{12}{4} = 3$$
$$y - 32 = 3(x - 5)$$
$$y - 32 = 3x - 15$$
$$y = 3x + 17$$

When $x = 20$, $y = 3(20) + 17$

$y = \$77$

19. $f(g(x)) = f(2x+3)$

$$= (2x+3)^2 - 2(2x+3) + 16$$
$$= 4x^2 + 12x + 9 - 4x - 6 + 16$$
$$= 4x^2 + 8x + 19$$

20.
$$f(x) = x^3 + 7$$
$$y = x^3 + 7$$
$$x = y^3 + 7$$
$$x - 7 = y^3$$
$$\sqrt[3]{x-7} = y$$
$$f^{-1}(x) = \sqrt[3]{x-7}$$

21. (a) $f(x) = |x-6|$ is not one-to-one.
For example, $f(0) = 6$ and $f(12) = 6$.

(b) $f(x) = ax + b,\ a \neq 0$ is one-to-one.

(c) $f(x) = x^3 - 19$ is one-to-one.

22.
$$f(x) = \sqrt{\frac{3-x}{x}},\quad 0 < x \leq 3$$
$$y = \sqrt{\frac{3-x}{x}}$$
$$x = \sqrt{\frac{3-y}{y}}$$
$$x^2 = \frac{3-y}{y}$$
$$x^2y = 3 - y$$
$$x^2y + y = 3$$
$$y(x^2+1) = 3$$
$$y = \frac{3}{x^2+1}$$
$$f^{-1}(x) = \frac{3}{x^2+1}$$

23. $y = kx$ and $y = 30$ when $x = 5$

$$30 = k(5)$$
$$6 = k$$
$$y = 6x$$

24. $y = k/x$ and $y = 0.5$ when $x = 14$

$$0.5 = k/14$$
$$7 = k$$
$$y = 7/x$$

25. $z = \dfrac{kx^2}{y}$ and $z = 3$ when $x = 3,\ y = -6$

$$3 = \frac{k(3)^2}{-6}$$
$$-18 = 9k$$
$$-2 = k$$
$$z = \frac{-2x^2}{y}$$

CHAPTER 4

Practice Test Solutions

1. x-intercepts: $(1, 0)$, $(5, 0)$

y-intercept: $(0, 5)$

Vertex: $(3, -4)$

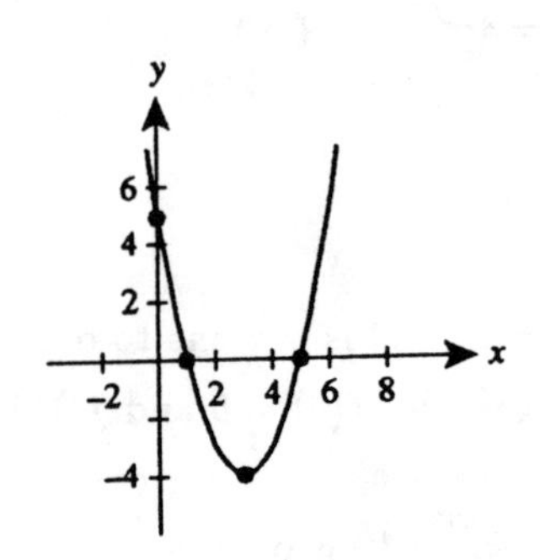

2. $a = 0.01,\ b = -90$

$$\frac{-b}{2a} = \frac{90}{2(.01)} = 4500 \text{ units}$$

3. Vertex $(1, 7)$ opening downward through $(2, 5)$

$y = a(x-1)^2 + 7$ Standard form

$5 = a(2-1)^2 + 7$

$5 = a + 7$

$a = -2$

$y = -2(x-1)^2 + 7$

$= -2(x^2 - 2x + 1) + 7$

$= -2x^2 + 4x + 5$

4. $y = \pm a(x-2)(3x-4)$ where a is any real number

$y = \pm a(3x^2 - 10x + 8)$

5. Leading coefficient: -3

Degree: 5

Moves down to the right and up to the left

6. $0 = x^5 - 5x^3 + 4x$

$= x(x^4 - 5x^2 + 4)$

$= x(x^2 - 1)(x^2 - 4)$

$= x(x+1)(x-1)(x+2)(x-2)$

$x = 0,\ x = \pm 1,\ x = \pm 2$

7. $f(x) = x(x-3)(x+2)$

$$= x(x^2 - x - 6)$$

$$= x^3 - x^2 - 6x$$

8. Intercepts: $(0, 0)$, $(\pm 2\sqrt{3}, 0)$

Origin symmetry

Moves up to the right.

Moves down to the left.

x	-2	-1	0	1	2
y	16	11	0	-11	-16

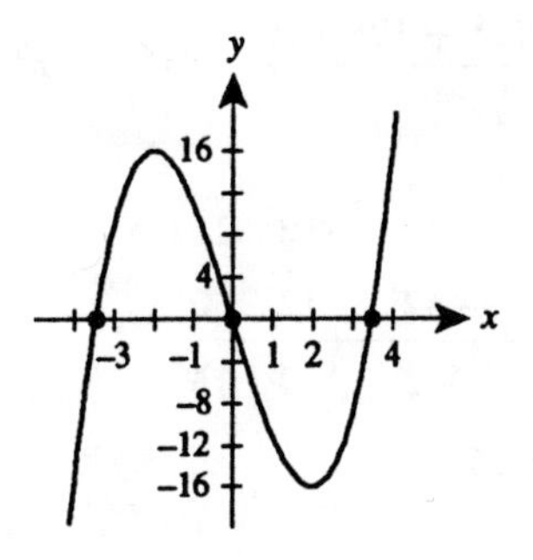

9.

$$\begin{array}{r|l}
 & 3x^3 + 9x^2 + 20x + 62 + \dfrac{176}{x-3} \\
\hline
x - 3 & 3x^4 + 0x^3 - 7x^2 + 2x - 10 \\
 & \underline{3x^4 - 9x^3} \\
 & \quad 9x^3 - 7x^2 \\
 & \quad \underline{9x^3 - 27x^2} \\
 & \qquad 20x^2 + 2x \\
 & \qquad \underline{20x^2 - 60x} \\
 & \qquad\quad 62x - 10 \\
 & \qquad\quad \underline{62x - 186} \\
 & \qquad\qquad 176
\end{array}$$

10.

$$\begin{array}{r|l}
 & x - 2 + \dfrac{5x - 13}{x^2 + 2x - 1} \\
\hline
x^2 + 2x - 1 & x^3 + 0x^2 + 0x - 11 \\
 & \underline{x^3 + 2x^2 - x} \\
 & \quad -2x^2 + x - 11 \\
 & \quad \underline{-2x^2 - 4x + 2} \\
 & \qquad 5x - 13
\end{array}$$

11. $$\begin{array}{r|rrrrrr} -5 & 3 & 13 & 0 & 0 & 12 & -1 \\ & & -15 & 10 & -50 & 250 & -1310 \\ \hline & 3 & -2 & 10 & -50 & 262 & -1311 \end{array}$$

$$\frac{3x^5 + 13x^4 + 12x - 1}{x+5} = 3x^4 - 2x^3 + 10x^2 - 50x + 262 - \frac{1311}{x+5}$$

12. $$\begin{array}{r|rrrr} -6 & 7 & 40 & -12 & 15 \\ & & -42 & 12 & 0 \\ \hline & 7 & -2 & 0 & 15 \end{array}$$

$f(-6) = 15$

13. $0 = x^3 - 19x - 30$

Possible Rational Roots: $\pm 1,\ \pm 2,\ \pm 3,\ \pm 5,\ \pm 6,\ \pm 10,\ \pm 15,\ \pm 30$

$$\begin{array}{r|rrrr} -2 & 1 & 0 & -19 & -30 \\ & & -2 & 4 & 30 \\ \hline & 1 & -2 & -15 & 0 \end{array} \quad -2 \text{ is a zero.}$$

$0 = (x+2)(x^2 - 2x - 15)$

$0 = (x+2)(x+3)(x-5)$

Zeros: $x = -2,\ x = -3,\ x = 5$

14. $0 = x^4 + x^3 - 8x^2 - 9x - 9$

Possible Rational Roots: $\pm 1,\ \pm 3,\ \pm 9$

$$\begin{array}{r|rrrrr} 3 & 1 & 1 & -8 & -9 & -9 \\ & & 3 & 12 & 12 & 9 \\ \hline & 1 & 4 & 4 & 3 & 0 \end{array} \quad x = 3 \text{ is a zero.}$$

$0 = (x-3)(x^3 + 4x^2 + 4x + 3)$

Possible Rational Roots of $x^3 + 4x^2 + 4x + 3:\ \pm 1,\ \pm 3$

$$\begin{array}{r|rrrr} -3 & 1 & 4 & 4 & 3 \\ & & -3 & -3 & -3 \\ \hline & 1 & 1 & 1 & 0 \end{array} \quad x = -3 \text{ is a zero.}$$

–CONTINUED ON NEXT PAGE–

14. –CONTINUED–

$0 = (x-3)(x+3)(x^2+x+1)$

The zeros of x^2+x+1 are $x = \dfrac{-1 \pm \sqrt{3}\,i}{2}$.

Zeros: $x = 3,\ x = -3,\ x = -\frac{1}{2} + \frac{\sqrt{3}}{2}i,\ x = -\frac{1}{2} - \frac{\sqrt{3}}{2}i$

15. $0 = 6x^3 - 5x^2 + 4x - 15$

Possible Rational Roots: $\pm 1,\ \pm 3,\ \pm 5,\ \pm 15,\ \pm\frac{1}{2},\ \pm\frac{3}{2},\ \pm\frac{5}{2},\ \pm\frac{15}{2},\ \pm\frac{1}{3},\ \pm\frac{5}{3},\ \pm\frac{1}{6},\ \pm\frac{5}{6}$

16. $0 = x^3 - \frac{20}{3}x^2 + 9x - \frac{10}{3}$

$0 = 3x^3 - 20x^2 + 27x - 10$

Possible Rational Roots: $\pm 1,\ \pm 2,\ \pm 5,\ \pm 10,\ \pm\frac{1}{3},\ \pm\frac{2}{3},\ \pm\frac{5}{3},\ \pm\frac{10}{3}$

$$\begin{array}{r|rrrr} 1 & 3 & -20 & 27 & -10 \\ & & 3 & -17 & 10 \\ \hline & 3 & -17 & 10 & 0 \end{array}$$

$0 = (x-1)(3x^2 - 17x + 10)$

$0 = (x-1)(3x-2)(x-5)$

Zeros: $x = 1,\ x = \frac{2}{3},\ x = 5$

17. Possible Rational Roots: $\pm 1,\ \pm 2,\ \pm 5,\ \pm 10$

$$\begin{array}{r|rrrrr} 1 & 1 & 1 & 3 & 5 & -10 \\ & & 1 & 2 & 5 & 10 \\ \hline & 1 & 2 & 5 & 10 & 0 \end{array} \quad x = 1 \text{ is a zero.}$$

$$\begin{array}{r|rrrr} -2 & 1 & 2 & 5 & 10 \\ & & -2 & 0 & -10 \\ \hline & 1 & 0 & 5 & 0 \end{array} \quad x = -2 \text{ is a zero.}$$

$f(x) = (x-1)(x+2)(x^2+5)$

$\quad = (x-1)(x+2)(x+5i)(x-5i)$

18. $f(x) = (x-2)[x-(3+i)][x-(3-i)]$

$= (x-2)[x^2 - x(3-i) - x(3+i) + (3+i)(3-i)]$

$= (x-2)[x^2 - 6x + 10]$

$= x^3 - 8x^2 + 22x - 20$

19.

$$\begin{array}{r|rrrr} 3i & 1 & 4 & 9 & 36 \\ & & 3i & 12i-9 & -36 \\ \hline & 1 & 4+3i & 12i & 0 \end{array}$$

20.

Iteration	*a*	*c*	*b*	*f(a)*	*f(c)*	*f(b)*	*Error*
1	0	0.5	1	−1	0.125	2	0.5
2	0	0.25	0.5	−1	−0.4844	0.125	0.25
3	0.25	0.375	0.5	−0.4844	−0.1973	0.125	0.125
4	0.375	0.4375	0.5	−0.1973	−0.0413	0.125	0.0625
5	0.4375	0.4688	0.5	−0.0413	0.0405	0.125	0.0313
6	0.4375	0.4531	0.4688	−0.0413	−0.0007	0.0405	0.0156
7	0.4531	0.4610	0.4688	−0.0008	0.0198	0.0405	0.0078
8	0.4531	0.4570	0.4610	−0.0008	0.0095	0.0198	0.0039
9	0.4531	0.4550	0.4570	−0.0008	0.0043	0.0095	0.0020
10	0.4531	0.4541	0.4550	−0.0008	0.0018	0.0043	0.0010
11	0.4531	0.4536	0.4541	−0.0008	0.0005	0.0018	0.0008
12	0.4531	0.4534	0.4536	−0.0008	0.0000	0.0005	0.0008

$x \approx 0.453$

CHAPTER 5

Practice Test Solutions

1. Vertical asymptote: $x = 0$

Horizontal asymptote: $y = \frac{1}{2}$

x-intercept: $(1, 0)$

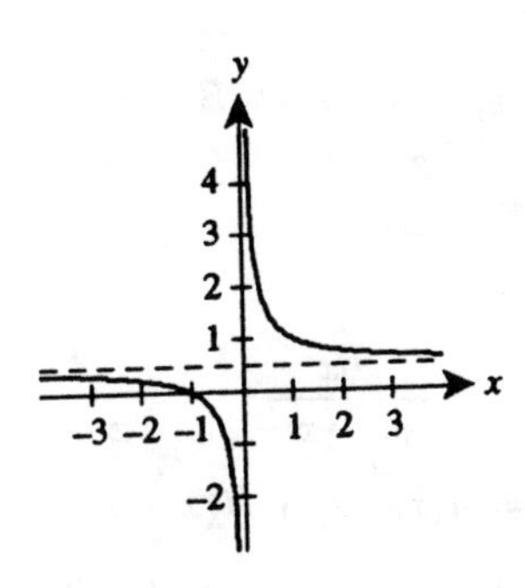

2. Vertical asymptote: $x = 0$

Slant asymptote: $y = 3x$

x-intercepts: $\left(\pm\frac{2}{\sqrt{3}}, 0\right)$

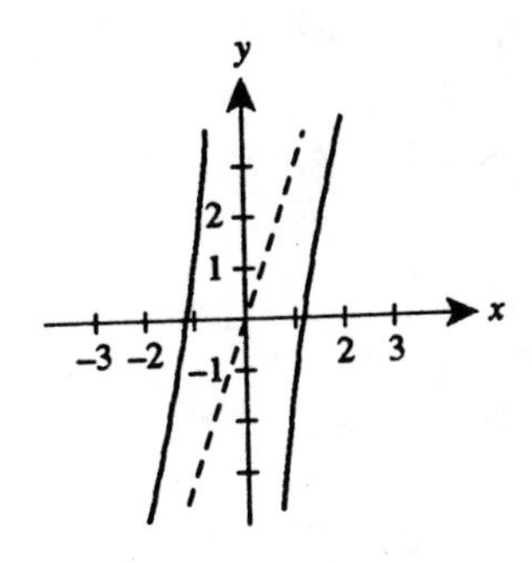

3. $y = 8$ is a horizontal asymptote since the degree of the numerator equals the degree of the denominator.

There are no vertical asymptotes.

4. $x = 1$ is a vertical asymptote.

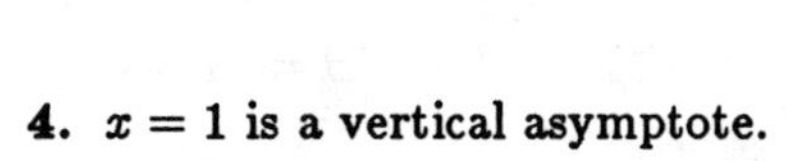

$$\frac{4x^2 - 2x + 7}{x - 1} = 4x + 2 + \frac{9}{x - 1}$$

so $y = 4x + 2$ is a slant asymptote.

5. $f(x) = \frac{x - 5}{(x - 5)^2} = \frac{1}{x - 5}$

Vertical asymptote: $x = 5$

Horizontal asymptote: $y = 0$

y-intercept: $\left(0, -\frac{1}{5}\right)$

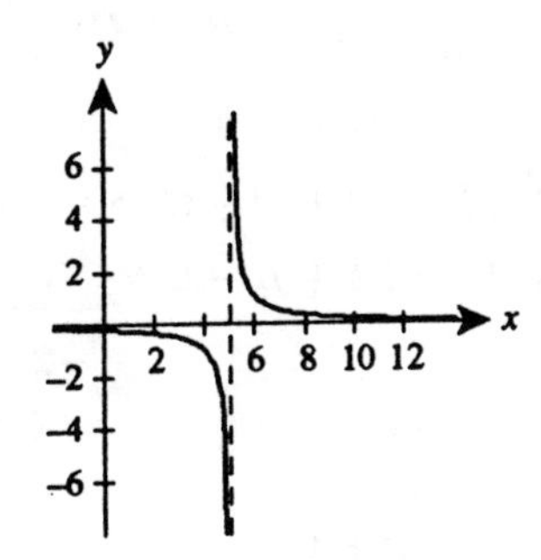

6. $\dfrac{1-2x}{x^2+x} = \dfrac{1-2x}{x(x+1)} = \dfrac{A}{x} + \dfrac{B}{x+1}$

$1 - 2x = A(x+1) + Bx$

When $x = 0,\ 1 = A$.

When $x = -1,\ 3 = -B \Rightarrow B = -3$.

$\dfrac{1-2x}{x^2+x} = \dfrac{1}{x} - \dfrac{3}{x+1}$

7. $\dfrac{6x}{x^2-x-2} = \dfrac{6x}{(x+1)(x-2)} = \dfrac{A}{x+1} + \dfrac{B}{x-2}$

$6x = A(x-2) + B(x+1)$

When $x = -1,\ -6 = -3A \Rightarrow A = 2$.

When $x = \ \ 2,\ 12 = \ \ 3B \Rightarrow B = 4$.

$\dfrac{6x}{x^2-x-2} = \dfrac{2}{x+1} + \dfrac{4}{x-2}$

8. $\dfrac{6x-17}{(x-3)^2} = \dfrac{A}{x-3} + \dfrac{B}{(x-3)^2}$

$6x - 17 = A(x-3) + B$

When $x = 3,\ 1 = B$.

When $x = 0,\ -17 = -3A + B \Rightarrow A = 6$.

$\dfrac{6x-17}{(x-3)^2} = \dfrac{6}{x-3} + \dfrac{1}{(x-3)^2}$

9. $\dfrac{3x^2-x+8}{x^3+2x} = \dfrac{3x^2-x+8}{x(x^2+2)} = \dfrac{A}{x} + \dfrac{Bx+C}{x^2+2}$

$3x^2 - x + 8 = A(x^2+2) + (Bx+C)x$

When $x = 0,\ 8 = 2A \Rightarrow A = 4$.

$$\begin{array}{ll}
\text{When } x = \ \ 1,\ 10 = 3A + B + C \Rightarrow & -2 = \ \ B + C. \\
\text{When } x = -1,\ 12 = 3A + B - C \Rightarrow & \underline{\ \ \ 0 = \ \ B - C\ \ } \\
 & -2 = 2B \qquad \Rightarrow B = -1 \\
 & \qquad\qquad\qquad\ \ C = -1
\end{array}$$

$\dfrac{3x^2-x+8}{x^3+2x} = \dfrac{4}{x} - \dfrac{x+1}{x^2+2}$

10. $(x-0)^2 = 4(5)(y-0)$

Vertex: $(0, 0)$

Focus: $(0, 5)$

Directrix: $y = -5$

11. $(y-0)^2 = 4(7)(x-0)$

$$y^2 = 28x$$

12. $a = 12,\ b = 5,\ h = k = 0,$

$c = \sqrt{144-25} = \sqrt{119}$

Center: $(0, 0)$

Foci: $(\pm\sqrt{119},\ 0)$

Vertices: $(\pm 12,\ 0)$

13. Center: $(0, 0)$

$c = 4,\ 2b = 6 \Rightarrow b = 3,$

$a = \sqrt{16+9} = 5$

$$\frac{x^2}{25} + \frac{y^2}{9} = 1$$

14. $a = 12,\ b = 13,\ c = \sqrt{144+169} = \sqrt{313}$

Center: $(0, 0)$

Foci: $(0,\ \pm\sqrt{313})$

Vertices: $(0,\ \pm 12)$

Asymptotes: $y = \pm\frac{12}{13}x$

15. Center: $(0, 0)$

$a = 4,\ \pm\frac{1}{2} = \pm\frac{b}{4} \Rightarrow b = 2$

$$\frac{x^2}{16} - \frac{y^2}{4} = 1$$

16. $p = 4$

$(x-6)^2 = 4(4)(y+1)$

$(x-6)^2 = 16(y+1)$

17.

$$\begin{aligned} 16x^2 - 96x + 9y^2 + 36y &= -36 \\ 16(x^2 - 6x + 9) + 9(y^2 + 4y + 4) &= -36 + 144 + 36 \\ 16(x-3)^2 + 9(y+2)^2 &= 144 \\ \frac{(x-3)^2}{9} + \frac{(y+2)^2}{16} &= 1 \end{aligned}$$

$a = 4,\ b = 3,\ c = \sqrt{16-9} = \sqrt{7}$

Center: $(3,\ -2)$

Foci: $(3,\ -2 \pm \sqrt{7})$

Vertices: $(3,\ -2 \pm 4)$ or $(3,\ 2)$ and $(3,\ -6)$

18. Center: $(3, 1)$

$a = 4,\ 2b = 2 \Rightarrow b = 1$

$$\frac{(x-3)^2}{16} + \frac{(y-1)^2}{1} = 1$$

19. Center: $(-3, 1)$

Vertices: $\left(-3 \pm \frac{1}{2},\ 1\right)$ or $\left(-\frac{5}{2},\ 1\right)$ and $\left(-\frac{7}{2},\ 1\right)$

Foci: $\left(-3 \pm \frac{\sqrt{13}}{6},\ 1\right)$

Asymptotes: $y = \pm\frac{1/3}{1/2}(x+3) + 1 = \pm\frac{2}{3}(x+3) + 1$

$a = \frac{1}{2},\ b = \frac{1}{3},\ c = \sqrt{\frac{1}{4} + \frac{1}{9}} = \frac{\sqrt{13}}{6}$

$$\frac{(x+3)^2}{1/4} - \frac{(y-1)^2}{1/9} = 1$$

20. Center: $(3, 0)$

$a = 4,\ c = 7,\ b = \sqrt{49 - 16} = \sqrt{33}$

$$\frac{y^2}{16} - \frac{(x-3)^2}{33} = 1$$

CHAPTER 6

Practice Test Solutions

1. $x^{3/5} = 8$

$$x = 8^{5/3} = (\sqrt[3]{8})^5 = 2^5 = 32$$

2. $3^{x-1} = \frac{1}{81}$

$$3^{x-1} = 3^{-4}$$

$$x - 1 = -4$$

$$x = -3$$

3. $f(x) = 2^{-x} = \left(\frac{1}{2}\right)^x$

x	-2	-1	0	1	2
$f(x)$	4	2	1	$\frac{1}{2}$	$\frac{1}{4}$

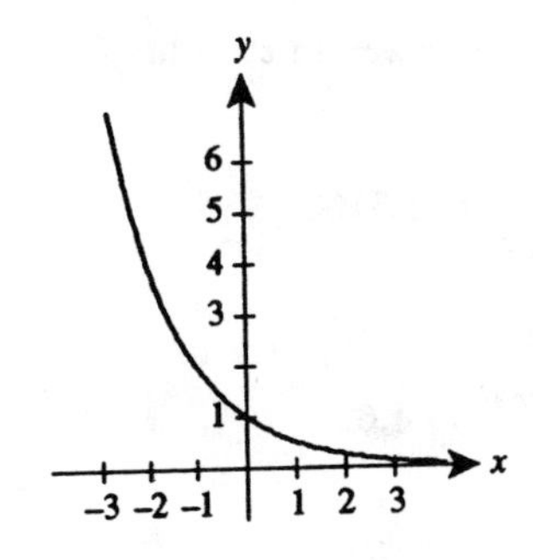

4. $g(x) = e^x + 1$

x	-2	-1	0	1	2
$g(x)$	1.14	1.37	2	3.72	8.39

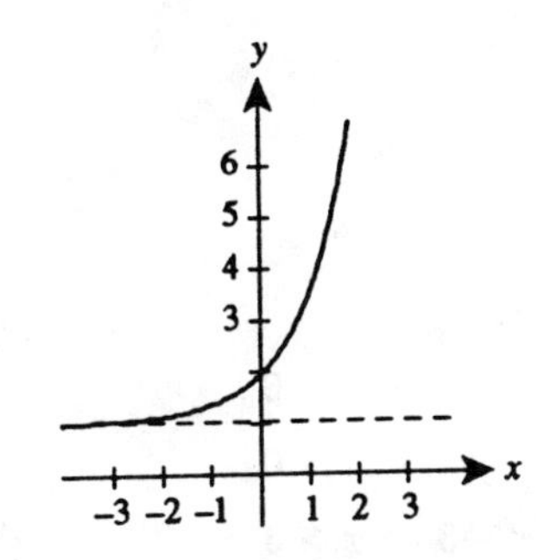

5. $A = P\left(1 + \frac{r}{n}\right)^{nt}$

(a) $A = 5000\left(1 + \frac{0.09}{12}\right)^{12(3)} \approx \$6543.23$

(b) $A = 5000\left(1 + \frac{0.09}{4}\right)^{4(3)} \approx \$6530.25$

(c) $A = 5000e^{(0.09)(3)} \approx \$6549.82$

6. $7^{-2} = \frac{1}{49}$

$$\log_7 \frac{1}{49} = -2$$

7. $x - 4 = \log_2 \frac{1}{64}$

$$2^{x-4} = \frac{1}{64}$$
$$2^{x-4} = 2^{-6}$$
$$x - 4 = -6$$
$$x = -2$$

8. $\log_b \sqrt[4]{8/25} = \frac{1}{4} \log_b \frac{8}{25}$

$$= \tfrac{1}{4}[\log_b 8 - \log_b 25]$$
$$= \tfrac{1}{4}[\log_b 2^3 - \log_b 5^2]$$
$$= \tfrac{1}{4}[3 \log_b 2 - 2 \log_b 5]$$
$$= \tfrac{1}{4}[3(0.3562) - 2(0.8271)]$$
$$= -0.1464$$

9. $5 \ln x - \dfrac{1}{2} \ln y + 6 \ln z = \ln x^5 - \ln \sqrt{y} + \ln z^6 = \ln\left(\dfrac{x^5 z^6}{\sqrt{y}}\right)$

10. $\log_9 28 = \dfrac{\log 28}{\log 9} \approx 1.5166$

11. $\log N = 0.6646$

$$N = 10^{0.6646} \approx 4.62$$

12.

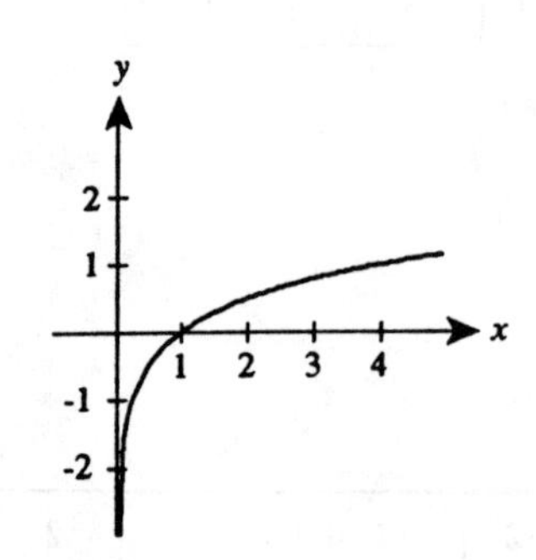

13. Domain:

$$x^2 - 9 > 0$$
$$(x + 3)(x - 3) > 0$$
$$x < -3 \text{ or } x > 3$$

14.

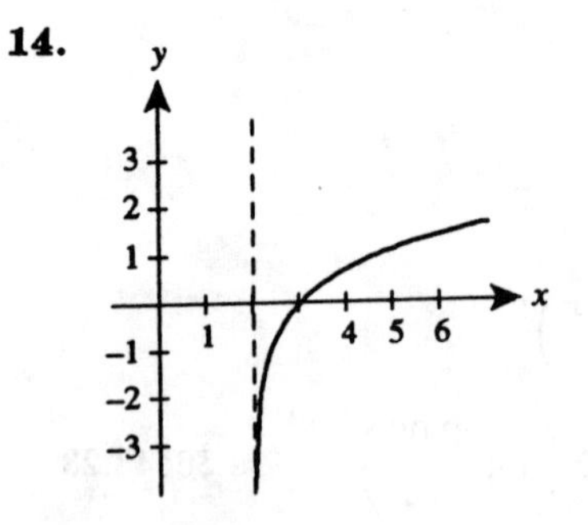

15. $\dfrac{\ln x}{\ln y} \neq \ln(x - y)$ since $\dfrac{\ln x}{\ln y} = \log_y x$

16. $5^x = 41$

$$x = \log_5 41 = \frac{\ln 41}{\ln 5} \approx 2.3074$$

17. $x - x^2 = \log_5 \frac{1}{25}$

$$5^{x-x^2} = \frac{1}{25}$$

$$5^{x-x^2} = 5^{-2}$$

$$x - x^2 = -2$$

$$0 = x^2 - x - 2$$

$$0 = (x+1)(x-2)$$

$$x = -1 \text{ or } x = 2$$

18. $\log_2 x + \log_2(x-3) = 2$

$$\log_2[x(x-3)] = 2$$

$$x(x-3) = 2^2$$

$$x^2 - 3x = 4$$

$$x^2 - 3x - 4 = 0$$

$$(x+1)(x-4) = 0$$

$$x = 4$$

$$x = -1$$

(extraneous solution)

19. $$\frac{e^x + e^{-x}}{3} = 4$$

$$e^x(e^x + e^{-x}) = 12e^x$$

$$e^{2x} + 1 = 12e^x$$

$$e^{2x} - 12e^x + 1 = 0$$

$$e^x = \frac{12 \pm \sqrt{144 - 4}}{2}$$

$e^x = 11.9161$ or $e^x = 0.0839$

$x = \ln 11.9161$ $\quad$ $x = \ln 0.0839$

$x \approx 2.4779$ $\quad$ $x \approx -2.4779$

20. $$A = Pe^{rt}$$

$$12{,}000 = 6000e^{0.13t}$$

$$2 = e^{0.13t}$$

$$0.13t = \ln 2$$

$$t = \frac{\ln 2}{0.13}$$

$t \approx 5.3319$ yr or 5 yr 4 mo

CHAPTER 7

Practice Test Solutions

1.
$$x + y = 1$$
$$3x - y = 15 \Rightarrow y = 3x - 15$$
$$x + (3x - 15) = 1$$
$$4x = 16$$
$$x = 4$$
$$y = -3$$

2.
$$x - 3y = -3 \Rightarrow x = 3y - 3$$
$$x^2 + 6y = 5$$
$$(3y - 3)^2 + 6y = 5$$
$$9y^2 - 18y + 9 + 6y = 5$$
$$9y^2 - 12y + 4 = 0$$
$$(3y - 2)^2 = 0$$
$$y = \frac{2}{3}$$
$$x = -1$$

3.
$$x + y + z = 6 \Rightarrow z = 6 - x - y$$
$$2x - y + 3z = 0 \qquad 2x - y + 3(6 - x - y) = 0 \quad \Rightarrow -x - 4y = -18$$
$$5x + 2y - z = -3 \qquad 5x + 2y - (6 - x - y) = -3 \Rightarrow 6x + 3y = 3$$
$$x = 18 - 4y$$
$$6(18 - 4y) + 3y = 3$$
$$-21y = -105$$
$$y = 5$$
$$x = 18 - 4y = -2$$
$$z = 6 - x - y = 3$$

4.
$$x + y = 110 \quad \Rightarrow \quad y = 110 - x$$
$$xy = 2800$$
$$x(110 - x) = 2800$$
$$0 = x^2 - 110x + 2800$$
$$0 = (x - 40)(x - 70)$$
$$x = 40 \quad \text{or} \quad x = 70$$
$$y = 70 \qquad\quad y = 40$$

5.
$$2x + 2y = 170 \Rightarrow y = \frac{170 - 2x}{2} = 85 - x$$
$$xy = 2800$$
$$x(85 - x) = 2800$$
$$0 = x^2 - 85x + 2800$$
$$0 = (x - 25)(x - 60)$$
$$x = 25 \quad \text{or} \quad x = 60$$
$$y = 60 \qquad\quad y = 25$$

Dimensions: $60' \times 25'$

6.
$$\begin{aligned} 2x + 15y &= 4 \quad\Rightarrow\quad & 2x + 15y &= 4 \\ x - 3y &= 23 \quad\Rightarrow\quad & 5x - 15y &= 115 \\ & & 7x &= 119 \\ & & x &= 17 \\ & & y &= \frac{x-23}{3} \\ & & &= -2 \end{aligned}$$

7.
$$\begin{aligned} x + y &= 2 \quad\Rightarrow\quad & 19x + 19y &= 38 \\ 38x - 19y &= 7 \quad\Rightarrow\quad & 38x - 19y &= 7 \\ & & 57x &= 45 \end{aligned}$$

$$x = \frac{45}{57} = \frac{15}{19}$$

$$y = 2 - x = \frac{38}{19} - \frac{15}{19} = \frac{23}{19}$$

8.
$$\begin{aligned} 0.4x + 0.5y &= 0.112 \quad\Rightarrow\quad & 0.28x + 0.35y &= 0.0784 \\ 0.3x - 0.7y &= -0.131 \quad\Rightarrow\quad & 0.15x - 0.35y &= -0.0655 \\ & & 0.43x &= 0.0129 \end{aligned}$$

$$x = \frac{0.0129}{0.43} = 0.03$$

$$y = \frac{0.112 - 0.4x}{0.5} = 0.20$$

9. Let $x =$ amount in 11% fund and $y =$ amount in 13% fund.

$$x + y = 17000 \Rightarrow y = 17000 - x$$

$$0.11x + 0.13y = 2080$$

$$\begin{aligned} 0.11x + 0.13(17000 - x) &= 2080 \\ -0.02x &= -130 \\ x &= \$6500 \\ y &= \$10,500 \end{aligned}$$

10. $(4, 3)$, $(1, 1)$, $(-1, -2)$, $(-2, -1)$

$$n = 4, \quad \sum_{i=1}^{4} x_i = 2, \quad \sum_{i=1}^{4} y_i = 1, \quad \sum_{i=1}^{4} x_i^2 = 22, \quad \sum_{i=1}^{4} x_i y_i = 17$$

$$\begin{aligned} 4b + 2a &= 1 \quad\Rightarrow\quad & 4b + 2a &= 1 \\ 2b + 22a &= 17 \quad\Rightarrow\quad & -4b - 44a &= -34 \\ & & -42a &= -33 \end{aligned}$$

$$a = \frac{33}{42} = \frac{11}{14}$$

$$b = \frac{1}{4}\left(1 - 2\left(\frac{33}{42}\right)\right) = -\frac{1}{7}$$

$$y = ax + b = \frac{11}{14}x - \frac{1}{7}$$

11. $$\begin{aligned} x + y \quad &= -2 \\ 2x - y + z &= 11 \\ 4y - 3z &= -20 \end{aligned} \Rightarrow \begin{aligned} -2x - 2y \quad &= 4 \\ 2x - y + z &= 11 \\ \hline -3y + z &= 15 \end{aligned} \qquad \begin{aligned} -9y + 3z &= 45 \\ 4y - 3z &= -20 \\ \hline -5y \quad &= 25 \\ y &= -5 \\ x &= 3 \\ z &= 0 \end{aligned}$$

12. $$\begin{aligned} 4x - y + 5z &= 4 \\ 2x + y - z &= 0 \\ 2x + 4y + 8z &= 0 \end{aligned} \begin{aligned} &\Rightarrow \\ &\Rightarrow \\ & \end{aligned} \begin{aligned} 4x - y + 5z &= 4 \\ -4x - 2y + 2z &= 0 \\ \hline -3y + 7z &= 4 \end{aligned}$$

$$\begin{aligned} 2x + 4y + 8z &= 0 \\ -2x - y + z &= 0 \\ \hline 3y + 9z &= 0 \\ -3y + 7z &= 4 \\ \hline 16z &= 4 \\ z &= \tfrac{1}{4} \\ y &= -\tfrac{3}{4} \\ x &= \tfrac{1}{2} \end{aligned}$$

13. $$\begin{aligned} 3x + 2y - z &= 5 \\ 6x - y + 5z &= 2 \end{aligned} \begin{aligned} &\Rightarrow \\ &\Rightarrow \end{aligned} \begin{aligned} 6x + 4y - 2z &= 10 \\ -6x + y - 5z &= -2 \\ \hline 5y - 7z &= 8 \\ y &= \frac{8 + 7z}{5} \end{aligned}$$

$$\begin{aligned} 3x + 2y - z &= 5 \\ 12x - 2y + 10z &= 4 \\ \hline 15x \quad + 9z &= 9 \\ x = \frac{9 - 9z}{15} &= \frac{3 - 3z}{5} \end{aligned}$$

Let $z = a$, then $x = \dfrac{3 - 3a}{5}$ and $y = \dfrac{8 + 7a}{5}$.

14. $y = ax^2 + bx + c$ passes through $(0, -1)$, $(1, 4)$, and $(2, 13)$.

At $(0, -1)$, $-1 = a(0)^2 + b(0) + c \Rightarrow c = -1$

At $(1, 4)$, $4 = a(1)^2 + b(1) - 1 \Rightarrow 5 = a + b \Rightarrow 5 = a + b$

At $(2, 13)$, $13 = a(2)^2 + b(2) - 1 \Rightarrow 14 = 4a + 2b \Rightarrow -7 = -2a - b$

$$\begin{aligned} -2 &= -a \\ a &= 2 \\ b &= 3 \end{aligned}$$

Thus, $y = 2x^2 + 3x - 1$.

15. $s = \frac{1}{2}at^2 + v_0t + s_0$ passes through $(1, 12)$, $(2, 5)$, and $(3, 4)$.

At $(1, 12)$, $12 = \frac{1}{2}a + v_0 + s_0 \Rightarrow 24 = a + 2v_0 + 2s_0$

At $(2, 5)$, $5 = 2a + 2v_0 + s_0 \Rightarrow -5 = -2a - 2v_0 - s_0$

At $(3, 4)$, $4 = \frac{9}{2}a + 3v_0 + s_0$

$$19 = -a + s_0$$

$$\begin{aligned} 15 &= 6a + 6v_0 + 3s_0 \\ -8 &= -9a - 6v_0 - 2s_0 \\ \hline 7 &= -3a + s_0 \\ -19 &= a - s_0 \\ \hline -12 &= -2a \\ a &= 6 \\ s_0 &= 25 \\ v_0 &= -16 \end{aligned}$$

Thus, $s = \frac{1}{2}(6)t^2 - 16t + 25 = 3t^2 - 16t + 25$.

16. $x^2 + y^2 \geq 9$

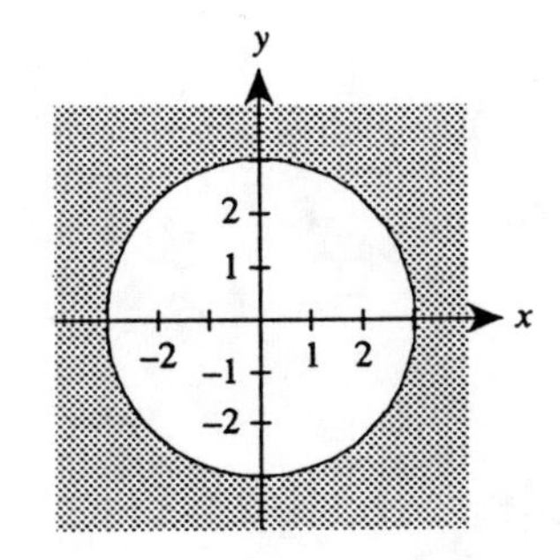

17. $x + y \leq 6$

$x \geq 2$

$y \geq 0$

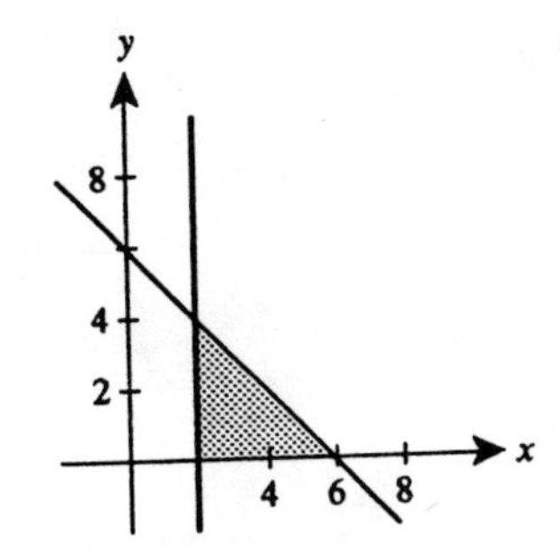

18. Line through (0, 0) and (0, 7):

$x = 0$

Line through (0, 0) and (2, 3):

$y = \frac{3}{2}x$ or $3x - 2y = 0$

Line through (0, 7) and (2, 3):

$y = -2x + 7$ or $2x + y = 7$

Inequalities: $x \geq 0$

$3x - 2y \leq 0$

$2x + y \leq 7$

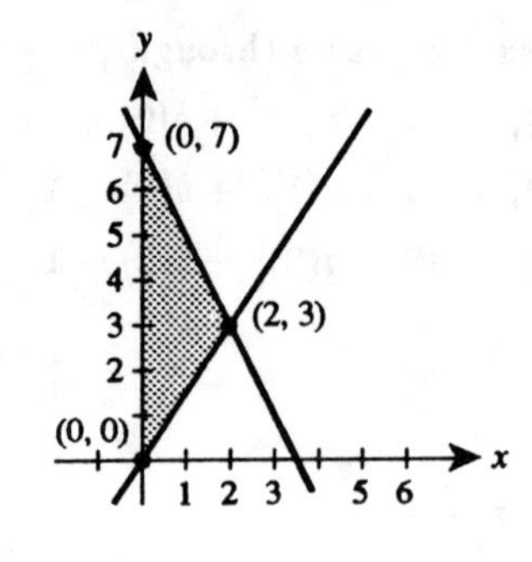

19. Vertices: (0, 0), (0, 7), (6, 0), (3, 5)

$z = 30x + 26y$

At (0, 0), $z = 0$

At (0, 7), $z = 182$

At (6, 0), $z = 180$

At (3, 5), $z = 220$

The maximum value of z is 220.

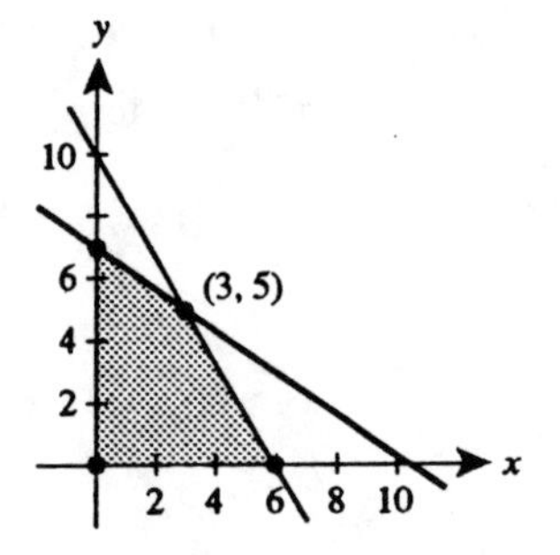

20.

$$x^2 + y^2 \leq 4$$
$$(x - 2)^2 + y^2 \geq 4$$

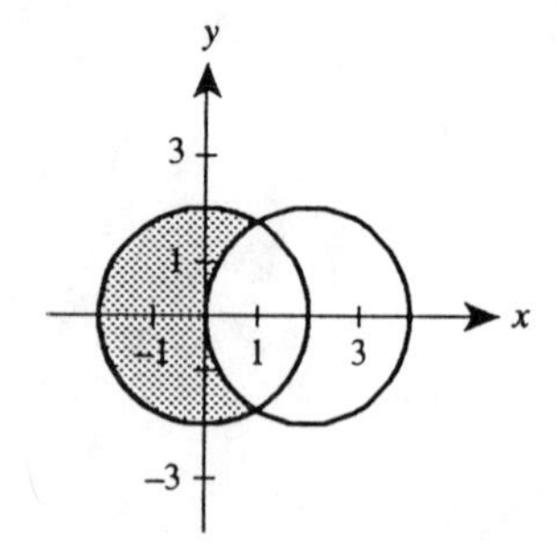

CHAPTER 8

Practice Test Solutions

1. $\begin{bmatrix} 1 & -2 & 4 \\ 3 & -5 & 9 \end{bmatrix} -3R_1 + R_2 \rightarrow \begin{bmatrix} 1 & -2 & 4 \\ 0 & 1 & -3 \end{bmatrix} 2R_2 + R_1 \rightarrow \begin{bmatrix} 1 & 0 & -2 \\ 0 & 1 & -3 \end{bmatrix}$

2. $3x + 5y = 3$

$2x - y = -11$

$$\left[\begin{array}{cc:c} 3 & 5 & 3 \\ 2 & -1 & -11 \end{array}\right] \quad -R_2 + R_1 \rightarrow \left[\begin{array}{cc:c} 1 & 6 & 14 \\ 2 & -1 & -11 \end{array}\right]$$

$$-2R_1 + R_2 \rightarrow \left[\begin{array}{cc:c} 1 & 6 & 14 \\ 0 & -13 & -39 \end{array}\right]$$

$$-\tfrac{1}{13}R_2 \rightarrow \left[\begin{array}{cc:c} 1 & 6 & 14 \\ 0 & 1 & 3 \end{array}\right]$$

$$-6R_2 + R_1 \rightarrow \left[\begin{array}{cc:c} 1 & 0 & -4 \\ 0 & 1 & 3 \end{array}\right]$$

Answer: $x = -4,\ y = 3$

3. $2x + 3y = -3$

$3x + 2y = 8$

$x + y = 1$

$$\left[\begin{array}{cc:c} 2 & 3 & -3 \\ 3 & 2 & 8 \\ 1 & 1 & 1 \end{array}\right] \quad R_3 \leftrightarrow R_1 \quad \left[\begin{array}{cc:c} 1 & 1 & 1 \\ 3 & 2 & 8 \\ 2 & 3 & -3 \end{array}\right]$$

$$\begin{array}{r} \\ -3R_1 + R_2 \rightarrow \\ -2R_1 + R_3 \rightarrow \end{array} \left[\begin{array}{cc:c} 1 & 1 & 1 \\ 0 & -1 & 5 \\ 0 & 1 & -5 \end{array}\right]$$

$$\begin{array}{r} R_2 + R_1 \rightarrow \\ -R_2 \rightarrow \\ -R_2 + R_3 \rightarrow \end{array} \left[\begin{array}{cc:c} 1 & 0 & 6 \\ 0 & 1 & -5 \\ 0 & 0 & 0 \end{array}\right]$$

Answer: $x = 6,\ y = -5$

4. $x + 3z = -5$

$2x + y = 0$

$3x + y - z = 3$

$$\left[\begin{array}{ccc|c} 1 & 0 & 3 & -5 \\ 2 & 1 & 0 & 0 \\ 3 & 1 & -1 & 3 \end{array}\right] \quad \begin{array}{r} \\ -2R_1 + R_2 \rightarrow \\ -3R_1 + R_3 \rightarrow \end{array} \left[\begin{array}{ccc|c} 1 & 0 & 3 & -5 \\ 0 & 1 & -6 & 10 \\ 0 & 1 & -10 & 18 \end{array}\right]$$

$$\begin{array}{r} \\ \\ -R_2 + R_3 \rightarrow \end{array} \left[\begin{array}{ccc|c} 1 & 0 & 3 & -5 \\ 0 & 1 & -6 & 10 \\ 0 & 0 & -4 & 8 \end{array}\right]$$

$$\begin{array}{r} -3R_3 + R_1 \rightarrow \\ 6R_3 + R_2 \rightarrow \\ -\frac{1}{4}R_4 \rightarrow \end{array} \left[\begin{array}{ccc|c} 1 & 0 & 0 & 1 \\ 0 & 1 & 0 & -2 \\ 0 & 0 & 1 & -2 \end{array}\right]$$

Answer: $x = 1,\ y = -2,\ z = -2$

5. $$\begin{bmatrix} 1 & 4 & 5 \\ 2 & 0 & -3 \end{bmatrix} \begin{bmatrix} 1 & 6 \\ 0 & -7 \\ -1 & 2 \end{bmatrix} = \begin{bmatrix} -4 & -12 \\ 5 & 6 \end{bmatrix}$$

6. $$3A - 5B = 3\begin{bmatrix} 9 & 1 \\ -4 & 8 \end{bmatrix} - 5\begin{bmatrix} 6 & -2 \\ 3 & 5 \end{bmatrix}$$
$$= \begin{bmatrix} 27 & 3 \\ -12 & 24 \end{bmatrix} - \begin{bmatrix} 30 & -10 \\ 15 & 25 \end{bmatrix}$$
$$= \begin{bmatrix} -3 & 13 \\ -27 & -1 \end{bmatrix}$$

7. $$f(A) = \begin{bmatrix} 3 & 0 \\ 7 & 1 \end{bmatrix}^2 - 7\begin{bmatrix} 3 & 0 \\ 7 & 1 \end{bmatrix} + 8\begin{bmatrix} 1 & 0 \\ 0 & 1 \end{bmatrix}$$
$$= \begin{bmatrix} 3 & 0 \\ 7 & 1 \end{bmatrix}\begin{bmatrix} 3 & 0 \\ 7 & 1 \end{bmatrix} - \begin{bmatrix} 21 & 0 \\ 49 & 7 \end{bmatrix} + \begin{bmatrix} 8 & 0 \\ 0 & 8 \end{bmatrix}$$
$$= \begin{bmatrix} 9 & 0 \\ 28 & 1 \end{bmatrix} - \begin{bmatrix} 21 & 0 \\ 49 & 7 \end{bmatrix} + \begin{bmatrix} 8 & 0 \\ 0 & 8 \end{bmatrix}$$
$$= \begin{bmatrix} -4 & 0 \\ -21 & 2 \end{bmatrix}$$

8. False since

$$(A + B)(A + 3B) = A(A + 3B) + B(A + 3B)$$
$$= A^2 + 3AB + BA + 3B^2$$

9. $\left[\begin{array}{cc|cc} 1 & 2 & 1 & 0 \\ 3 & 5 & 0 & 1 \end{array}\right]$ $\begin{array}{r} \\ -3R_1 + R_2 \rightarrow \end{array} \left[\begin{array}{cc|cc} 1 & 2 & 1 & 0 \\ 0 & -1 & -3 & 1 \end{array}\right]$

$$\begin{array}{r} 2R_2 + R_1 \rightarrow \\ -R_2 \rightarrow \end{array} \left[\begin{array}{cc|cc} 1 & 0 & -5 & 2 \\ 0 & 1 & 3 & -1 \end{array}\right]$$

$$A^{-1} = \begin{bmatrix} -5 & 2 \\ 3 & -1 \end{bmatrix}$$

10. $\left[\begin{array}{ccc|ccc} 1 & 1 & 1 & 1 & 0 & 0 \\ 3 & 6 & 5 & 0 & 1 & 0 \\ 6 & 10 & 8 & 0 & 0 & 1 \end{array}\right]$ $\begin{array}{r} \\ -3R_1 + R_2 \rightarrow \\ -6R_1 + R_3 \rightarrow \end{array} \left[\begin{array}{ccc|ccc} 1 & 1 & 1 & 1 & 0 & 0 \\ 0 & 3 & 2 & -3 & 1 & 0 \\ 0 & 4 & 2 & -6 & 0 & 1 \end{array}\right]$

$$\begin{array}{r} -R_2 + R_1 \rightarrow \\ \frac{1}{3}R_2 \rightarrow \\ -4R_2 + R_3 \rightarrow \end{array} \left[\begin{array}{ccc|ccc} 1 & 0 & \frac{1}{3} & 2 & -\frac{1}{3} & 0 \\ 0 & 1 & \frac{2}{3} & -1 & \frac{1}{3} & 0 \\ 0 & 0 & -\frac{2}{3} & -2 & -\frac{4}{3} & 1 \end{array}\right]$$

$$\begin{array}{r} \frac{1}{2}R_3 + R_1 \rightarrow \\ R_3 + R_2 \rightarrow \\ -\frac{3}{2}R_3 \rightarrow \end{array} \left[\begin{array}{ccc|ccc} 1 & 0 & 0 & 1 & -1 & \frac{1}{2} \\ 0 & 1 & 0 & -3 & -1 & 1 \\ 0 & 0 & 1 & 3 & 2 & -\frac{3}{2} \end{array}\right]$$

$$A^{-1} = \begin{bmatrix} 1 & -1 & \frac{1}{2} \\ -3 & -1 & 1 \\ 3 & 2 & -\frac{3}{2} \end{bmatrix}$$

11. (a) $x + 2y = 4$

$3x + 5y = 1$

$\left[\begin{array}{cc|cc} 1 & 2 & 1 & 0 \\ 3 & 5 & 0 & 1 \end{array}\right]$ $\begin{array}{r} \\ -3R_1 + R_2 \rightarrow \end{array} \left[\begin{array}{cc|cc} 1 & 2 & 1 & 0 \\ 0 & -1 & -3 & 1 \end{array}\right]$

$$\begin{array}{r} -2R_2 + R_1 \rightarrow \\ -R_2 \rightarrow \end{array} \left[\begin{array}{cc|cc} 1 & 0 & -5 & 2 \\ 0 & 1 & 3 & -1 \end{array}\right]$$

$$X = A^{-1}B = \begin{bmatrix} -5 & 2 \\ 3 & -1 \end{bmatrix} \begin{bmatrix} 4 \\ 1 \end{bmatrix} = \begin{bmatrix} -18 \\ 11 \end{bmatrix}$$

$x = -18,\ y = 11$

(b) $x + 2y = 3$

$3x + 5y = -2$

$$X = A^{-1}B = \begin{bmatrix} -5 & 2 \\ 3 & -1 \end{bmatrix} \begin{bmatrix} 3 \\ -2 \end{bmatrix} = \begin{bmatrix} -19 \\ 11 \end{bmatrix}$$

$x = -19,\ y = 11$

12. $\begin{vmatrix} 6 & -1 \\ 3 & 4 \end{vmatrix} = 24 - (-3) = 27$

13. $\begin{vmatrix} 1 & 3 & -1 \\ 5 & 9 & 0 \\ 6 & 2 & -5 \end{vmatrix}\begin{matrix} 1 & 3 \\ 5 & 9 \\ 6 & 2 \end{matrix} = (-45 + 0 - 10) - (-54 + 0 - 75) = 74$

14. $\begin{vmatrix} 1 & 4 & 2 & 3 \\ 0 & 1 & -2 & 0 \\ 3 & 5 & -1 & 1 \\ 2 & 0 & 6 & 1 \end{vmatrix} = \begin{vmatrix} 1 & 2 & 3 \\ 3 & -1 & 1 \\ 2 & 6 & 1 \end{vmatrix} + 2\begin{vmatrix} 1 & 4 & 3 \\ 3 & 5 & 1 \\ 2 & 0 & 1 \end{vmatrix}$

$= 51 + 2(-29) = -7$ Expansion along Row 2.

15. $\begin{vmatrix} 3 & 0 & 0 \\ 0 & 3 & 0 \\ 0 & 0 & 3 \end{vmatrix} = 3(3)(3)\begin{vmatrix} 1 & 0 & 0 \\ 0 & 1 & 0 \\ 0 & 0 & 1 \end{vmatrix} = -3^3\begin{vmatrix} 1 & 0 & 0 \\ 0 & 0 & 1 \\ 0 & 1 & 0 \end{vmatrix}$

True

16. $\begin{vmatrix} 6 & 4 & 3 & 0 & 6 \\ 0 & 5 & 1 & 4 & 8 \\ 0 & 0 & 2 & 7 & 3 \\ 0 & 0 & 0 & 9 & 2 \\ 0 & 0 & 0 & 0 & 1 \end{vmatrix} = 6(5)(2)(9)(1) = 540$

17. Area $= \dfrac{1}{2}\begin{vmatrix} 0 & 7 & 1 \\ 5 & 0 & 1 \\ 3 & 9 & 1 \end{vmatrix}$

$= \dfrac{1}{2}(31)$

$= 15.5$ square units

18. $x = \dfrac{\begin{vmatrix} 4 & -7 \\ 11 & 5 \end{vmatrix}}{\begin{vmatrix} 6 & -7 \\ 2 & 5 \end{vmatrix}} = \dfrac{97}{44}$

19. $z = \dfrac{\begin{vmatrix} 3 & 0 & 1 \\ 0 & 1 & 3 \\ 1 & -1 & 2 \end{vmatrix}}{\begin{vmatrix} 3 & 0 & 1 \\ 0 & 1 & 4 \\ 1 & -1 & 0 \end{vmatrix}} = \dfrac{14}{11}$

20. $y = \dfrac{\begin{vmatrix} 721.4 & 33.77 \\ 45.9 & 19.85 \end{vmatrix}}{\begin{vmatrix} 721.4 & -29.1 \\ 45.9 & 105.6 \end{vmatrix}} = \dfrac{12{,}769.747}{77{,}515.530} \approx 0.1647$

CHAPTER 9

Practice Test Solutions

1. $a_n = \dfrac{2n}{(n+2)!}$

$a_1 = \dfrac{2(1)}{3!} = \dfrac{2}{6} = \dfrac{1}{3}$

$a_2 = \dfrac{2(2)}{4!} = \dfrac{4}{24} = \dfrac{1}{6}$

$a_3 = \dfrac{2(3)}{5!} = \dfrac{6}{120} = \dfrac{1}{20}$

$a_4 = \dfrac{2(4)}{6!} = \dfrac{8}{720} = \dfrac{1}{90}$

$a_5 = \dfrac{2(5)}{7!} = \dfrac{10}{5040} = \dfrac{1}{504}$

Terms: $\dfrac{1}{3}, \dfrac{1}{6}, \dfrac{1}{20}, \dfrac{1}{90}, \dfrac{1}{504}$

2. $a_n = \dfrac{n+3}{3^n}$

3. $\displaystyle\sum_{i=1}^{6}(2i-1) = 1+3+5+7+9+11$

$= 36$

4. $a_1 = 23,\ d = -2$

$a_2 = a_1 + d = 21$

$a_3 = a_2 + d = 19$

$a_4 = a_3 + d = 17$

$a_5 = a_4 + d = 15$

Terms: 23, 21, 19, 17, 15

5. $a_1 = 12,\ d = 3,\ n = 50$

$a_n = a_1 + (n-1)d$

$a_{50} = 12 + (50-1)3 = 159$

6. $a_1 = 1$

$a_{200} = 200$

$S_n = \dfrac{n}{2}(a_1 + a_n)$

$S_{200} = \dfrac{200}{2}(1 + 200) = 20{,}100$

7. $a_1 = 7,\ r = 2$

$a_2 = a_1 r = 14$

$a_3 = a_1 r^2 = 28$

$a_4 = a_1 r^3 = 56$

$a_5 = a_1 r^4 = 112$

Terms: 7, 14, 28, 56, 112

8. $\sum_{n=0}^{9} 6\left(\frac{2}{3}\right)^n,\ a_1 = 6,\ r = \frac{2}{3},\ n = 10$

$$S_n = \frac{a_1(1-r^n)}{1-r}$$

$$= \frac{6(1-(\frac{2}{3})^{10})}{1-\frac{2}{3}} \approx 17.6879$$

9. $\sum_{n=0}^{\infty} (0.03)^n,\ a_1 = 1,\ r = 0.03$

$$S = \frac{a_1}{1-r} = \frac{1}{1-0.03} = \frac{1}{0.97} = \frac{100}{97} \approx 1.0309$$

10. For $n = 1$, $1 = \frac{1(1+1)}{2}$.

Assume that $1 + 2 + 3 + 4 + \cdots + k = \frac{k(k+1)}{2}$.

Now for $n = k + 1$,

$$1 + 2 + 3 + 4 + \cdots + k + (k+1) = \frac{k(k+1)}{2} + k + 1$$

$$= \frac{k(k+1)}{2} + \frac{2(k+1)}{2}$$

$$= \frac{(k+1)(k+2)}{2}.$$

Thus, $1 + 2 + 3 + 4 + \cdots + n = \frac{n(n+1)}{2}$ for all integers $n \geq 1$.

11. For $n = 4$, $4! > 2^4$.

Assume that $k! > 2^k$.

Then $(k+1)! = (k+1)(k!) > (k+1)2^k > 2 \cdot 2^k = 2^{k+1}$.

Thus, $n! > 2^n$ for all integers $n \geq 4$.

12. ${}_{13}C_4 = \frac{13!}{(13-4)!4!} = 715$

13. $(x+3)^5 = x^5 + 5x^4(3) + 10x^3(3)^2 + 10x^2(3)^3 + 5x(3)^4 + (3)^5$

$= x^5 + 15x^4 + 90x^3 + 270x^2 + 405x + 243$

14. ${}_{12}C_5 x^7(-2)^5 = -25{,}344x^7$

15. ${}_{30}P_4 = \dfrac{30!}{(30-4)!} = 657{,}720$

16. $6! = 720$ ways

17. ${}_{12}P_3 = 1320$

18. $P(2) + P(3) + P(4) = \dfrac{1}{36} + \dfrac{2}{36} + \dfrac{3}{36}$

$= \dfrac{6}{36} = \dfrac{1}{6}$

19. $P(K,\ B10) = \dfrac{4}{52} \cdot \dfrac{2}{51} = \dfrac{2}{663}$

20. Let $A =$ probability of no faulty units.

$P(A) = \left(\dfrac{997}{1000}\right)^{50} \approx 0.8605$

$P(A') = 1 - P(A) \approx 0.1395$